SELF-ORGANIZATION IN BIOLOGICAL SYSTEMS

PRINCETON STUDIES IN COMPLEXITY

Self-Organization in Biological Systems

SCOTT CAMAZINE

JEAN-LOUIS DENEUBOURG

NIGEL R. FRANKS

JAMES SNEYD

GUY THERAULAZ

ERIC BONABEAU

ORIGINAL LINE DRAWINGS BY WILLIAM RISTINE AND MARY ELLEN DIDION

STARLOGO PROGRAMMING BY WILLIAM THIES

PRINCETON UNIVERSITY PRESS

PRINCETON AND OXFORD

Published by Princeton University Press, 41 William Street, Princeton, New Jersey 08540
In the United Kingdom: Princeton University Press, 3 Market Place, Woodstock,
Oxfordshire OX20 1SY

Library of Congress Cataloging-in-Publication Data

Self-organization in biological systems / Scott Camazine ... [et al.].
p. cm. — (Princeton studies in complexity)
Includes bibliographical references (p.).
ISBN 0-691-01211-3 (cloth : alk. paper)
1. Biological systems. 2. Self-organizing systems. I. Camazine, Scott. II. Series.

QH313 .S477 2001
570'.1'1 — dc21

00-045329

This book has been composed in Times Roman

The paper used in this publication meets the minimum requirements of ANSI/NISO
Z39.48-1992 (R1997) (*Permanence of Paper*)
www.pup.princeton.edu

Printed in the United States of America

10 9 8 7 6 5 4 3 2 1

Contents

Explanation of Color Plates

Similar patterns and structures appear throughout the natural world. The wind-blown ripples of a sand dune, the alternating black and white stripes of a zebra, and the neuronal projections from one eye or the other to the brain all share the same characteristic striped pattern. Is this merely a coincidence, perhaps a lack of creativity on the part of Nature? Certainly not. Though these structures bear no obvious relation to one another—the constituent components being lifeless grains of sand, pigment-producing cells, or neurons—there is an explanation for the commonality of form. All these striped structures share the underlying mechanism by which patterns emerge through self-organization. Nature's stripes, spirals, spots, and blotches often develop without reference to a blueprint, a master plan, or set of explicitly coded genetic instructions for where each stripe or spot occurs. Instead these patterns emerge spontaneously by means of simple, local interactions among the constituent components.

Some of the earliest studies of self-organization dealt with physical and chemical systems (Plate 1). This trend continues with recent research examining many examples of self-organization in other nonliving systems—sand dune stripes, paint shrinkage, surface buckling, and convection patterns in tundra soils, clouds, and the sun's photosphere.

When we turn to living systems, particularly cellular systems involved in developmental processes, self-organization has been invoked to explain the skin and surface markings of mammalian coats, seashells, butterfly wings, beetle elytra, fish, reptiles, and amphibians (Plate 2). In these systems, the development of patterns involves "societies" of cells whose local interactions, at a microscopic scale, result in colorful patches of scales or pigments. In contrast, most of this book is devoted to self-organization in living systems composed of multicellular animals (fireflies, flocks of birds, schools of fish, and especially the social insects). Here we emphasize one crucial difference between the nonliving and the biological systems. In the latter, self-organized patterns have almost certainly been molded by natural selection (see Chapter 7). Nonetheless, a comparison of living with nonliving patterns reveals some remarkable similarities that are no mere coincidence, but rather an indication of the underlying similarity in pattern formation mechanisms.

In addition to animal systems, self-organized pattern formation frequently reveals itself among plants and fungi—the branching patterns of slime molds and the movement of the individual amoebae; the ornamentation on the surface of pollen grains and mushroom caps; the growth of lichens; and the spiral arrangement of flower parts in daisies and pine cones (Plates 3 and 8).

Among the social insects, self-organization has been shown to play a role in building behavior, decision making, synchronization of activities and trail formation (Plates 4, 5, 6, and 7). The emergent patterns and decision-making processes of these systems are often nonintuitive due to the large numbers of nonlinear interactions involved. For this reason, mathematical models and simulations provide useful techniques for studying self-organizing systems and exploring the consequences of the myriad interactions among component subunits. Many of these simulations are agent-based, that is, each individual subunit is followed throughout the simulation, and its behavior over time is determined through local interactions with other subunits and local cues from the environment. In order to follow the behavior of an individual, it is usually necessary to mark the individual with a spot of paint or a numbered tag (as in Plate 4f). Detailed experimental observations of individual bees, ants, and other living subunits provide the necessary information for developing realistic cellular automaton (CA) models and Monte Carlo simulations (Plates 6, 7, and 8).

SELF-ORGANIZATION IN BIOLOGICAL SYSTEMS

Prologue

Aims and Scope of the Book

This book examines biological structures built through mechanisms involving self-organization. The structures of interest are those that develop through interactions among organisms, hence we focus on objects built by (or of) groups of organisms. Moreover, we focus our attention on those products of group activity which are group-level adaptations, not merely incidental by-products of the behaviors of a group's members (Williams 1966).

A prime example of an adaptive structure is a nest built by a colony of the fungus-growing termite *Macrotermes* (Figure 18.1). With its thick protective walls and labyrinth of ventilation ducts, this air-conditioned castle of clay confers large positive fitness effects on the genes of its termite builders, by providing them with a safe and stable environment. One aim of this book is to understand how such structures are built, and the role in their construction played by mechanisms involving self-organization.

The book is divided into three parts. Part I is an introduction to self-organizaton as it relates to the biological systems that are the subject matter of this book. It provides both the conceptual basis and tools for understanding the examples of self-organization that constitute the remainder of the book. In Part II, we present certain examples that show how self-organized structures arise in groups of organisms that are gregarious for at least a portion of their lives. The structures built by these groups are generally less sophisticated than the highly adaptive structures that are built by insect societies—also the subject matter of Part II. In Part III we summarize the lessons learned from self-organization, identify new avenues of research, and suggest how the self-organization approach will improve our understanding of the building of biological structures in general.

Even though the study of self-organization is a relatively new field, there is already a large literature on numerous topics. However, most concern the fields of physics, chemistry, biochemistry, and developmental biology (Prigogine and Glansdorf 1971; Haken 1977; Nicolis and Prigogine 1977, 1989; Murray 1988, 1989; Kauffman 1993; Kapral and Showalter 1994; Nicolis 1995; Goldbeter 1996; Bak 1996) rather than organismal biology, our focus in this book. Although we devote one chapter of the book to the well-studied aggregation patterns formed by unicellular slime molds, most of the book concerns itself with groups of more complex multicellular organisms which utilize self-organizing mechanisms of pattern formation, decision-making, and collective behavior. Some examples include the coordinated movements of fish in a school, the synchronized flashing of fireflies, and the collective foraging and building behavior of social insects.

What is of special interest to us are the *mechanisms* by which such structures develop and are maintained. Recent research has begun to reveal that even the most sophisticated structures that we will consider, such as the nests of termite colonies, are self-organized structures built through the iteration of surprisingly simple behaviors performed by large numbers of individuals that rely only on local information. Our primary goal in writing this book is to demonstrate, for a wide range of examples, the link between the rather simple behavioral programs of the individuals in a group and the sophisticated structures and patterns that emerge from their collective activity. This goal raises a number of questions to be addressed throughout the book: (1) To what extent can mechanisms of pattern formation based upon self-organization account for biological structure? (2) What are the alternative mechanisms of biological pattern formation? (3) Under what circumstances do organisms use self-organization versus these alternatives? (4) What level of complexity at the individual level is required to generate the observed complexity at the group (collective) level? (5) How much of the observed complexity at the group level is a reflection of complexity of the environment rather than complexity at the level of the individual? (6) To what extent have widely differing organisms adopted similar, convergent strategies of pattern formation?

We wish to emphasize one more important idea at the start of this book: Much of the complexity of self-organized structures seen in biology arises because the rules governing the interactions among the components of biological systems have evolved through natural selection. The process of evolution has generated an enormous diversity of behavioral and physiological interactions, far surpassing the diversity of interactions possible in chemical and physical systems. This makes the study of biological self-organization particularly exciting and challenging. Furthermore, it guarantees that the study of biological self-organization will not simply be a reworking of chemical and physical models of self-organization using the same equations with the variables simply carrying different names.

Acknowledgments

We have many people to thank for their contributions to our thoughts and writings on self-organization. Foremost among these are Rüdiger Wehner who initiated this project when he invited Scott Camazine, Jean-Louis Deneubourg, Nigel Franks and Tom Seeley to spend a year at the Wissenschaftskolleg in Berlin, Germany. Although our interactions were far from self-organized, the Wissenschaftskolleg provided us with an ideal milieu for working closely, and sharing our thoughts about the role of self-organization in biological systems. We greatly appreciate the opportunity afforded by the Wissenschaftskolleg, and the generous assistance of the staff at the institute.

For their support at various times during this project or for certain aspects of the experimental work reported here, Nigel Franks wishes to thank the Wissenschaftskolleg, Rüdiger Wehner, Tom Seeley, Ana Sendova-Franks, The Department of Biology and Biochemistry of the University of Bath, Iain Couzin, The Leverhulme Trust and the Smithsonian Tropical Research Institute.

Jean-Louis Deneubourg would like to thank Claire Detrain, Arnaud Lioni, Jesus Millor, Grégoire Nicolis, Jacques Pasteels, Ilya Prigogine and Philippe Rasse for the fruitful discussions, and The Solvay Institutes and the FNRS for their support.

Guy Theraulaz and Eric Bonabeau acknowledge grants from the GIS (Groupement d'Intérêt Scientifique) Sciences de la Cognition and from the Conseil Régional Midi-Pyrénées. Guy Theraulaz wishes to thank Stéphane Blanco, Vincent Fourcassié, Richard Fournier, Bob Jeanne and Jean-Louis Joly for their friendship and many fruitful discussions. Eric Bonabeau wishes to thank the Santa Fe Institute for support through the Interval Research Fellowship.

Scott Camazine would like to thank Bob Jeanne for comments on an early version of the manuscript, and for the helpful feedback during discussions with several of his students. In addition, he would like to thank Tom Seeley for countless hours of advice and stimulating discussions as his Ph.D. advisor and colleague. At Penn State, Camazine thanks Albert Rozo for assistance with many tasks related to preparing the manuscript. I offer special thanks to my wife, Susan Trainor, for technical and emotional support during the writing of this book.

At Princeton University Press, many people participated in bringing this book to fruition. Among them are Sam Elworthy, Brigitte Pelner, Jack Repcheck, Michelle McKenna, Malcolm DeBevoise, and Heidi Sheehan. Bob Bernhard did an excellent job of reading through the entire manuscript, and providing valuable suggestions for improving the readability, grammar and technical aspects of our writing.

We would like to thank Bill Thies for his skills in developing the StarLogo simulations mentioned in the book. Scott Camazine learned a great deal from Bill, who enthusiastically worked on these programs as a high school student, the year before entering M.I.T. Among the artists who contributed illustrations for the book, we thank William Ristine, Mary Ellen Didion, Mark A. Klingler, Anne-Catherine Mailleux and Stephane Portha.

There are undoubtedly many others who we have inadvertently omitted from our acknowledgments. We thank everyone who contributed to this book directly, and to those who provided research, data, and insight into the ideas presented here.

Part I

Introduction to Biological Self-Organization

Figure 1.1 Self-organized pattern of wind-blown ripples on the surface of a sand dune.
(Photo © 1994 Bob Barber/ColorBytes)

1

What Is Self-Organization?

Technological systems become organized by commands
from outside, as when human intentions lead to the building
of structures or machines. But many natural systems
become structured by their own internal processes: these are
the self-organizing systems, and the emergence of order
within them is a complex phenomenon that intrigues
scientists from all disciplines.

> —F. E. Yates et al., *Self-Organizing Systems:*
> *The Emergence of Order*

Self-Organization Defined

Self-organization refers to a broad range of pattern-formation processes in both physical and biological systems, such as sand grains assembling into rippled dunes (Figure 1.1), chemical reactants forming swirling spirals (Figure 1.3a), cells making up highly structured tissues, and fish joining together in schools. A basic feature of these diverse systems is the means by which they acquire their order and structure. In self-organizing systems, pattern formation occurs through interactions internal to the system, without intervention by external directing influences. Haken (1977, p. 191) illustrated this crucial distinction with an example based on human activity: "Consider, for example, a group of workers. We then speak of organization or, more exactly, of organized behavior if each worker acts in a well-defined way on given external orders, i.e., by the boss. We would call the same process as being self-organized if there are no external orders given but the workers work together by some kind of mutual understanding." (Because the "boss" does not contribute directly to the pattern formation, it is considered external to the system that actually builds the pattern.)

Systems lacking self-organization can have order imposed on them in many different ways, not only through instructions from a supervisory leader but also through various directives such as blueprints or recipes, or through pre-existing patterns in the environment (templates).

To express as clearly as possible what we mean by self-organization in the context of pattern formation in biological systems, we provide the following definition: *Self-organization is a process in which pattern at the global level of a system emerges solely from numerous interactions among the lower-level components of the system. Moreover, the rules specifying interactions among the system's components are executed using only local information, without reference to the global pattern.* In short, the pattern is an emergent property of the system, rather than a property imposed on the system by an external ordering influence. Emergent properties will be defined in later chapters, but for now suffice to say that emergent properties are features of a system that arise unexpectedly from interactions among the system's components. An emergent property cannot be understood simply by examining in isolation the properties of the system's components, but requires a consideration of the interactions among the system's components. It is important to point out that system components do not necessarily have to interact directly. As described in Chapter 2, and Figure 2.4, individuals may interact indirectly if the behavior of one individual modifies the environment and thus affects the behavior of other individuals.

Pattern in Group Activities

Critical to understanding our definition of self-organization is the meaning of the term *pattern*. As used here, pattern is a particular, organized arrangement of objects in space or time. Examples of biological pattern include a school of fish, a raiding column of army ants, the synchronous flashing of fireflies, and the complex architecture of a termite mound. Examples of other biological patterns include lichen growth (Figure 1.2a), pigmentation patterns on shells, fish and mammals (Murray 1988, Meinhardt 1995), (Figures 1.2b,c,d) and the ocular dominance stripes in the visual cortex of the macaque monkey brain (Hubel and Wiesel 1977) (Figure 1.2e).

To understand how such patterns are built, it is important to note that in some cases the building blocks are living units — fish, ants, nerve cells, etc. — and in others they are inanimate objects such as bits of dirt and fecal cement that make up the termite mound. In each case, however, a system of living cells or organisms builds a pattern and succeeds in doing so with no external directing influence, such as a template in the environment or directions from a leader. Instead, the system's components interact to produce the pattern, and these interactions are based on local, not global, information. In a school of fish, for instance, each individual bases its behavior on its perception of the position and velocity of its *nearest* neighbors, rather than knowledge of the global behavior of the whole school. Similarly, an army ant within a raiding column bases its activity on *local* concentrations of pheromone laid down by other ants rather than on a global overview of the pattern of the raid.

The literature on nonlinear systems often mentions self-organization, emergent properties, and complexity as well as dissipative structures and chaos

Figure 1.2

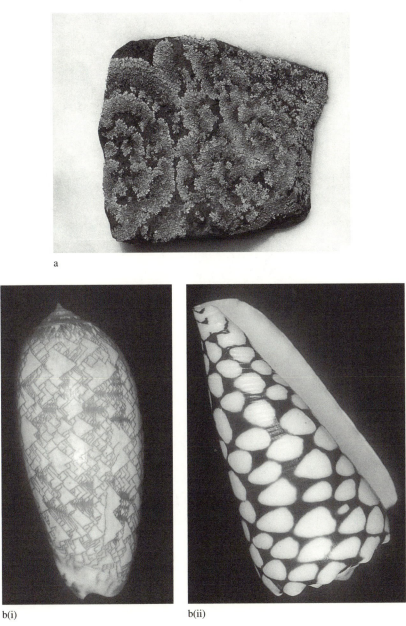

a

b(i)   b(ii)

Figure 1.2 Self-organized patterns in biological systems include: (a) lichen growth; (b) pigmentation of a porphyry olive shell (*Olivia porphyria*) (i) and a marble cone shell (*Conus marmoreus*) (ii); (*Figure 1.2 continued next page*)

Figure 1.2 continued

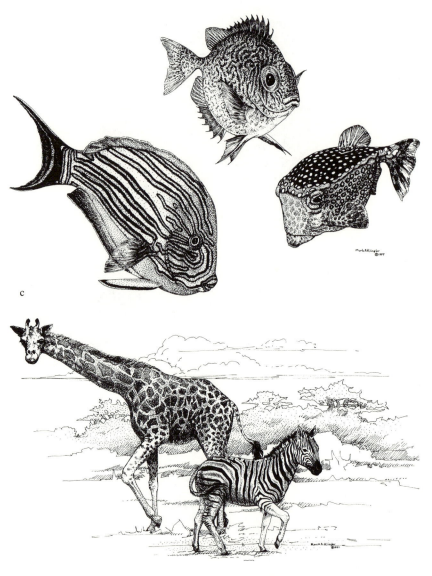

c

d

(c) skin pigmentation on fish (clockwise from top—vermiculated rabbitfish (*Siganus vermiculatus*), male boxfish (*Ostracion solorensis*), and surgeonfish (*Acanthurus lineatus*)); (d) zebra and giraffe coat patterns. (*Figure 1.2 continued next page*)

Figure 1.2 continued

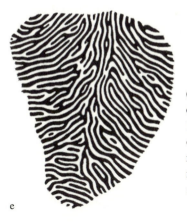

(e) ocular dominance stripes in the visual cortex of the macaque monkey (from Hubel and Wiesel 1977). Cortical regions receiving inputs from one of the monkey's eyes are shown in black while regions receiving inputs from the other eye are represented by white regions between the black stripes.

e

(Prigogine and Glansdorf 1971; Nicolis and Prigogine 1989). The terms *chaos* and *dissipative structures* have precise scientific meanings that may differ from popularized definitions, so it is important to discuss these terms at this point. To begin with, the term *complex* is a relative one. Individual organisms may use relatively simple behavioral rules to generate structures and patterns at the collective level that are relatively more complex than the components and processes from which they emerge. As discussed in Chapter 6 (see Box 1), systems are complex not because they involve many behavioral rules and large numbers of different components but because of the nature of the system's global response. Complexity and complex systems, on the other hand, generally refer to a system of interacting units that displays global properties not present at the lower level. These systems may show diverse responses that are often sensitively dependent on both the initial state of the system and nonlinear interactions among its components. Since these nonlinear interactions involve amplification or cooperativity, complex behaviors may emerge even though the system components may be similar and follow simple rules.

Complexity in a system does not require complicated components or numerous complicated rules of interaction.

Self-Organization in Biology

The concept of self-organization in biological systems can be conveyed through counterexamples. A marching band forming immense letters on a football field provides one such example. Here the band's members are guided in their behavior by a set of externally imposed instructions for the movements of each band member that specify in fine detail the final configuration of the whole band. A particular member of the band may know that the instructions are to march to the 50-yard line, turn left 90 degrees and march 10 paces. To the extent that the band member follows this recipe for contributing to the pattern and ignores local information, such as position relative to neighbors, this pattern formation would not be considered self-organized.

Similarly, a team of carpenters building a house is a pattern-formation process that functions without self-organization. Here members of the construction crew are guided in their collective behavior by predetermined externally imposed instructions expressed as blueprints, that precisely specify the final structure of the house. Letter formation by a marching band and house construction by a construction crew both involve pattern building in space.

Let us also consider two counterexamples to self-organization that involve pattern building over time. One such example is oarsmen in a rowing team pulling on their oars in perfect synchrony with one another and with appropriate adjustments of their stroke frequency. This pattern arises when each oarsman responds to the coxswain's shouted instructions indicating when to begin each stroke. Clearly, this is an example of a group generating a pattern by following explicit orders from a leader based on the overall state of the group members. The rhythmic contractions of muscle fibers in the heart are also a counterexample to self-organization. Here the pattern arises as the component building blocks (the muscle fibers), follow instructions from special excitable cells that act as an external pacemaker and send a rhythmic electrical signal to the fibers.[1]

We can easily see how a system can form a precise pattern if it receives instructions from outside—such as a blueprint, recipe, or signals from a pacemaker—but it is less obvious how a definite pattern can be produced in the absence of such instructions. A general answer to this puzzle is provided in the next chapter, while specific answers for particular biological patterns constitute the main body of this book. For now, it is merely asserted that pattern formation often is achieved by systems without external guidance.

The mechanisms of self-organization in biological systems differ from those in physical systems in two basic ways. The first is the greater complexity of the subunits in biological systems. The interacting subunits in physical systems are inanimate objects such as grains of sand or chemical reactants. In biological systems there is greater inherent complexity when the subunits are living organisms such as fish or ants or neurons.

The second difference concerns the nature of the rules governing interactions among system components. In chemical and physical systems, pattern is created through interactions based solely on physical laws. For example, heat applied evenly to the bottom of a tray filled with a thin sheet of viscous oil transforms the smooth surface of the oil into an array of hexagonal cells of moving fluid called *Bénard convection cells* (Figure 1.3) (Velarde and Nor-

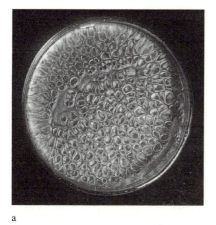 

a   b

Figure 1.3 Further examples of self-organized patterns in physical and chemical systems: (a) hexagonal Bénard convection cells created when a thin sheet of viscous oil is heated uniformly from below. Aluminum powder was added to the oil to show the convection pattern; and (b) spiral patterns produced by the Belousov-Zhabotinski reaction. The chemistry of the reaction is explained by Winfree (1972, 1984). (Image courtesy of Stefan C. Müller)

mand 1980). The molecules of oil obey physical laws related to surface tension, viscosity, and other forces governing the motion of molecules in a heated fluid. Likewise, when wind blows over a uniform expanse of sand a pattern of regularly spaced ridges is formed (Figure 1.1) through a set of forces attributable to gravity and wind acting on the sand particles (Anderson 1990; Forrest and Haff 1992).

Of course, biological systems obey the laws of physics, but in addition to these laws the physiological and behavioral interactions among the living components are influenced by the genetically controlled properties of the components. In particular, the subunits in biological systems acquire information about the local properties of the system and *behave* according to particular genetic programs that have been subjected to natural selection. This adds an extra dimension to self-organization in biological systems, because in these systems selection can finely tune the rules of interaction. By tuning the rules, selection shapes the patterns that are formed and thus the products of group activity can be adaptive. What is also intriguing about pattern formation in biological systems and lends excitement to studies of self-organization in animal groups is the recent realization that interactions among system components can be surprisingly simple, even when extremely sophisticated patterns are built, such as the labyrinthine nests of termites, the spatial patterns of army ant raids, and the coordinated movements of fish in a school.

Figure 2.1 The upper portion of the figure is a top view of the polygonal pattern of male *Tilapia mossambica* nest territories. Each territory is a pit dug in the sandy bottom of the water. The rims of the pits form the boundaries of the territories and create a pattern of polygons that results from a combination of positive feedback and negative feedback (from Barlow 1974). Similarly shaped nesting territories of male bluegills are seen in the lower portion of the figure. Each colonial male defends a territory bordered by the nest sites of other males. Predators, such as bass (above), bullhead catfish (left), and pumpkinseed sunfish (right foreground) roam the colony in search of eggs. (Drawing © Matt Gross; used with permission)

2

How Self-Organization Works

Positive feedback isn't always negative.
                              —M. Resnick, *Learning about Life*

Self-organizing systems typically are comprised of a large number of frequently similar components or events. The principal challenge is to understand how the components interact to produce a complex pattern. The best approach is to first understand the two basic modes of interaction among the components of self-organizing systems: positive and negative feedback. The next step is to discuss information, since each component or individual must acquire and process information to determine its actions.

Feedback: Positive and Negative

Most self-organizing systems use positive feedback. This may be surprising since most biologists probably are more familiar with negative feedback, a mechanism commonly used to stabilize physiological processes (homeostasis) and avoid undesirable fluctuations. We are probably all familiar with the regulation of blood sugar levels, a process that proceeds smoothly in most people but functions abnormally in diabetics. Blood sugar levels are regulated by a negative feedback mechanism involving the release of insulin (Figure 2.2a) (Mountcastle 1974). An increase in blood glucose following ingestion of a sugary meal quickly triggers the release of insulin from the pancreas. Insulin has a number of physiological effects including the conversion of glucose to glycogen, an energy storage compound deposited in the liver. This negative feedback mechanism, counteracts increases in blood-sugar level. In this case, we see negative feedback acting to maintain the status quo by damping large fluctuations in blood glucose level. In diabetics, however, a failure of adequate insulin secretion results in elevated blood sugar levels.

A similar example involves the homeostatic regulation of body temperature. In warm-blooded mammals, internal body temperature is monitored by sensitive temperature receptors in the hypothalamic area of the brain (Brooks and Koizumi 1974). Arterial blood from throughout the body is monitored in the hypothalamus and if blood temperature is within a narrow range around the thermal setpoint, the organism feels comfortable. If, however, the organism

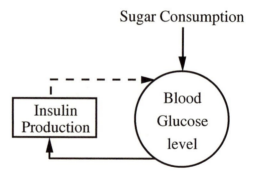

a

Sugar Consumption

Insulin Production

Blood Glucose level

b

Drop in ambient temperature

Bundling up or shivering

Body Temperature

Figure 2.2 Examples of negative feedback regulation that compensates for a perturbation of the system and results in homeostasis: (a) a rise in blood glucose level (solid line) following the ingestion of food triggers the release of insulin and results in a drop (dotted line) in the glucose level; and (b) a drop in body temperature (dotted line) caused by exposure to cold weather, may cause a person to shiver or dress more warmly and experience a rise (solid line) in body temperature.

experiences a thermal stress, as would arise if a person stepped outdoors on a winter day without enough clothing (Figure 2.2b), then a discrepancy arises between the organism's actual body temperature and the thermal setpoint. The organism feels cold. The discrepancy triggers various behavioral and physiological responses, such as putting on a warm coat or shivering, that counteract the drop in body temperature.

In both instances, the individual acquires and processes information that elicits a negative feedback response: A small perturbation applied to the system triggers an opposing response that counteracts the perturbation. In the first case an *increase* in blood glucose triggers a compensating response leading to a *decrease* in blood glucose. In the second case, a *decrease* in body temperature results in responses that *increase* body temperature. In these classical examples of regulatory systems, a single negative feedback loop performs the important role of counteracting changes imposed on a system.

Positive Feedback and the Creation of Pattern

In contrast to negative feedback, positive feedback generally promotes changes in a system. The explosive growth of the human population provides a familiar example of the effects of positive feedback (Figure 2.3). For the past several centuries, each generation has more than reproduced itself, so more births occur with each successive generation, which further increases the population and results in still more births and a yet greater population. The snowballing effect of positive feedback takes an initial change in a system and reinforces that change in the *same* direction as the initial deviation. Self-enhancement, amplification, facilitation, and autocatalysis are all terms used to describe positive feedback. The growth of the human population will also eventually be stabilized by negative feedback in the form of a lower birth rate (or a higher death rate) when the population becomes extremely large (Figure 2.3).

Consider another example of positive feedback—the clustering or aggregation of individuals. Many birds, such as seagulls (Kruuk 1964), herons (Krebs 1974), and blackbirds (Horn 1968) nest in large colonies. Group nesting evidently provides individuals with certain benefits, such as better detection of

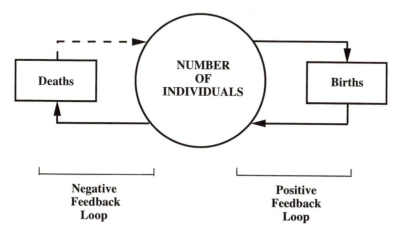

Figure 2.3 A simple model of population growth may involve a positive feedback loop of increased births and a negative feedback loop of increased deaths.

predators or greater ease in finding food. The mechanism by which colonial nesting arises is apparently that birds preparing to nest are attracted to sites where other birds are already nesting. This imitative behavior is a positive feedback process in which one individual follows the behavioral rule, "I nest close where you nest." As more birds nest at a particular site, the attractive stimulus becomes greater and the probability that newly arriving birds join the colony becomes greater, leading to an even larger aggregation. The key point is that the aggregation of nesting birds at a particular site is *not* purely a consequence of each bird being attracted to the site per se. Rather, the aggregation evidently arises primarily because each bird is attracted to *others*, hence because of interactions between individual birds in the system. In an environment with numerous, equally suitable nesting sites such as an array of islands, the birds aggregate on only a few of the many possible sites (see, for example, Krebs 1974).

Throughout this book, various forms of positive feedback are encountered that play a major role in building group activity. Fireflies flashing in synchrony follow the rule, "I signal when you signal," fish traveling in schools abide by the rule, "I go where you go," and so forth. In humans, the "infectious" quality of a yawn or laughter (Provine 1986, 1996) is a familiar example of positive feedback of the form, "I do what you do." Seeing a person yawning, or even just thinking of yawning, can trigger a yawn.

Positive Feedback and the Amplification of Fluctuations

Consider another example of colonial nesting. Male bluegill sunfish (*Lepomis macrochirus*) nest in colonies of up to 150 individuals (Figure 2.1b). This behavior presumably evolved as a defense against brood predators (Gross and MacMillan 1981), with each individual a member of a "selfish herd" (Hamilton 1971). By joining the group, and so surrounding its nest with the nests of others, each fish reduces the exposure of its brood to predators. But what is the mechanism that led to these aggregations? It may simply be that each male bluegill follows the behavioral rule, "I nest where others nest."

Now envision how the nesting pattern will appear in a large lake with a perfectly homogeneous bottom, hence an abundance of identical sites available for nesting. First, consider a scenario where the density of bluegills is very low. For lack of behavioral interactions among the males, the bluegills may end up nesting far apart over the lake bottom. A male may not find another nesting male in his vicinity, or even if it does find another male, it may be that a single adjacent male may not provide sufficient stimulus to nest nearby. Positive feedback under these conditions is insufficient to initiate the aggregation of nest sites, so a stable state will be reached in which nesting sites are randomly distributed. Next consider a scenario with a higher density of bluegills. Through a random process, several nesting sites occasionally will be close enough to provide a sufficiently strong attraction to stimulate other bluegills to nest nearby, and form an aggregation nucleus. The random pattern of nest sites will now be unstable and a cluster of nest sites will grow. In the terminology of nonlinear dynamics, a stationary state became unstable through the amplification of random fluctuations. At a critical density of bluegills, a pattern arises within the system—a homogeneous, random array of nesting sites becomes a cluster. In fact, the nucleation of a cluster of nest sites in one area can lead to a self-organized regular spacing of clusters throughout the lake bottom.

Keeping Positive Feedback Under Control

The amplifying nature of positive feedback means that it has the potential to produce destructive explosions or implosions in any process where it plays a role. How can such snowballing be kept under control? This is where negative feedback plays a critical role, providing inhibition to offset the amplification and helping to shape it into a particular pattern. Given that male bluegills try to nest near one another, does one not find overcrowded colonies of fish? Although the details are unknown, it seems certain that negative feedback is involved. The fish have some limits in their behavioral tendency to nest where others nest. If too many male fish congregate within a given area, additional fish may be inhibited from nesting in this same area. Thus the behavioral rule may be more complicated than initially suggested, possessing both an autocatalytic as well as an antagonistic aspect: "I nest where others nest, *unless the area is overcrowded*." In this case both the positive and negative feedback may be coded into the behavioral rules of the fish.

In other cases one finds that the inhibition arises automatically, often simply from physical constraints. For instance, in a lake that contains only a small number of bluegills the buildup of males at a site is self-limiting. In this situation there is no need for the fish to employ a mechanism of negative feedback to avoid overcrowding because once all the fish have clustered in an area, the positive feedback ceases on its own. Exhaustion or consumption of the building blocks is often an important mechanism for limiting positive feedback.

There is nothing deeply thought-provoking about negative feedback that simply forces a process to a complete stop. But in the case of the bluegills (and other fish such as *Tilapia*) that assemble into a colony, negative feedback actually shapes the process and creates a striking pattern in the spatial array of nests. Through longer-range interactions, the fish aggregate in a positive feedback manner, resulting in colonial nesting. But in their short-range interactions the fish are influenced more by negative feedback: "Keep away! Don't nest where I am nesting." This occurs as each male fish builds a nest consisting of a depression in the sandy lake bottom, and diligently defends the nest area from intrusions by neighboring males. As a result of the interplay between the opposing tendencies to squeeze together yet maintain a personal territory, the breeding ground becomes a beautiful closely packed, polygonal array of nests (Figure 2.1a). It is likely, however, that a polygonal pattern itself serves no function and has no adaptive significance. Instead, the regular geometric spacing of nests probably is an epiphenomenon, an incidental consequence of each individual striving to be close, but not too close, to a neighbor. Mechanistically, it arises automatically through a self-organizing process similar to the hexagonal close-packing of round marbles placed in a dish.

A second example neatly illustrates the interplay between positive feedback, which is behaviorally coded, and negative feedback, which arises merely as a physical constraint. Consider a child making sand castles at the beach. Suppose the child wants to build a tall spire using dry sand. Naively, he employs a positive feedback rule: "Add more sand to where the sand pile is tallest." The pile starts out flat and initially gets steeper and steeper as it grows in size, but to the child's frustration he can never build the very steep tall tower he wants. Once the slope of the pile has reached a certain critical angle, the addition of more sand triggers a series of avalanches that brings the pile's profile back to the same angle (Bak et al. 1987; Jaeger and Nagel 1992). In this example, the positive feedback arises from the child's behavioral rule—"Add sand to sand."—but negative feedback arises automatically from the physical constraints of gravity and friction between the sand particles. On a particular beach with sand grains of a particular size, shape, and wetness, no matter how hard the child tries the slope of the pile always returns to the same self-organized angle of repose.

Self-enhancing positive feedback coupled with antagonistic negative feedback provides a powerful mechanism for creating structure and pattern in many physical and biological systems involving large numbers of components: the regular spacing of ridges and furrows in the surface of a sand dune (Figure 1.1), the pattern of ocular dominance stripes in the visual cortex of the brain (Figure 1.2e) or of zebra stripes (Figure 1.2d), and the skin pigmentation patterns of fish (Figure 1.2c) (Anderson 1990; DeAngelis et al. 1986; Forrest and Haff 1992; Meinhardt 1982; Miller et al. 1989; Murray 1981, 1988; Swindale 1980).

How Organisms Acquire and Act upon Information

The defining characteristic of self-organizing systems is that their organization arises entirely from multiple interactions among their components. In the case of animal groups, these internal interactions typically involve information transfers between individuals. Biologists have recently recognized that information can flow within groups via two distinct pathways—signals and cues (Lloyd 1983; Seeley 1989b). Signals are stimuli shaped by natural selection specifically to convey information, whereas cues are stimuli that convey information only incidentally. The distinction between signals and cues is illustrated by the difference between ant and deer trails. The chemical trail deposited by ants as they return from a desirable food source is a signal. Over evolutionary time such trails have been molded by natural selection for the purpose of sharing with nestmates information about the location of rich food sources. In contrast, the rutted trail made by deer walking through the woods is a cue. Almost certainly, deer trails have not been shaped by natural selection for communication among deer but are a simple by-product of animals walking along the same path. Nonetheless, these trails may provide useful information to deer.

Interactions within self-organized systems are based on both signals and cues. But whereas information transfer via signals tends to be conspicuous, since natural selection has shaped signals to be strong and effective displays, information transfer via cues is often more subtle and based on incidental stimuli in an organism's social environment (Seeley 1989b). The lack of prominence of cues means that many interactions within animal groups are easily overlooked, a fact that contributes to the seemingly mysterious origins of the emergent properties of self-organized groups.

Information Gathered from One's Neighbors

Chapter 11 discusses in some detail how fish coordinate their movements as they travel in a school. The following example of self-organization is briefly described here to illustrate that sometimes the most important information comes directly from an individual's neighbors, often its nearest neighbors. In coordinating their movements in a school, fish use both positive and negative feedback mechanisms (Huth and Wissel 1992; Partridge 1982; Partridge and Pitcher 1980; Partridge et al. 1980; Pitcher et al. 1976). Positive feedback operates as it does in the nesting colonies of seabirds; individuals are attracted to the presence of other individuals, resulting in the fish assembling into schools. Negative feedback functions in spacing fish within the school. A schooling fish that gets too close to a neighbor moves away to avoid a collision. It is as if each member were connected to its neighbor by a rubber band that pulls the individual toward a neighbor that gets too far away, but is also connected by a spring that pushes the individual away from neighbors that are too close. Here again, the behavioral rule, "Be most attracted to the largest group of fish," provides positive feedback to create a cluster of individuals. At close range, negative

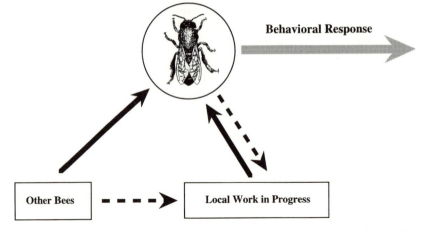

Figure 2.4 Information within an animal group may flow from one individual to another directly (generally via communication signals) or indirectly (generally via cues arising from work in process). Indirect flow is particularly important during the joint building of nests, for here individuals can efficiently coordinate their activities through information embodied in the structure of the partially completed nest. In the figure, the solid arrows show information from the environment or other individuals to the focal individual. The dotted arrows correspond to behavioral actions that may modify the environment and lead to a particular behavioral response by the focal individual.

feedback, "If too close, move away," imposes shape and pattern on the cluster.

Exactly how does each fish maintain the preferred spacing? Studies have shown that a school has no leader (Partridge 1982; Partridge et al. 1980). Indeed, in schools containing thousands of fish it is inconceivable either that one supervisory individual could monitor everybody's position and broadcast the moment-by-moment instructions needed to maintain the school's spatial structure, or that individual fish within the school could monitor the movements of the leader and follow accordingly. Coherence is achieved, instead, by each fish gathering information only about its nearest neighbors and responding accordingly. Many examples presented in this book show how individuals acquire only limited information, so that each individual's perception of group activity is myopic, not at all synoptic.

Such limited information acquisition presumably reflects the tremendous difficulty of acquiring a more complete knowledge of the group. Each fish traveling in a school, for instance, must constantly adjust its speed and direction of travel. Fish do not have time to gather more than momentary impressions of the movement patterns of their nearest neighbors before they must act. Any delay in gathering, processing, and acting upon information will result in the fish bumping into others or being left behind. A fish must also contend with the sheer physical constraint of being surrounded by nearby fish, which surely blocks most stimuli from distant members of the school.

Members of a self-organized group often rely on simple behavioral rules of thumb to guide their actions (Krebs and Davies 1984; Stephens and Krebs 1986). The reason is that it is usually difficult, if not impossible, for an organism to obtain complete global information in a reasonable amount of time. These rules necessarily are often based upon local (hence incomplete) information, but this is generally sufficient. A member of a fish school does not need to know the long-range direction taken by the school or even the precise trajectories of all or any of its neighbors. It needs only apply a few simple rules of thumb, such as these: Approach neighbors if neighbors are too far away; Avoid collisions with nearby fish; If the first two rules have been obeyed and neighbors are at the "preferred" distance, then continue to move in the same direction.

The concept that individuals can function effectively with information acquired by monitoring just their nearest neighbors applies not only to fish schools, but to many self-organized systems. Another example of this is the coordinated rhythm of flashing fireflies that is discussed in Chapter 10. In this case each firefly primarily detects the flashes of immediately neighboring fireflies to adjust the timing of its own flash.

Information Gathered from Work in Progress (Stigmergy)

Information acquired directly from other individuals is only one source of information used by organisms in self-organizing systems. In situations where many individuals contribute to a collective effort, such as a colony of termites building a nest, stimuli provided by the emerging structure itself can be a rich source of information for the individual (Figure 2.4). In other words, information from the local environment and work-in-progress can guide further activity. As a structure such as a termite mound develops, the state of the building process continually provide new information for the builders.

In the study of social insects, the term *stigmergy* (Grassé 1959, 1967) has been used to describe such recursive building activity (See Chapter 4 for a more detailed discussion). "In stigmergic labor it is the product of work previously accomplished, rather than direct communication among nestmates, that induces the insects to perform additional labor" (Wilson 1971, p. 229).

There is good reason for many students of social insects to emphasize that workers rely on information derived from the environment rather than from fellow workers. Large colonies of social insects pose perplexing puzzles when viewed anthropomorphically. Maurice Maeterlinck (1927, p. 137), speaking of termites, wrote, "What is it that governs here? What is it that issues orders, foresees the future, elaborates plans and preserves equilibrium?" The mystery is how thousands of termites coordinate their activities during the building of a nest, a mound many thousands of times larger than a single individual. It seems certain that no termite possesses knowledge of the ultimate form of the structure, much less maintains an overview of the emerging structure as it takes shape. Furthermore, the duration of the building process spans several termite lifetimes, making it all the more mysterious how the work can progress smoothly over time.

For the reasons just mentioned, it seems clear that coordinated building activity does not depend on supervisor termites monitoring the construction progress and issuing instructions. The more likely explanation is a process of decentralized coordination based on stigmergic activity where individuals respond to stimuli provided through the common medium of the emerging nest. Instead of coordination through direct communication among nestmates, each individual can adjust its building behavior to fit with that of its nestmates through the medium of the work in progress. Each termite can communicate indirectly with its nestmates (or with itself) across both space and time by means of the small changes each one makes in the shared nest structure. More will be said about stigmergy in the next section, and in Chapter 4.

Positive Feedback, Stigmergy, and the Amplification of Fluctuations

During the initial stages of stigmergic activity, random fluctuations and chance heterogeneities may arise and become amplified by positive feedback to create the required structures. For example, consider the building activity of the termites described by Grassé (1959). When removed from their nest and introduced into a novel environment, such as a petri dish lined with a thin layer of soil, the termites will construct a series of pillars and arches made from pellets of earth and excrement (Figures 18.1 and 18.2). Grassé noticed a distinct period during the building process that he called "*la phase d'incoordination*," in which the workers randomly deposit pellets on the substrate. At this point they appear to work essentially independently and incoherently, with total indifference to the activities of their nestmates. In particular, the work appears incoherent because many individuals remove pellets of material from a particular location while others just as excitedly put them back in the same location.

Moreover, the same individual appears undecided about whether to build up or tear down its own work.

Eventually, however, small fluctuations in the deposition of pellets on the ground can generate a critical density of pellets in places, and the appearance of these incipient pellet piles induces an abrupt change in the building behavior of the termites. What was initially vague and diffuse activity suddenly becomes transformed into behavior infinitely more precise. But how exactly did this precipitous transition to *la phase de coordination* occur?

In the initial uncoordinated stage of activity, the relatively homogeneous substrate fails to provide a sufficient stimulus for deposition of material. Workers are stimulated both to deposit and remove pellets of earth and excrement. Therefore, nothing much gets done. Over time, however, and merely by chance heterogeneities in the substrate arise. A pellet placed haphazardly atop another pellet creates an inhomogeneity in the surface that provides an attractive stimulus for the deposition of yet another pellet. Once a critical density of pellets rises from the initial featureless plain, a snowballing effect takes control and the coordinated phase of activity ensues with many workers all building in the same place. Positive feedback serves to raise a pillar suddenly into the air. But even during the coordinated stage of activity, it is unlikely that the workers' actions are coordinated by means of direct communication between individuals. Rather, they are more likely coordinated through stigmergy creating positive feedback, such that many workers are attracted to build in the same location.

As discussed in greater detail in Chapter 18, the stigmergic mechanism may also account for an orderly spacing of pillars in the termite mound. Positive feedback acts over the short range to stimulate the deposition of more pellets, but outside this zone of attraction negative feedback evidently operates. Around each growing column the supply of earth is rapidly consumed as workers excavate material to add to the pillar. In so doing, they inhibit the initiation of nearby building activity. Overall, the building process starts with a homogeneous plain, then becomes focussed through a process of random fluctuations and nucleation, and then proceeds through an orderly array of positive and negative feedback zones. The sequence is entirely analogous to the positive and negative feedback that created the polygonal array of nests for the bluegill sunfish.

It is useful to consider the mechanistic origins of the positive and negative feedback in the termites' nest construction. The positive feedback process—expressed as the rule, "Build where there is already some building"—presumably is based on the termites' innate behavioral repertoire encoded in the termites' genes. However, there is no need to genetically encode the inhibitory feedback process that brakes and shapes the building. Instead, this inhibition is supplied by depletion of the number of termites in the vicinity.

There is no need to explicitly code for the behavioral rule, "Don't build one pillar near another." The diameter of the pillars and their spacing is probably determined through an interplay of many factors, including: strength of the positive feedback stimuli, which may include attraction pheromones added to structures as they are built, the amount and consistency of the soil, and the number and density of termites engaged in building. How column-building activity might progress to the eventual creation of precise and complex galleries and passageways will be discussed in greater detail in Chapter 18.

Summary

The preceding examples indicate that positive feedback is a powerful mechanism for building structure in biological systems. Without an antagonizing inhibitory mechanism, however, the process may become uncontrollable. Negative feedback brakes and shapes what could otherwise become an amorphous, overgrown structure. Positive and negative feedback mechanisms are set into motion when an individual acquires and acts on information gathered from other individuals, work in progress, or the initial state of the environment.

Where large numbers of individuals act simultaneously, a system can suddenly break out of an amorphous state and begin to exhibit order and pattern. All that is required sometimes for this transition is the implementation of a few simple rules based on positive feedback. Relatively little needs to be coded at the behavioral level and the information required for action by the individual is often local rather than global. In place of explicitly coding for a pattern by means of a blueprint or recipe, self-organized pattern-formation relies on positive feedback, negative feedback, and a dynamic system involving large numbers of actions and interactions.

With such self-organization, environmental randomness can act as the "imagination of the system," the raw material from which structures arise. Fluctuations can act as seeds from which patterns and structures are nucleated and grow. The precise patterns that emerge are often the result of negative feedback provided by these random features of environment and the physical constraints they impose, not by behaviors explicitly coded within the individual's genome.

Box 2.1 Negative Feedback, Positive Feedback, and the Amplification of Fluctuations

Consider two model systems consisting of interacting subunits, in one case positively charged particles and in another a group of male sunfish. Such a system yields different behaviors and spatial patterns of the subunits depending on whether the interactions between subunits include negative or positive feedback. In the case where all the particles are positively charged, the particles repel one another. When the particles are initially forced into one end of the chamber the interactions among subunits are governed by negative feedback; the more particles in a region, the greater the repulsive force on a particle. Such a system actively opposes any unequal distribution of particles in a region. In the second case, in which the subunits are male sunfish governed by a behavioral positive feedback, the greater the number of fish in a region the more attractive that region is for nesting. Here, even if the system initially starts out with a uniform distribution of fish, through fluctuations an instability may be reached in which one region experiences a moment in which it has significantly more fish than other regions, thus, permitting a symmetry-breaking amplification to occur. As the difference in numbers of fish between regions increases, the positive feedback becomes even stronger, allowing one region to capture all the fish. As discussed in the next chapter, whether all the subunits in a system cluster in one location will depend on the strength of the positive feedback as well as the initial density of system particles. (See the program Particles to explore systems of particles governed by negative or positive feedback. It can be downloaded at the following website address: http://beelab.cas.psu.edu.)

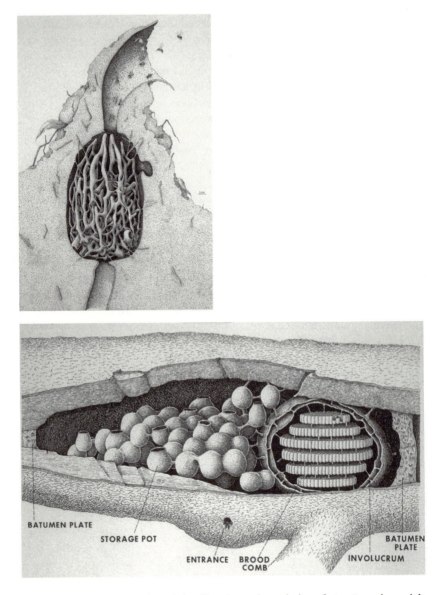

BATUMEN PLATE

STORAGE POT

ENTRANCE   BROOD COMB

BATUMEN PLATE

INVOLUCRUM

Figure 3.1 Typical examples of the diversity and regularity of structures in social-insect nests are the internal nest structure of the stingless bees. *Trigona testacea*, with its funnel-shaped entrance (top), and *Melipona interrupta grandis* in a hollow branch. The anastomosing rods (top) presumably provide resting space for the colony's defense force. Self-organization probably plays a role, at least partially, in the building of the nests. (Original drawings, courtesy J. M. F. de Camargo, from Michener 1974)

3

Characteristics of Self-Organizing Systems

Simple and complex systems exhibit... the spontaneous
emergence of order, the occurrence of self-organization.
—S. A. Kauffman, *The Origins of Order:*
Self-Organization and Selection in Evolution

We have defined self-organization and briefly discussed how it works. Now, we will describe some of the characteristics of self-organizing systems. What general features do these systems possess?

Self-Organizing Systems Are Dynamic

The multiplicity of interactions that characterizes self-organizing systems emphasizes that such systems are dynamic and require continual interactions among lower-level components to produce and maintain structure. This point is made more clearly by contrasting a dynamic process of pattern formation with an alternative, essentially static process illustrated by the assembly of a jigsaw puzzle. A jigsaw puzzle is a global structure with an intricate pattern constructed from lower-level subunits, the pieces of the puzzle. The pieces are put together in a precise manner to create a pattern. Each piece of the puzzle has a particular shape and set of markings that complements the shape and markings of the pieces to which it fits. To create the global pattern, one carefully matches the pieces together, and once the pieces of the puzzle are fit together the action stops. The pattern and structure are locked into place.

Edelman (1984, p. 120) provides a lovely metaphor for such a static mechanism of pattern formation in his discussion of the role of cell-adhesion molecules in regulating cell movements and morphogenetic processes during embryological development:

> There are two alternative ways patterns might be formed at the cellular level without the direct intervention of some kind of "little architect" or "construction demon." The first way would require prelabeling all cells with molecular markers (presumably proteins), each one spatially complementary to some other marker on a cell to be placed next to it in the pattern. This is essentially how parts of the great offshore abbey of Mont-Saint-Michel were built. Stones were cut and shaped on the mainland,

marked by their makers and reassembled on the island according to a plan. The Mont-Saint-Michel model is a metaphor for various "chemoaffinity" theories of cell adhesion. The major difficulty with such theories is that if the pattern to be formed is complex, has much variation in shape or has many elements and much local detail (as, for example, in the brain), then the number of specific surface markers determining each cell's location must be enormous. Inasmuch as such markers are most likely to be specific proteins, each encoded by a different gene, the number of genes would be correspondingly large.... Moreover, a pattern made this way is prefigured and essentially static: once the right markers come together, no further dynamism is necessary.

Several alternative, more dynamic ways of generating patterns have been described. Turing (1952), in what has been called "one of the most important papers in theoretical biology" (Murray 1988, p. 80), postulated a mechanism of generating animal coat patterns based on reacting and diffusing chemicals that he called morphogens (see Figure 1.2). The "Brussels school" extended these ideas of self-organized pattern formation to a variety of chemical and biochemical systems (Nicolis and Prigogine 1977, 1989). In these open systems, in which there is a continual influx of energy or matter, reactions occur far from chemical equilibrium, and structures emerge through interactions obeying nonlinear kinetics. Such structures are called *dissipative*. At about the same time, Haken (1978) introduced the concept of synergetics as a unifying approach to pattern formation in various disciplines. (See also reviews by Levin and Segel 1985 and Schöner and Kelso 1988.) Without going into a technical discussion of the similarities and differences among these different explanations of self-organized pattern formation, we again refer to Edelman (1984, p. 120) who provides a useful visual metaphor of this process:

There is an alternative and more dynamic way of generating patterns, akin to what might be observed in a mountain stream. In this kinetic, far-from-equilibrium situation, pattern results from the play of energy as it is dissipated into the environment against various constraints. To make the simplest case for this mountain-stream example, imagine a stream of water running down a mountainside and striking a submerged boulder whose temperature is below freezing. At first the flow of water will be influenced only slightly by the boulder and the stream will remain a single stream. In time, however, as water freezes onto the boulder, the enlarging structure may suddenly become a barrier causing the stream to split into two and assume a new shape as it runs down the mountain. All subsequent shapings of the stream will be influenced by the effect of the original freezing. Rivulets downstream may break into a variety of new and intricate patterns as they meet different constraints at lower levels. Seen from above, the entire stream will nonetheless have a definite shape.

As described in the next section, this dynamic process of pattern formation gives rise to emergent properties, an example of which is the sudden bifurcation of the stream as it courses down the mountain.

Self-Organizing Systems Exhibit Emergent Properties

In the first chapter self-organizing systems were shown to possess emergent properties. Emergence refers to a process by which a system of interacting subunits acquires qualitatively new properties that cannot be understood as the simple addition of their individual contributions. Since these system-level properties arise unexpectedly from nonlinear interactions among a system's components, the term *emergent property* may suggest to some a mysterious property that materializes magically. To dispel this notion, two examples are given that illustrate an emergent property of the system. The first is the biological phenomenon of clustering by larvae of the bark beetle, *Dendroctonus micans*. (We will consider this example in greater detail in Chapter 9.) The second is a physical phenomenon—Bénard convection—which was mentioned earlier (Figure 1.3a) but will be discussed again in more detail.

The eggs of *Dendroctonus* beetles are laid in batches beneath the bark of spruce trees. Larvae hatch from the eggs and feed as a group, side by side, on the phloem tissues just inside the tree bark (Deneubourg et al. 1990a). Previous studies have shown that the larvae emit an attractive pheromone (Grégoire et al. 1982). In a series of experiments (described in Chapter 9), the larvae were randomly placed on a circular sheet of filter paper 24 cm in diameter between two glass plates separated by 3 mm to allow the larvae free movement. The subsequent positions of the larvae were observed over time. The degree of clustering exhibited by the larvae was found to depend strongly on the initial larval density. At low density (0.04 larvae/cm^2), a loose cluster appeared, but it did so only slowly, in approximately 1 hour, and comprised only 25 percent of the population (Figure 9.4). In contrast, at high density (0.17 larva/cm^2) a single, tight cluster rapidly assembled (Figure 9.3). Within 5 min about 50% of the larvae were clustered in the arena's center and after 20 min some 90 percent of the larvae joined this cluster. The experiments demonstrated a simple emergent property—a cluster—in a group where the individuals initially were homogeneously distributed. At a certain density of larvae, the system spontaneously organizes itself.[1]

An even more dramatic example of spontaneous emergence of pattern is the well-known phenomenon of Bénard convection cells described in Box 3.1. Here an initially homogeneous layer of fluid becomes organized into a regular array of hexagonal cells of moving fluid (Figure 1.3a). The striking pattern of convection does not appear gradually but arises suddenly. At a certain moment determined by the amount of heat applied to the bottom of the fluid layer, the initially homogeneous regime becomes unstable and changes to a new pattern.

In the terminology of dynamic systems, this emergent pattern or property is called an *attractor* of the system. Under a particular set of initial conditions and parameter values, an attractor is the state toward which the system converges over time. In the Bénard convection system, one attractor (seen under conditions of a small temperature gradient) is the random motion of the fluid molecules. A different attractor appears when the temperature gradient is increased to a critical value.

The mere detection of a pattern in nature, however, is inadequate to distinguish self-organizing mechanisms from other mechanisms of pattern formation. Observing a pattern at a moment or even over a period of time does not enable one to identify the mechanism of that pattern's formation. One must understand the pattern-formation machinery inside the system and be able to observe its operation to know whether the pattern is self-organized. Most importantly, one needs to devise means of experimentally perturbing the pattern-formation system and to obtain evidence that supports models based on self-organization as opposed to other models based on greater degrees of centralized or external control.

Compare an aggregation of *Dendroctonus* larvae to a group of people huddled under a bus stop to get out of the rain. In both cases one observes a cluster of individuals, but in the first the cluster arises through a self-organized process involving interactions among the individuals, whereas individuals in the second case are independently attracted to a preexisting focus of aggregation.

A striking feature of self-organized systems is the occurrence of a bifurcation—a sudden transition from one pattern to another following even a small change in a parameter of the system. One speaks of "tuning" a parameter in the system to invoke the onset of a different pattern. In the *Dendroctonus* example, one tunable parameter is the initial density of the setup. In the Bénard convection system, a tunable parameter is the amount of heat applied to the lower surface of the dish. By making small adjustments in such parameters, one can induce large changes in the state of the system, since the system may now be on a trajectory that flows to a quite different attractor. Most self-organized systems have many tunable parameters. Let us explore this phenomenon of parameter tuning more closely.

Parameter Tuning

A mathematical model popularized by Robert May (1974; 1976) has become one of the premier examples in the field of chaos theory, and it is a classic example of population growth for a hypothetical organism with nonoverlapping generations. (Also refer to the less technical presentations by Crutchfield et al. 1986; Gleick 1987; Dewdney 1991.) This system is a useful one for gaining an intuitive understanding of how a system can undergo dramatic transitions between two ordered states, or from an ordered state to a chaotic state or vice versa.[2]

As shown in Box 3.2, we can model the growth of a certain population with the logistic difference equation: $N_{t+1} = rN_t(1 - N_t)$. This equation has the single variable, N_t, which is the population size in the current generation, and varies between 0 and 1, where 0 is extinction and 1 is the population at the carrying capacity of the environment. The single parameter, r represents the intrinsic reproductive rate of the species. The population size in the next generation is given by N_{t+1}. To determine the population size over time, the equation is solved iteratively, starting with an arbitrary population value for N_t in the range of 0 to 1. The result of each iteration is the new population value N_{t+1}, which is then substituted in the equation as the population size in the current generation. The process can be repeated *ad infinitum*. One finds that if r is within a certain range ($0 < r < 1$), then repeated iterations result in extinction of the population regardless of the initial population size, N_t. This is to be expected for the obvious reason that each individual does not replace itself in the next generation. Similarly, if $1 \leq r < 3$, then the population again shows simple behavior, approaching a constant size after several generations, as one might intuitively expect for a population living under constant environmental conditions. Regardless of its initial size the population approaches the same final size, an *attractor*. But if we increase r slightly beyond 3, the population suddenly develops a new pattern; it enters a regime where it oscillates between two values. If we continue to increase r to more than 3.4, the system undergoes another abrupt transition where the oscillations between two population sizes change to oscillations between four population sizes. The system is now behaving in an unanticipated way. If r is raised yet again, beyond 3.57, the population exhibits deterministic chaos, changing erratically between generations with no regular pattern.

The appearance of a qualitative change in behavior when a parameter-value changes quantitatively is called a bifurcation. At the bifurcation between a single, stable population and oscillations between two different population values, r provides sufficient positive feedback in the system for the population size to undergo a cyclic rise and fall. The population size overshoots and then crashes.

Self-organizing systems, with nonlinear positive feedback interactions characteristically show bifurcations. In the *Dendroctonus* system, experiments reveal a bifurcation at a particular density of larvae; in the Bénard convection system, a bifurcation occurs as the amount of heat applied to the bottom of the fluid layer reaches a certain level. In many real-world systems, especially those in biology, it is difficult to control parameter values precisely enough to reveal such abrupt bifurcations, but in our hypothetical population we have complete control of the system. So in this situation it is easy to demonstrate the sudden emergence of novel behavior as one gradually tunes a system parameter.

Multistable Systems

Another type of behavior often exhibited by self-organizing systems is multi-stability, in which multiple possible stable states, or attractors may occur. This raises the question of what determines which of the various alternatives the system will exhibit. In the case of our hypothetical population, the behavior of the system is sensitive both to the parameter-value r and to the initial value of the population N_t. Figure 3.2 portrays the behavior of the system over the entire range of r, for all initial conditions. When different patterns arise in systems with multiple regimes there is usually no way of knowing a priori which particular regimes ultimately will be chosen. The final states attained by such systems usually depend on the initial conditions and a range of initial conditions that act as a *basin of attraction* for a particular attractor.

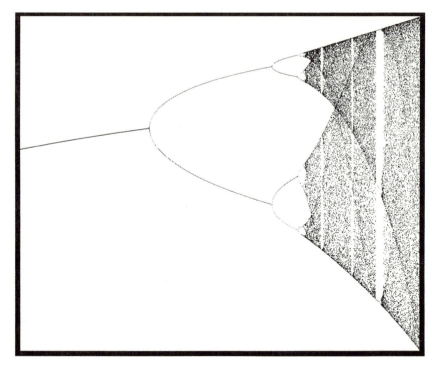

Figure 3.2 The bifurcation diagram above is for the logistic difference equation. Values along the x-axis are the tuning parameter, r, from 0 to 4. The corresponding population values (ranging from 0 to 1) are shown along the y axis. The diagram was generated with the program, Bifurcation Diagram, which can be downloaded at http://beelab.cas.psu.edu.

Biological and Physical Parameters

For the biological systems discussed in this book—what is meant by a "parameter"? In systems that can be described by a simple mathematical equation, we have no difficulty identifying the parameters, such as r in the above population equation. But of course, the situation is more complicated in actual living systems. As biologists interested in the behavior of living organisms, we may distinguish between two basic types of parameters in self-organized systems: those intrinsic to the organisms (the biological parameters) and those that arise from the environment or by physical constraints (the physical parameters). However, this distinction is irrelevant to the pattern formation process. For example, a parameter can affect the rules of thumb that describe the probability of performing a certain behavior under specified circumstances. The execution of a rule of thumb depends on information (about itself and the environment) that an organism acquires moment by moment, and on genetically encoded information that an organism possesses intrinsically. In formulating models of pattern formation in the examples in this book, we generally start out with a presentation of the behavioral rules of thumb used by the individuals in the system. For example, a rule of thumb for an army ant, "The more pheromone detected, the quicker the running speed," might translate to a differential equation such as $dx/dt = vC$, where dx/dt is the distance moved by the ant per unit time (its speed) as a function of the pheromone concentration, C. The parameter, v, relates the walking speed to the pheromone concentration.

Physical parameters also play important roles in biological systems. Consider the example of trail-following by ants. We need to know how the concentration of the trail pheromone varies over time, not only as a function of the behavior of the ant depositing the pheromone, but also as a function of the evaporation rate of the pheromone. The evaporation rate is determined by the pheromone's chemical structure and the physical conditions, such as temperature and air flow. Natural selection can influence the evaporation rate by determining the chemical structure of the pheromone, but the volatility of a particular pheromone compound is governed by the laws of physics.

Consequences of Emergent Properties in Self-Organization

The striking phenomenon of tunable emergent properties can have important evolutionary consequences for self-organized systems. We have shown how a small change in a system parameter can result in a large change in the overall behavior of the system. Is it possible that such properties could provide self-organized systems with adaptive, flexible responses to changing conditions in the environment and to changing needs of the system? If so, how might this flexibility arise?

Let us assume that natural selection can tune a particular behavioral parameter to a range of values close to a bifurcation point. Then, small adjustments in the parameter for each individual within a group may induce large changes in the collective properties of the group, thereby endowing the group with a wide range of responses and the ability to switch from one behavioral response to another.

Flexibility of this type may operate on a day to day basis, or over a longer time span, such as throughout the seasons. To take a hypothetical example: in the spring honey bee colonies undergo a transition from one mode of behavior to another as colonies switch from a nonswarming to a swarming state. The colony initially produces only worker bees, but eventually it begins to produce a batch of queens. The colony then divides, with approximately half the workers leaving (swarming) with the original queen to establish a new colony and the other half remaining at home with one of the new queens to continue the original colony. In some cases, however, the colony produces multiple swarms. What is responsible for these different swarming responses?

Rather than assuming that different behavioral rules determine the type of swarming outcome, let us suppose that the bees respond with the *same* set of behavioral rules to slightly *different* circumstances such as the initial colony size. In such a system, there may be an economy of behavioral complexity at the individual level required to switch from one kind of behavior to another. Thus, tunable parameters and bifurcations might provide an efficient mechanism for producing flexibility in biological systems.

Role of Environmental Factors

Environmental parameters may play a crucial role in shaping self-organized systems. The environment specifies some of the initial conditions, and positive feedback results in great sensitivity to these conditions. In particular, positive feedback can amplify initial random fluctuations or heterogeneities in the environment, and as a result the system may exhibit a number of different outcomes. A clear example is the raiding patterns of army ant colonies, analyzed in detail in Chapter 14. Deneubourg et al. (1989) examined models in which the *same* set of behavioral and physiological rules apply to *different* army ant species, but under different environmental conditions. Their striking finding was that distinct morphological patterns of army ant raids emerged merely by varying the initial distribution of food in the environment. It was not at all obvious, a priori, that the system would display multiple stable regimes as a function of variation in environmental parameters.

Biologists are accustomed to considering differences between species' behavior patterns as the phenotypic expressions of underlying genotypic differences that evolved over an evolutionary time-scale in response to environmental conditions. We suggest that certain species-specific patterns may be self-organized expressions of differences in environmental variables. Not surprisingly, differences in the raiding patterns of *Eciton burchelli* versus *E. rapax* may also reflect genetically based differences in pheromones or behaviors. Undeniably, such differences in biological parameters probably do exist, but a remarkable fact is that models of the raiding patterns demonstrate that differences between species in raiding patterns could arise simply from differences in the spatial distribution of each species' prey.

Self-Organization Can Promote Stable Patterns

We have emphasized transitions from one pattern to another as one or more of a system's parameters changes value. This may have given the impression that self-organizing systems are rather fragile, erratic, and susceptible to perturbations. However, most of the self-organizing systems described in this book are extremely robust, by which we mean they are stable over a wide range of parameter values. Although we have pointed out that natural selection may tune a particular parameter to the vicinity of a bifurcation point, it appears that most systems operate in a parameter range far from bifurcation points and, therefore, stubbornly resist transition from one pattern to another. The reason for this seems clear. Most of the patterns discussed in this book are adaptive and in most cases would be highly maladaptive if the behavior of the builders did not consistently produce the typical species-specific superstructure or pattern.

Consider again the example of the bark beetles (*Dendroctonus*). Although experimental situations can be contrived in which the larvae do not aggregate, under normal conditions the larvae almost always operate in a parameter range where strong aggregation occurs. This makes sense, for if this clustered feeding is an important adaptation for countering the tree's defensive production of sticky resin, then that clustering is expected to be consistently observed under natural conditions. Natural selection is expected to tune the larvae's behavioral and physiological parameters so that clustering occurs under the full range of conditions that the larvae would be expected to encounter.

Given that natural selection frequently favors a particular pattern for a system, many self-organizing systems are expected to resist perturbations and operate with great stability within a single regime. In such circumstances, the pattern exhibits the property of self-repair. In other words, the pattern is an attractor of the system. This phenomenon is discussed further in Chapter 16 with analysis of pattern formation on the comb of honey bee colonies.

Self-Organization and the Evolution of Pattern and Structure

Intuitively, it would seem easier for natural selection to make adjustments in the processes underlying an existing structure than to evolve a fundamentally new structure. It seems likely, therefore, that natural selection generates new adaptive structures and patterns by tuning system parameters in self-organized systems rather than by developing new mechanisms for each new structure.

Consider the wide range of different coat patterns of mammals or shell patterns of mollusks. Self-organized pattern formation mechanisms have been hypothesized for these systems (Ermentrout et al. 1986; Fowler et al. 1992; Lindsay 1977; Murray 1981; Murray 1988). An intriguing feature of this hypothesis is that fundamentally similar mechanisms may account for a wide variety of different patterns. In the past most biologists probably would have thought that the strikingly different color patterns on different shells or mammals arise through qualitatively different mechanisms. However, the concept of self-organization alerts us to the possibility that strikingly *different patterns* may result from the *same mechanism* operating in a different parameter range. This underscores the possibility that in evolution important changes in the properties of organisms and groups of organisms might result from slight changes in the tuning of parameters of the underlying developmental systems.

Simple Rules, Complex Patterns—The Solution to a Paradox

Biologists have been puzzled by the fact that the amount of information stored in the genes is much smaller than the amount of information needed to describe the structure of the adult individual. The puzzle now may be solved by noticing that the genes are not required to specify all the information regarding adult structure, but need only carry a set of rules to generate that information (Maruyama 1963, p. 171).

Most self-organizing systems, like biological systems in general, are highly complex and probably use multiple rules. Termites, for example, probably use more than a single simple set of rules for constructing their intricate mounds (Figures 18.1 and 18.2). Nonetheless, it should be stressed that simple nonlinear interactions between large numbers of individuals can lead to surprisingly complex patterns at the group level, patterns that often are unexpected even if detailed knowledge exists of the group's members and their interactions. An important goal of this book is to explore the question of how much—more to the point, how little—complexity must be built into the components of a self-organized system to generate the observed complexity at the group level. This question has important evolutionary implications.

As Maruyama (1963) points out, we know that it is impossible for each detail of an organism to be *explicitly coded* in the genome. For example, the human body can produce antibodies against a nearly unlimited number of foreign substances, yet the body has only about 100,000 different genes. An individual can synthesize more than 100 million distinct antibody proteins at any one time, so each protein obviously cannot be specified by a separate gene. Instead, natural selection led to a clever device for economizing on the information that needs to be genetically coded for the immune system. The device is a combinatorial scheme for generating diversity (Janeway 1993). Antibody genes are not distinct units; they are inherited as fragments joined together to form a complete gene within individual B lymphocytes, of which the body has about 10 trillion. One cannot improve on François Jacob's (1982, p. 38) elegant description of the process:

> [A] mammal can produce several millions or tens of millions of different antibodies, a number far greater than the number of structural genes in the mammalian genome. Actually a small number of genetic segments is used, but the diversity is generated during the development of the embryo by the cumulative effect of different mechanisms operating at three levels. First, at the *cell* level, every antibody-forming cell produces only one type of antibody, the total repertoire of antibodies in the organism being formed by the whole population of such cells. Second at the *protein* level, every antibody is formed by the association of two types of protein chains, heavy and light; each of these chains can be sampled from a pool of several thousand and their combinatorial association generates a diversity of several million types. Third, at the *gene* level, every gene coding for an antibody chain, heavy or light, is prepared during embryonic development by joining several DNA segments, each one sampled from a pool of similar but not identical sequences. This combinatorial systems allows a limited amount of genetic information in the germ line to produce an enormous number of protein structures with different binding capacities in the soma. This process clearly illustrates the way nature operates to create diversity: by endlessly combining bits and pieces.

This example shows that various mechanisms exist for economizing on the information that needs to be coded in a system. Self-organization is one type of mechanism for creating structure with a minimal amount of genetic coding. The antibody combinatorial mechanism is another. Let us now return to the original question: How much behavioral information needs to be coded explicitly in the genome of a self-organized system? Athough we cannot provide a precise answer, we suggest that it is far less information than might have been assumed in the past.

Box 3.1 Bénard Convection

The Bénard convection system has become a classic example of a self-organizing system. It is simple to demonstrate and displays a complex emergent pattern (Figure 1.3a). Convection is a process of fluid flow that occurs when a liquid or gas is heated (DeAngelis et al. 1986; Velarde and Normand 1980). When a fluid is heated from below, the bottom layer expands and becomes less dense. This lighter, warmer layer tends to rise and the heavier, cooler layer above tends to sink. The result is convective fluid transport.

At the turn of the last century the French investigator Henri Bénard studied a convective system that revealed unusual patterns. The system comprised a thin layer of spermaceti oil heated uniformly from below while its upper surface was kept relatively cool by contact with the atmosphere. As long as the vertical temperature gradient in the fluid was sufficiently small, no special pattern appeared; but when a gradual increase occurred in a system parameter—the amount of heat applied to the lower surface—the previously uniform surface of the fluid suddenly became a tesselated mosaic of polygons. The critical value of the tuning parameter depended on the fluid's viscosity, surface tension, depth, and other factors. Indeed, any of these factors can be taken as the tuning parameter, although it is most convenient to adjust the amount of heat. The following description of the Bénard convection system provides a good explanation of the phenomenon (Velarde and Normand 1980, p. 94–95):

> Consider a small parcel of fluid near the bottom of the layer. Because of the elevated temperature at the bottom, the parcel has a density that is less than the average density of the entire layer. As long as the parcel remains in place, however, it is surrounded by fluid of the same density, and so has neutral buoyancy. All the forces acting on it are in balance, and it neither rises nor sinks.
>
> Suppose now that through some random perturbation the parcel of fluid is given a slight upward motion. What effect does the displacement have on the balances of forces? The parcel now is surrounded by cooler and denser fluid. As a result it has positive buoyancy, so it tends to rise. The net upward force is proportional to the density difference and to the volume of the parcel. Thus an initial upward displacement of the warm fluid is amplified by the density gradient, and the amplification gives rise to forces that cause further upward movement. A similar analysis could be made for a slight downward displacement of a parcel of cool, dense fluid near the top of the layer. On moving downward the parcel would enter an environment of lower average density, and so the parcel would become heavier than its surroundings. It would therefore tend to sink, amplifying the initial perturbation. Natural convection is the result of these combined upward and downward flows, and it tends to overturn the entire layer of fluid.

The next description of the onset of instability is expressed in the terminology of self-organization and dynamic systems (Prigogine and Stengers 1984, p. 142):

> The "Bénard instability" is another striking example of the instability of a stationary state giving rise to a phenomenon of spontaneous self-organization. The instability is due to a vertical temperature gradient set up in a horizontal liquid layer. The lower surface of the latter is heated to a given temperature, which is higher than that of the upper surface. As a result of these boundary conditions, a permanent heat flux is set up, moving from the bottom to the top. When the imposed gradient reaches a threshold value, the fluid's state of rest—the stationary state in which heat is conveyed by conduction alone, without convection—becomes unstable. . . . The Bénard instability is a spectacular phenomenon. The convection motion produced actually consists of the complex spatial organization of the system. Millions of molecules move coherently, forming hexagonal convection cells of a characteristic size.

The development of hexagonal cells in the Bénard convection system may be comparable to the development of clustering in the *Dendroctonus* beetle system. In Bénard convection, random motion of the molecules (in which heat transfer is by conduction) competes with coherent motion of the molecules (when convection occurs). With the beetle larvae, random motion of the larvae occurs in the absence of any chemical cues, but motion becomes oriented when a pheromonal gradient is established by the larvae themselves. In both cases, the system exhibits a spontaneous transition to a more ordered state as a particular parameter is gradually increased beyond its bifurcation point.

It is easy to demonstrate Bénard convection in the classroom. Bénard used spermaceti (sperm whale) oil, and Velarde and Normand (1980) used silicone oil to which flakes of aluminum were added to make the flow visible. We can use ordinary vegetable oil to which we add very fine aluminum powder (called bronzing powder, available at art supply stores). A pinch of powder in a cup of oil is sufficient. We place a layer of this oil about 1 cm thick in a 10-cm-diameter glass petri dish. Rather than use an apparatus with which we could gradually and uniformly increase the temperature at the bottom of the dish, we simply place the dish briefly on a laboratory hot plate, then carefully remove the dish and place it on a table top. If the bottom of the dish is heated sufficiently, the characteristic pattern will suddenly appear, and remain for several minutes. It is a temporarily stable, swirling pattern of polygonal Bénard convection cells.

Box 3.2 Tuning the Growth Rate Parameter in the Logistic Difference Equation

In many species of insect, such as certain butterflies, generations are nonoverlapping. Such species have eggs that hatch in the spring after overwintering. The adults live through the summer and then die after laying eggs in the fall. To describe the growth of such a population, one can use the so-called logistic difference equation:

$$N_{t+1} = rN_t(1 - N_t)$$

We can think of this equation as describing the population size in the next generation (N_{t+1}) as a function of the current population size (N_t) and a parameter, r. Population size is scaled to vary between 0 (no individuals) and 1 (the maximum number of individuals). Here the subscripts, t and $t + 1$, refer respectively to the current time and the time of the next generation. This equation tells us that the population reached in the next generation (N_{t+1}) depends on the number of individuals in the current generation (N_t), which makes sense since the current individuals are those that will be laying the eggs for the next generation. The parameter, r, corresponds to an intrinsic reproductive rate indicating the average fecundity (number of offspring surviving to adulthood) of an individual. The equation supposes that, in the absence of a limiting factor such as overcrowding, the population in the next generation will be rN_t. Greater values of r result in greater numbers of individuals in the next generation. The parameter r specifies the strength of the positive feedback in the system. The factor $(1 - N_t)$ plays an important role in the system: it provides the negative feedback. It also makes the equation nonlinear, giving it many of its unusual properties. In this model the population is scaled between the limits of zero (extinct) and one (the maximum carrying capacity of the population). Thus the factor $(1 - N_t)$ limits population growth as it nears its carrying capacity, because as N_t approaches 1, the factor $(1 - N_t)$ approaches zero. Expanding the right hand side of the equation yields $rN_t - rN_t^2$, which is the equation of a parabolic curve.

Without the factor $(1 - N_t)$, the right hand side of the equation is simply $r N_t$, the equation of a line with slope r.

In exploring this simple model of population growth, we wish to demonstrate how a system can suddenly go from one state to another through the gradual tuning of a parameter.[2] As described in the main text, the final population size reached after several iterations of the equation depends on the value of the reproductive parameter, r. Over certain ranges of r the population size reaches a single value, but as the parameter is increased the population size oscillates first between two values (so-called "period two behavior"), then four values, eight values and so on. In the terminology of nonlinear dynamics, the system exhibits a sequence of period-doubling bifurcations.

Another feature of nonlinear systems that can be explored with this equation is the transition to chaos. As r is increased beyond 3.57, the system not only fails to reach a stable value but also does not oscillate among a number of fixed values. Instead, no pattern occurs in the sequence of population levels from generation to generation. The system is said to be chaotic. Prior to Robert May's work, it is likely that such unpredictable behavior in an insect population would have been attributed to random external influences or noise in the measurements of population size. But in our hypothetical population governed by this simple deterministic equation, noise is not provided by the environment, the model, or by random errors in data collection that so often plague field studies. The chaotic behavior is called deterministic chaos. Here, the term *chaos* has a precise mathematical meaning that should not be confused with randomness or noise. Deterministic chaos is the unpredictable behavior of a nonlinear system within a certain parameter range. Deterministic means that subsequent population values are determined precisely by its equation. What is so unexpected, however, is that a deterministic equation can yield unpredictable results.

Chaos is not a topic emphasized in this book, largely because the systems dealt with here do not normally exhibit chaotic behavior. No doubt this is because natural selection tunes the parameters of living systems to avoid chaos. In most situations, it would probably be grossly maladaptive for a living system to exhibit chaotic, disorganized patterns.

Even a simple nonlinear equation can exhibit complex behavior, and so researchers have developed a graphical method for showing in a single figure the behavior of the system over a range of parameter values. This is called a bifurcation diagram (Figure 3.2). For the logistic difference equation, the diagram displays values of r on the x axis. The corresponding population sizes appear on the y axis. For each value of r, many initial values of the population size, N_t, are iterated one by one, and after a large number of iterations, the population size for each N_t is plotted on the graph. We see that for certain values of r, a single steady state is reached regardless of the initial population size. Other regions show period two behavior and the transition to chaos, as well as parameter zones within the chaotic region where the population level suddenly switches to fixed population sizes.

Box 3.3 Toppling Dominoes: A Mechanical Example of Tuning a Parameter

The example of clustering by *Dendroctonus* beetle larvae illustrated how the initial density of the larvae was a parameter that could be tuned to generate sufficient positive feedback to initiate clustering. A simple mechanical analog of this system uses dominoes. Instead of tuning the initial density of larvae to affect a clustering process, the density of dominoes can be changed to affect a chain reaction of toppling.

Consider an arena of fixed size, say 1 m^2, seeded with a variable number of dominoes, each standing on its narrow end. If the density of dominoes is sufficiently low they will be sufficiently separated so that toppling a single one results in few, if any, subsequent topples. As the density of dominoes is gradually increased a density is eventually reached at which the fall of a single domino triggers a chain reaction of topples throughout the system. Experiments can be performed in which one counts the number of topples initiated by knocking over a single randomly chosen domino. If the experiment is repeated many times with different domino-densities, one can plot the average number of topples per experiment as a function of the domino density. The plot would be nonlinear, with a low number of topples up to a certain domino density, and then a rapidly rising portion of the curve at higher domino densities where the fall of a single domino results in a chain reaction of many topples.

Domino density is not the only parameter that can be tuned. The probability that a chain reaction propagates through the system is also a function of the height of a domino and the area of its base. The taller a domino, the more likely it will hit another domino when it falls. The smaller the area on which a domino stands, the more likely it will topple when hit by another domino. Thus a parameter equal to (density × height)/area of base also can be tuned to specify the probability that a chain reaction propagates throughout the system.

In Chapter 6, this domino system is shown to be a mechanical analog of a cellular-automaton simulation of an epidemic. The spread of disease in a population resembles the propagation of topples in the domino system.

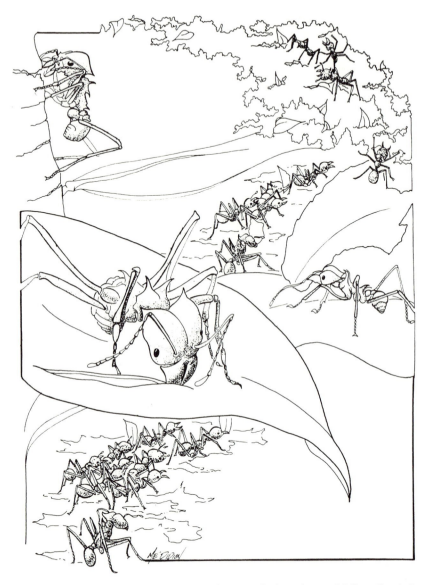

Figure 4.1 Leaf-cutter ants (*Atta* sp.) are shown gathering pieces of foliage for their fungal gardens. An ant will often cut out a disk-shaped section of leaf by rotating its body as it gnaws through the leaf, in the same way we use a compass to draw a circle. In this sense, the ant's body serves as a template for cutting the proper shape. (See Figure 4.4 for a similar example.) (Illustration by Mary Ellen Didion)

4

Alternatives to Self-Organization

But who is the architect who designs the plans which the
workers execute.... No reasonable person can imagine for
one moment that every small worker is conscious of the
purpose of its work, that it carries in its mind the plan, or
even part of the plan of the building operations.
　　　　　　　　　—E. Marais, *The Soul of the White Ant*

Alternate Paths to Order

When studying the mechanisms of pattern formation in a living system
it is important to keep in mind not only the ideas of self-organization, but
also the alternative mechanisms of pattern formation. Certainly not all patterns arise through self-organization, and even those based primarily on self-
organization may involve other mechanisms as well. Order within a system
can arise through an interplay between internal interactions among the components and external forces imposing order from the outside. Discussions of
self-organized systems must clearly define the level of organization under investigation. In discussing the brain for instance, the neurons and their local
interactions (synaptic signaling) may be considered the source of the emergent
self-organized property of thought; at a different level of analysis, however,
the brain may be viewed as a centralized leader with external control over the
outputs of the motor nerves.

The alternatives to self-organization described in this chapter are particularly relevant to systems composed of organisms. These alternatives have been
unable to explain certain collective behaviors in groups of organisms and have
led, historically, to the development of theories that are precursors of the self-
organization approach advanced in the book.

Leaders, Blueprints, Recipes, and Templates

We recognize four ways in which a group of organisms can build a well
ordered structure without self-organization—without interactions among the
system's components. We will illustrate each of these alternative mechanisms
with an example of pattern formation by humans, since the pattern-formation
processes within human groups are easily perceived. Each of these four mech-

anisms can be characterized by the means through which order is imposed on the group's pattern-formation activities: (1) leader, (2) blueprint, (3) recipe, and (4) template.

The Well-Informed Leader

The first alternative to self-organization is a well informed *leader* that directs the building activity of the group, providing each group member with detailed instructions about what to do to contribute to building the overall pattern. This is exemplified by the performance of a rowing crew, where stroke synchronization and coordinated shifts in stroke frequency are largely responses to the coxwain's commands. By contrast, a group that builds a pattern by means of self-organization needs no supervision by a leader.

Note, however, that the patterned behavior of a rowing crew is not based entirely on instructions from a leader; the oarsmen also receive input from one another as each pays close attention to the oar movements of fellow crew members. The coordinated behavior of a rowing crew reflects not only commands from a leader but also interactions among the oarsmen. This suggests that self-organization also plays a role in the formation of this pattern.

In analyzing the mechanism of pattern formation in a particular system, one should always look for key individuals that may play a leadership role in the organization and patterning of the system. The discovery of a leader provides some evidence that the system is not self-organized. Likewise, the discovery of interactions among components of the system provides support for the notion that the system is self-organized. Most importantly, an experimental approach to distinguish among these alternatives would involve perturbations of the system, either by removing the putative leader, or by blocking interactions among individuals, to test whether the pattern can be generated under these altered conditions.

Building by Blueprint

The second alternative to self-organization is for the group pattern-building to be directed by a *blueprint*, a compact representation of the spatial or temporal relationships of the parts of a pattern. A crew of construction workers, for example, receives a detailed and virtually complete description of the bridge, house, or whatever it builds from a set of blueprints. Likewise, each musician within a symphony orchestra receives a musical score, that fully specifies the pattern of notes within the composition and the tonal and temporal relationships among them. We note again, however, that neither of these two cooperative groups receives all of its instructions from the blueprint. A construction crew typically possesses a foreman who provides further specifications about the object under construction and additional instructions about the building process. Likewise, an orchestra has a conductor who sets the tempo and provides signals for group synchronization. Clearly a pattern-building group steered by external instructions may receive these instructions via multiple mechanisms, such as both a leader and a blueprint.

Following a Recipe

Each member of the group may possess a *recipe*, sequential instructions that precisely specify the spatial and temporal actions of the individual's contribution to the whole pattern. Those of us who are timid cooks may not bother to taste the dish as we go along, but merely follow the recipe to the letter. This lack of adjustment of the building process through feedback from the work in progress epitomizes our definition of a recipe. Most cooks, however, taste as they go along, modifying the process based on gustatorial feedback from the developing dish. In general, behavioral adjustments based on feedback from the emerging pattern (stigmergy) can be superimposed on the use of a recipe. As discussed in Chapter 2, stigmergy reflects interactions internal to the system, so it is a fundamentally different building mechanism than following a recipe.

Blueprints and recipes are similar in some respects. The distinction is that a blueprint does not specify *how* something is to be built, only *what* is to be built. In contrast, a recipe provides the step-by-step instructions for the pattern. This distinction becomes blurred in the case of a musical score, which in one sense can be seen as the blueprint for the composition, but can also be viewed as the step-by-step recipe to be followed to generate the score.

Examples of recipes used by a group of builders are the instructions underlying the plays in an American football game, since each member of the team executes instructions that specify where to go and what to do throughout the play. Here again, the group's instructions do not come in just one form, for each member of the team is guided not only by his personal recipe but also by instructions from a leader, the quarterback, that enable teammates to begin executing their recipes simultaneously. Also, once a play is underway each player may independently modify his behavior in light of unpredictable events unfolding around him, so the freedom to deviate from the recipe is essential. Similarly, an intricate ballet results largely from each dancer following a predetermined recipe from a choreographer. However, here too the graceful coordination among dancers arises not only through each dancer following her or his personal set of instructions, but also through subtle interactions among the dancers as they execute their parts.

Templates

The fourth alternative to self-organization is the use of a *template*—a full-size guide or mold that specifies the final pattern and strongly steers the pattern-formation process. A seamstress making a dress uses a sewing pattern, a paper template on which the precise dimensions of each piece of the garment are outlined. The template is pinned to the fabric and used as a guide to cut pieces of the proper size and shape. Other familiar examples of templates are candle molds, cookie cutters, or the dies used to mint coins. All such devices are templates for the production of a consistent, reproducible pattern.

Biological Examples of Alternatives to Self-Organization

Having defined four alternatives to self-organization and considered several relevant examples of pattern formation by human groups, one can question the extent to which these alternatives account for structures built by animal groups.

Leadership has often been invoked to explain the organized activities of animal groups, especially where it seems that one group member has more experience or is better informed than other individuals. A simple pattern that evidently does arise through leadership is the classical example of queuing behavior by ducklings or chicks, each following the mother, who makes sure that each duckling is following her (Figure 4.2a). As discussed in Chapter 11, it has also been suggested that leadership plays a major role in the coordinated movements of individuals in large groups, such as a school of fish or a flock of birds, but no evidence for leaders has been found in these groups.

Similarly, many people believe that in insect societies the mother queen—as the name suggests—functions as a ruler, a central authority that directs the activity of her offspring, the workers. To some extent, this may be true. In the honey bee colony, queen substance, a pheromone blend produced by the queen (Winston 1987) is an important regulator of several aspects of colony organization. The queen may play an even more important and central role in social insects with small colonies, such as primitively eusocial bees (Breed and Gamboa 1977) and some wasps (Reeve 1991; Reeve and Gamboa 1983, 1987). In Reeve's and Gamboa's studies of *Polistes* paper wasp colonies that typically consist of only a dozen or so individuals, they concluded that the members are "dynamically linked in a social feedback control system" in which the queen acts as a "central pacemaker and coordinator of colony activity." Queen aggression, or simply increased activity by the queen, was found to stimulate worker activity, leading to an increased rate of workers leaving the nest to forage. Presumably the queen of a paper wasp colony can function as a leader because her colony's nest is so small that she can examine the entire nest, gather complete information about the colony's needs, and provide accurate feedback to the workers.

Comparative studies by Jeanne (1991, 1998) show that wasps with small colony sizes are characterized by a centralized organization, while those species with large colonies are largely decentralized and probably governed much more by self-organizing processes. As the colony size increases, the regulatory activity is shifted from the queen to the workers themselves.

Figure 4.2

a

b

Figure 4.2 Alternatives to self-organization in biology include: leadership, where young ducklings follow the mother duck (a); or geese that have been hand-reared and trained to follow an ultralight airplane along their ancestral migratory route (b) (illustration by Mark A. Klingler); (*continued*)

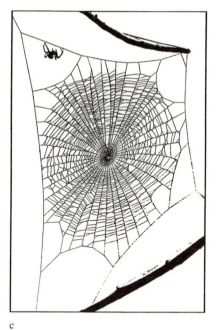

c

figure 4.2 continued
an innate recipe, such as spiders may use to build a web (c) (illustration by S. Camazine and A. Rozo); or a DNA template for its own duplication and production of RNA (d). (Illustration by Mark A. Klingler).

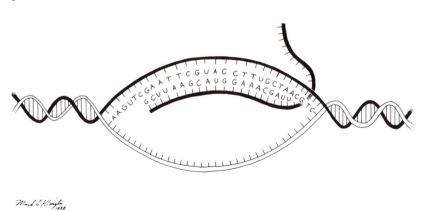

d

Reeve (1992) also provided evidence that among the eusocial naked mole rat (*Heterocephalus glaber*), the queen plays a central role in regulating activity levels of workers in the colony.

It is difficult to find clear examples of the use of blueprints by animal builders. Hansell (1984, p. 201) reviewed animal architecture and building behavior and discussed whether animals have some form of mental blueprint, an "internal schema or mental image of the desired condition, against which the external stimulus [the work in progress] is compared." One difficulty is that it is hard to distinguish the use of a mental blueprint from the application of a rich repertoire of responses to different contingencies that may arise in the course of building, since both can account for the flexible building behavior exhibited by many animals.

In the case of a series of stimulus-response relations, a particular structure stimulates certain building behavior that gives rise to a new aspect of the structure that provides a stimulus for the next phase of construction. Through a chain of such linked stimulus-response relations, the structure arises, following a fairly rigid construction sequence. Deviations from this sequence caused by accidents, or whatever, are managed by backup stimulus-response relations. With a blueprint, however, "the animal on comparing the present with the desired situation can progress toward the goal along any number of alternative pathways" (Hansell 1984, p. 201). Supporting the idea that building activities are guided by a mental blueprint, Hansell reviewed many interesting examples among vertebrates and invertebrates demonstrating the variability and flexibility one might expect if the organism were using alternative paths to build and repair structures.

Nonetheless, it remains difficult to prove that such activity involves an actual mental blueprint. A clear counter-example to the concept of a mental blueprint guiding building behavior is an experimental study of the mud wasp (*Paralastor* sp.), which constructs a smooth-walled funnel over the opening of its nest entrance to deter parasites (Smith 1978). Experimental alterations of a funnel under construction result in bizarre, totally nonfunctional funnels (Figure 4.3). This demonstrates unambiguously that these particular wasps are not steering their building behavior through reference to a mental blueprint of the final structure. This example is discussed in more detail in Chapter 19.

There is better evidence that animals build by following recipes—instructions that specify a sequence of behaviors sometimes referred to as a genetically programmed behavior sequence (Eibl-Eibesfeldt 1970). Classic examples are solitary bees and wasps that must accomplish an entire sequence of acts in a fixed order, described more than a century ago by Fabre (1889). Another illustration is Eibl-Eibesfeldt's (1970, p. 18) description of a spider's construction of an egg case:

> While constructing a cocoon the spider *Cupiennius salei* first produces a base plate, then a raised rim that provides the opening into which the eggs are deposited. Having laid the eggs, the female closes the opening. If she is interrupted while spinning her cocoon, after the base plate has been completed, she will not produce a new base plate half an hour later

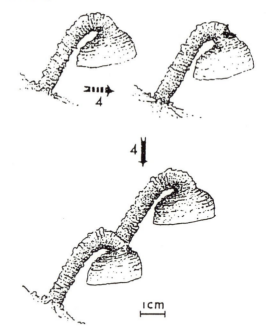

Figure 4.3 Illustrated is experimental proof that a mud wasp (*Paralastor* sp.) does not guide its building of a nest funnel through reference to a mental blueprint. When a hole is made in the top of the curved section (top), the wasp constructs a bizarre double funnel (bottom). If the individual possessed an innate blueprint of the structure, it would not build such an abnormal nest (from Smith 1978).

when she builds a new cocoon, but instead spins only a few threads and then continues with construction of the rim, so that the bottom of the cocoon remains open. If one adds the number of spinning movements she performed for the previous base plate and for the new substitute cocoon, the number roughly equals the number normally used to build a complete cocoon. She has available, so to speak, a limited number of spinning movements—approximately 6400 dabbing movements. This number of movements is performed, even if, under abnormal circumstances, she is no longer able to secrete any threads. This has happened when the glands dry up as a result of the hot lights used during filming. In such circumstances, the spider still produces her behavior program. After the appropriate number of ineffectual dabbing movements she will lay her eggs, which will then drop to the ground.

Cocoon-building by the cecropia silkworm moth, *Hyalophora cecropia*, also involves a stereotyped sequence of behaviors rigidly followed step by step (Van der Kloot and Williams 1953a, b). First, the caterpillar attaches anchoring threads to the substrate twigs and vegetation. Then, it lays down a tough outer envelope of silk. Inside the outer envelope, the caterpillar builds a loose filamentous layer and, finally, a closely woven inner envelope. The inner and outer envelopes have valve-like openings built at the same ends to provide the adult moth with an exit from the cocoon. Van der Kloot and Williams performed a number of experimental manipulations to determine whether the sequential steps in this behavioral pattern are rigidly stereotyped. By removing larvae from their partially finished cocoons and placing them inside envelopes from other cocoons, they found that each caterpillar always constructs its own cocoon, complete with an inner and outer envelope even if placed inside a complete inner envelope of another larva. Caterpillars removed from their cocoon and placed in a cylindrical container, never build a replacement envelope but simply follow a fixed behavioral recipe by continuing where they left off before being removed.

Nevertheless, cocoon-building by moths is not always so completely stereotyped that a larva is unable to modify its normal behavior in response to unusual—but not totally unexpected—events such as a change in the orientation of the cocoon during building. The moth *Dictyoploca japonica* normally spins its cocoon with the escape valves facing upward, and then pupates with its head toward the valve (Yagi 1926). If a cocoon is rotated 180° halfway through construction so that the larva and opening are facing downwards, the larva can turn itself around inside the cocoon and cut a new aperture in the top. These observations suggest that the basic plan of construction follows a genetically determined program or recipe that includes some contingency subroutines. Of course, such complexity in the recipe is adaptive when the animal must cope with an array of possible circumstances.

The principal argument against the concept of recipes is that the difficulty in a recipe-driven process of coordinating the activity of many different workers. Such rigidly deterministic behavior would probably cause problems in large colonies of social insects where individuals might inadvertently interrupt one another. With a recipe it is difficult for a worker that is involved in a particular step of a sequence to be replaced by another. It is partly to solve this problem that Grassé (1959) introduced the idea of stigmergy.

Templates are also used to create structure and pattern in biological systems. The classic example is seen at the subcellular level where DNA acts as a template for RNA, or messenger RNA acts as a template to specify a particular sequence of amino acids in a protein chain (Figure 4.2d). We cannot understand the primary structure of a particular protein — the linear sequence of amino acids — by examining the constituent amino acids themselves. We must refer to the messenger RNA template. As Rosen (1981) points out, "Since the information specifying the structure is not inherent in the subunits themselves, we cannot hope to understand the generation of the structure by studying the subunits in isolation." We must refer to the template.

Do organisms also use templates to build structures? One might not expect an animal builder to employ a physical template the way human builders do when constructing multiple copies of an object. However, an animal may use its body as a template. The male village weaverbird (*Ploceus cucullatus*) builds a saclike nest with long strips cut from giant grasses or palm leaves (Collias and Collias 1962) (Figure 4.4). The bird appears to follow a stereotyped building recipe involving itself as a kind of template for the nest. It begins by building a ring of strands attached to a branch. Then, once the ring is complete, the bird sits on the bottom of the ring and extends the ring into a hollow chamber by stretching as far in front of him as possible while sitting on the perch. As the bird weaves strips of material into the nest with its beak, the distance it can stretch defines the size and shape of the resulting hemispherical egg chamber.

Ideas Related to Self-Organization

As it gradually became apparent that leaders, blueprints, recipes, and templates cannot account for many instances of pattern formation in biological systems, various other ideas were suggested, many of which have affinities with the concept of self-organization. It is useful to review these ideas and examine their relationship to self-organization.

Stigmergy

Stigmergy was briefly discussed in Chapter 2. Here, a more detailed discussion is given to explain how stigmergy was an important antecedent to concepts of self-organization. Grassé (1959) originated the term stigmergy to refer to the mechanism by which the members of a termite colony coordinate their nest building activities. He believed that important determinants of an individual's building behavior were stimuli from work previously accomplished. "The stimulation of the workers by the very performances they have achieved is a significant one inducing accurate and adaptable responses, and has been named stigmergy" (Grassé 1959, p. 79). Clearly, this was an important insight. It explained how individual builders could act independently on a structure without direct interactions or sophisticated communications. Instead of the term *stigmergy*, Michener (1974, p. 33) used the expression "indirect social interactions" to describe the same mechanism of group integration and stress the importance of interindividual interactions, even though they were indirect:

> There are many activities in bee colonies that result in nest structures (or conditions of brood or stored food) to which other bees respond. The results are indirect social interactions—indirect because information transfer is not between two bees but between a construct made by bees and another bee. The construct (or brood or food) is made or cared for with

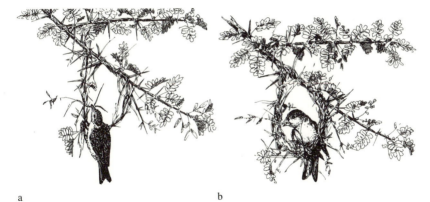

a   b

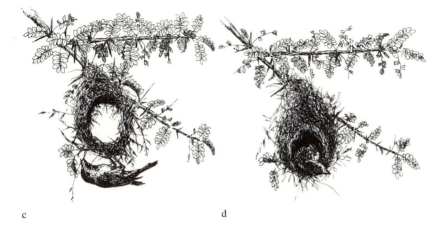

c   d

e

Figure 4.4 Sequence of nest-building in the village weaver bird: (a) construction of a vertical ring; (b) and (c) construction of the more or less hemispherical egg chamber through the use of the bird's body as a template; (d) and (e) the building of the antechamber and entrance. Based on a drawing in Collias and Collias (1962). (Illustration by Mark A. Klingler)

other primary objectives, not for signaling, although the information content may be essential for colony integration.

Grassé (1959, p. 79) realized that through stigmergy individuals might appear to be interacting directly but, in fact, they "do not constitute a working team," for they can be completely "indifferent to the behavior of their companions." Indeed, each builder contributes to the structure without communicating, and its contributions provide further cues to guide its own subsequent work as well as the work of others. This recursive control system uses stimuli from the work in progress to elicit a particular building response, which in turn acts as a stimulus for a further response, and so on in sequence.

Grasse imagined stigmergy to consist of a sequence of qualitatively different stimulus-response behaviors. A stimulus is transformed into another, qualitatively different, stimulus as a result of the building activity of the insect. The new stimulus may in turn release a different response from other insects. For example, a small defect such as a hole in the external wall of a nest may stimulate an insect to fill the hole with building material. Repairing the hole transforms the area into a flat surface. The flat surface is qualitatively different from the defective surface and may now stimulate the initiation of a new row of cells. An example of this type of stigmergic building behavior—termed *qualitative stigmergy*—will be discussed in a model of wasp nest construction in Chapter 19 (see also Theraulaz and Bonabeau, 1995a, b).

Another type of stigmergic building behavior is described in Chapter 18, in the context of termite-mound building. Here a stimulus of a certain intensity is transformed into the same stimulus with a higher intensity. For example, when a termite deposits a soil pellet soaked with pheromone, the pheromone contained in the pellet stimulates other termites to deposit other pellets nearby. The stimulus—a single pheromone-soaked pellet—is transformed into several pheromone-soaked pellets, that offer more intense and attractive stimuli than the original stimulus. This type of stigmergy has the flavor of self-organization since its most important component is the amplification of existing stimuli. Here stigmergy is merely an ingredient of self-organization and mediates interactions among workers. Self-organization is made possible by the intensity of the stigmergic interactions among termites that can adopt a continuum of values: The stronger the interactions (the greater the number of pheromone-impregnated pellets), the more likely the self-organizing effect is to intensify. Because the self-organized coordination of individual activities is facilitated by quantitative variations in the intensity of the interactions, this type of stigmergy is referred to as quantitative stigmergy.

Although it is conceptually easy to discriminate between qualitative and quantitative stigmergy, most situations in which stigmergy can be shown to play a role are likely to involve both types. A hole is qualitatively different from a flat surface, but when a wasp deposits building material such as chewed wood pulp in a hole, the newly deposited material may stimulate other wasps to help fill the hole because the wood pulp inevitably contains pheromone-laden salivary secretions. The same holds for the construction of a new comb: When a wasp starts a new comb she may indirectly stimulate other wasps to contribute to the new comb through pheromones released in the building material. More generally, if stimuli from qualitatively different classes have variable intensity and if the probability of releasing a certain behavior increases with stimulus-intensity, then the result is a combination of qualitative and quantitative stigmergy.

Stigmergy is effective for coordinating building activity over great spans of space and time. The structures built by social insects, for example, are often thousands of times larger than the builders and require many lifetimes to complete. Stigmergy also eliminates the need for individuals to be equipped with an inherited image or mental blueprint of the nest.

The elegant simplicity of stigmergy probably accounts for Grassé's (1967, p. 99) overly enthusiastic claim that "the building of these works [the sophisticated nests of *Macrotermes* colonies] and its coherence are perfectly explained by the theory of stigmergy." Subsequent researchers have approached stigmergy with a more critical eye, questioning the adequacy of stigmergy as an all-inclusive theory to explain nest construction by social insects. Harris and Sands (1965, p. 128) believe that stigmergy can explain the simpler aspects of building, but not "the more complicated termite nest structures as, for example, the peripheral tubes, exterior pores and spiral ramps of an *Apicotermes* nest." Stuart (1967) points out that stigmergy fails to explain how construction ends. Downing and Jeanne (1988) review other pitfalls of the theory, such as its inability to account for removal of material during construction or for adjustments to correct for construction errors.

These are valid criticisms only if one expects—as Grassé himself seems to have unrealistically expected—that stigmergy explains all aspects of nest construction. Instead, stigmergy may be regarded not as a complete theory of building activity but as an important concept that helps explain the flow of information among builders. Stigmergy is but one element in an overall set of mechanisms by which a group builds structures. It explains the source of much of the information—the stimuli—that each builder uses to determine what it should do.

Decentralized Control

Decentralized control is another concept that has been used to help explain social organization in insect colonies. Like stigmergy, decentralized control can be viewed as a general feature of self-organizing systems. It refers to a particular "architecture of information flow" (Seeley 1989, p. 549):

> Coordination of the activities in a honey bee colony arises without any centralized decision making. There is no evidence of an information and

control hierarchy, with some individuals taking in information about the colony, deciding what needs to be done, and issuing commands to other individuals who then perform necessary tasks.... [The] fact of decentralized control in honey bee colonies tells us that understanding how colonial coordination arises follows from understanding how each worker acquires information about her colony's needs.

Most of the mechanisms underlying self-organization are based on a decentralized architecture of information flow among a system's components. Individuals in self-organized social groups do not rely on instructions from well-informed individuals (leaders) in the upper echelons of a control hierarchy to know what to do.

Recent research in artificial life has generated great interest in decentralized control mechanisms. In this growing field of multi-agent systems a large literature now exists dealing with theoretical issues, the development of efficient algorithms, and robotics, as well as models inspired by insect societies (see, for example, Meyer and Wilson 1989, 1991; Ferber 1995; Bonabeau et al. 1998, 1999).

The relationships between stigmergy, decentralized control, and self-organization should now be more apparent. In a decentralized system, each individual gathers information on its own and decides for itself what to do. Stigmergy is one means of information flow within a decentralized system that involves gathering information from the shared environment. Another means of information flow within a decentralized system is from one individual to another, or from the group to an individual by means of cues or signals. These decentralized paths of information flow provide the essential means of interaction among the components in a self-organizing system.

Dense Heterarchies

The term *dense heterarchy* (Wilson and Hölldobler 1988) has been used to describe the basis of organization in social insect colonies, especially ant colonies:

[An] ant colony is a special kind of hierarchy, which can usefully be called a heterarchy. This means that the properties of the higher levels affect the lower levels to some degree, but induced activity in the lower units feeds back to influence the higher levels.

The concept of a dense heterarchy is related to the picture of information flow in stigmergy. But in addition to the idea of stigmergic-information flow, there is also the idea that direct communication between individuals or between groups and individuals is important: "The heterarchy is also highly connected or 'dense' in the sense that each individual member is likely to communicate with any other" (Wilson and Hölldobler 1988, p. 65). If a network is densely connected, this implies that the network is *not* set up in a hierarchical manner like "the partitioned hierarchies of human armies and factories, in which instructions flow down parallel independent groups of members through two or more levels of command" (Wilson and Hölldobler 1988). Instead, a dense heterarchy implies decentralized control since "the highest level of the heterarchy is the whole membership, rather than a particular set of 'bosses' who direct the nestmates. In particular, the queen is not at the head of the heterarchy." Wilson and Hölldobler (1988, p. 67) argue that through such a system of organization,

> The individual ant need operate only with "rules of thumb," elementary decisions based upon local stimuli that contain relatively small amounts of information.... Each of these rules is easily handled by individual workers. Each action is performed in a probabilistic manner with limited precision. But when put together in the form of heterarchies involving large numbers of workers and mass communication, a whole pattern emerges that is strikingly different and more complicated in form, as well as more precise in execution.

This is a good description of social system based on decentralized control where individuals respond not only to (stigmergic) stimuli from work-in-progress but also to stimuli received from neighbors.

Summary

The concepts proposed by Grassé, Michener, Seeley, and Wilson and Hölldobler use different terminology to describe the mechanisms of social organization of insect colonies but their ideas are somewhat similar. Many of these ideas are used in later chapters to describe self-organization in insect societies and living systems in general. Indeed, the concepts of self-organization proposed here are not completely new but grew from a number of related ideas that have been emerging for many years.

Perhaps the most important advance provided by the conceptual framework of self-organization is its introduction of a quantitative and dynamical approach to understanding social organization. Along with this comes the necessity of using certain tools such as mathematical modeling and simulation techniques. As Wilson (1971, p. 231) remarked:

> The total simulation of construction of complex nests from a knowledge of the summed behaviors of the individual insects has not been accomplished and stands as a challenge to both biologists and mathematicians. The eventual achievement of such a simulation will be the evidence of a fairly high level of sophistication in our understanding of social behavior.

We believe that self-organization applied to the study of social insects is the latest addition to theories of how social organization comes about. The self-organization approach with its mathematical tools may finally enable us to meet Wilson's challenge.

Figure 5.1 Beavers are well known for their "skill" in constructing lodges and dams from branches, mud, and other debris. However, the obvious analogy with human building activity is probably misleading. It is unlikely that beavers rely on an innate concept or blueprint of the structures they build. Instead, their building activity probably proceeds from an interplay among several mechanisms that may include: a genetically programmed sequence of behavior triggered by cues arising during construction, and stimulus-response behaviors that lead to self-organization through interactions among branches that are carried and moved about by the water currents and by the beavers themselves. The sound of water rushing over the dam or through holes in its structure is a key stimulus guiding certain aspects of the building behavior (Richard 1968, 1980). (Illustration by Mary Ellen Didion)

5

Why Self-Organization?

It is all but impossible to conceive how any one colony
member can oversee more than a minute fraction of the
construction work or envision in its entirety the plan of such
a finished product. Some of these nests require many
worker lifetimes to complete, and each new addition must
somehow be brought into a proper relationship with the
previous parts.
                                              —Edward O. Wilson

Self-Organization Versus Alternatives

This chapter addresses the question: "Why is self-organization sometimes
favored over other means of pattern formation in biological systems?"

Generally speaking, the rules in self-organizing systems can be quite eco-
nomical in the physiological and behavioral machinery needed to implement
them. They are more likely, therefore, to arise through evolutionary processes and more likely to carry smaller costs than more complicated rules.
Thus, we expect evolution by natural selection to favor mechanisms based on
self-organization whenever the alternative mechanisms—leaders, blueprints,
recipes, and templates—are unworkable or costly to implement in terms of genetic coding. For a more telling answer, however, we must consider exactly
why these alternatives are apt to work so poorly in particular circumstances.

Drawbacks of Alternatives

Why do biological systems often rely on interactions among their components rather than guidance from an external source to direct pattern-formation
processes? We believe that the answers to such questions reflect the limited
communication and cognitive abilities of individuals in a system; the problems
of making and using blueprints; the need for components to flexibly coordinate
their contributions to the desired pattern; and the lack of naturally occurring
templates.

Central Authority

The quotation from E. O. Wilson points up the engrossing enigma of how large complicated nests of social insects such as hornets and fungus-growing termites are built. Certainly one of the major problems associated with having a complex system run by a central authority is that it requires both an effective communication network among individuals and sophisticated cognitive abilities by the central planner.

Consider what, in essence, is required for an animal group to be directed by a leader as it builds a particular pattern. For this to happen, the leader of the group must have thorough knowledge of the desired pattern, must be able to maintain a synoptic view of the emerging structure and to devise and communicate instructions to all the other group's members. Obviously, such centralized control by a leader places formidable, if not impossible, burdens of information acquisition, processing, and transmission on the leader, especially if the group is large and the pattern being built is far larger than any individual group member. Such is the case for our favorite example of group-level pattern formation—nest construction by a colony of *Macrotermes* termites—where group size can exceed a half million individuals and the construction is some ten million times more massive than any of its builders (Figures 18.1 and 18.2).

Even in smaller groups of animals such as beavers (Figure 5.1), extensive complex habitats may be modified over many generations. In such cases it is not at all surprising that natural selection has favored a decentralized, self-organizing approach to pattern formation rather than relying on the "direct intervention of some kind of 'little architect' or 'construction demon'" (Edelman 1984, p. 120).

Problems with Blueprints

The problems of information collection, processing, and dissemination by an omniscient and clever supervisor are avoided if each member of a group has a personal copy of a blueprint, either mental or external, indicating the pattern to be built. If the blueprints possessed by the group's members are similar, then all the individuals should work toward a common goal and the final pattern should be coherent. It seems clear, however, that blueprints are not a widespread mechanism for guiding pattern formation in groups, with the obvious exception of human groups (Figure 5.2). Why is this? Perhaps one important reason is that it would be extremely costly to encode genetically the vast quantity of information that would need to be expressed in a mental blueprint for a complex structure, such as a termite nest.

Perhaps, too, it would be extremely difficult to transform information that is ultimately stored in an individual's genes into a detailed mental blueprint during development. Finally, it also seems that a serious problem is associated

Figure 5.2 The regularity seen in man-made structures, such as these, seldom involves self-organized processes. Instead pattern generally arises by means of preconceived plans and the use of devices such as templates, recipes, and blueprints. (a) A wall whose component stones have been fit together precisely. (b) Intricate crochet work, the result of careful planning and the implementation of a step-by-step recipe.

with executing the information expressed in a blueprint, because a blueprint does not specify *how* something is to be built, only *what* is to be built. This means that a blueprint does not provide a complete set of instructions for building a structure. Any animal guided by a blueprint must be able to figure out the actual building operations needed to produce a structure, and this may be extremely difficult. The difficulty will be especially great if the pattern is large and complex and requires a specific sequence of stages in the pattern-formation process. Thus it seems that reliance on blueprints is generally impossible for many animal groups, since this would require each group member to have an immense investment in genetically coded information and an unrealistically high level of mental sophistication.

Rigid Recipes Cannot Guide Flexible Building Behavior

Pattern-formation based on following a recipe circumvents the problem of instructions that are difficult to execute, for by its very nature a recipe provides step-by-step instructions for pattern-formation. It also skirts the problems associated with a leader, since it is inherently decentralized with each individual independently contributing to the pattern. Recipes, therefore, seem to be an excellent way to provide the instructions for pattern-formation. The instructions for structures built by single organisms, such as the cocoons of silkworm moths or the webs and egg cases of spiders (mentioned in Chapter 4), often take the form of a recipe.

A serious problem arises, however, when members of a group must use a recipe to work together to build a collective structure. Although the sequential instructions in a recipe are well-suited to a solitary builder, they are poorly suited to a group-building operation. A solitary builder such as a spider can be expected to encounter circumstances in a predictable sequence that would allow it reliably to execute its behavioral recipe. In contrast, an individual working within a collaborative group generally will not perform stereotyped sequences of building activities, since what it needs to do at each stage of building a pattern depends little on what it personally has done most recently. Rather, it depends primarily on what its fellow group members have recently accomplished. Thus individuals in a cooperative group need instructions that confer the extreme flexibility of behavior needed for coordinating the activities of many individuals. This need runs contrary to the basic nature of a recipe.

Templates Are Not Always Available

In most instances of pattern formation by biological systems the pattern is built from scratch in an environment that lacks anything remotely resembling a template for the pattern. Consider, for example, the example of pattern formation by a social insect colony—the air-conditioned termite mound (Figures 18.1 and 18.2). Mound construction starts underground and eventually extends above ground to form an impressive edifice, but in neither the subterranean nor the aerial construction phase are the termites guided by a three-dimensional template. As discussed in Chapter 18, however, one part of a termite mound's construction where a template may play a role is the building of the royal chamber, a thick-walled, bun-shaped cubicle completely enclosing the queen, and providing extra protection from intruders. The royal chamber's inner walls follow closely the contours of the queen's body, so it seems reasonable to hypothesize that the queen's body provides a template for chamber construction. Where such naturally occurring templates exist, they are expected to guide pattern-formation, but one suspects that such templates will prove very much the exception rather than the rule.

Summary

A reasonable suggestion is that pattern-formation by cooperative groups usually arises through self-organization rather than external guidance because the latter mechanisms generally are exceedingly difficult to implement. This seems especially true for pattern-formation by large groups. For large groups the high complexity and large scale of pattern-formation makes it virtually impossible for a leader to provide group members with detailed building instructions, leaves blueprints an insufficient source of instructions, renders fixed recipes of behavior inappropriate for the flexible building behavior that is required, and makes the occurrence of naturally occurring templates highly unlikely. Group-level pattern-formation through self-organization, in contrast, is based on rather simple instructions—which we perceive as rules of behavior—that are easily implemented by each member of the group.

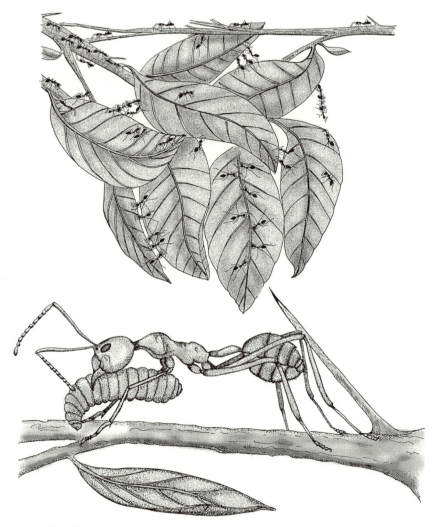

Figure 6.1 Weaver ants at work (*Oecophylla* sp.) (Illustration by Anne-Catherine Mailleux)

6

Investigation of Self-Organization

To bring a quality within the grasp of exact science, we
must conceive it as depending on the values of one or more
variable quantities, and the first step in our scientific
progress is to determine the number of these variables
which are necessary and sufficient to determine the quality.
—James Clerk Maxwell (quoted in Winfree, A.T.,
The Geometry of Biological Time)

One of the principal objects of theoretical research in any
department of knowledge is to find the point of view from
which the subject appears in its greatest simplicity.
—J. Willard Gibbs (quoted in Winfree, A.T.,
The Geometry of Biological Time)

Experimental Investigations

The study of self-organization logically starts with empirical studies that characterize the group-level pattern in detail so that we have a clear picture of the basic phenomenon to be explained (Figure 6.2). Then we look inside the group to identify its subunits and determine through empirical study the nature of their interactions. As previously outlined in Figure 2.4, we need to determine the information flow among the subunits and their behavioral rules of thumb. Ideally, one then proceeds to study the behavior of the intact group while delicately interfering with particular properties of its members, to determine what role these properties play in shaping the group's performance. This experimental stage of the investigation is often essential to unraveling the mechanisms of group functioning, since it can reveal subtle but important consequences of particular properties of the group's members.

Formulation of Rigorous Models

The approaches just described are likely to yield strong suggestions about how a group works. However, testing the accuracy and completeness of one's understanding requires a further stage, the formulation of a rigorous model that embodies current understanding of how the group works. Here a bottom-up approach is taken to model building, using empirical findings rather than intuition to shape the model. In the top-down approach, intuition rather than empirical findings often drives development of the model.

Modeling frequently requires translating a verbal understanding of the interactions among group members into a mathematical form such as a simulation or a set of equations. This exercise is itself useful as it requires the verbal postulates to be stated more precisely than they usually are in the initial conception of the mechanism. However, the principal aim of modeling is to check whether the processes identified through empirical analyses do produce the actual performance of the intact group. Unfortunately, simple qualitative reasoning often fails to predict the properties of systems, such as animal groups, with many components and complex dynamic interactions. Mathematical equations and computers, on the other hand, enable prediction of the properties of complex systems and provide a means of evaluating a model of a group's internal machinery. (The recent explosion in interest in self-organization is due, in part, to easier access to computers.)

If the predictions of the model fail to agree with the observations of the group's actual performance, then, clearly, at least one important aspect of the behavior of the group's members is poorly understood. In this situation, additional empirical investigations are needed to improve knowledge of how the group works. At this point we can again evaluate our understanding by formulating and testing a refined model of the group. Each repetition of the cycle of observation, experiment, model building, computer simulation, and reality check yields a better understanding of the subject. This approach to studying a self-organized system is summarized in Figure 6.2.

It may be useful here to clarify a common misunderstanding. If we proceed as outlined in Figure 6.2, the model will not automatically generate the global (group-level) pattern that was the impetus for the study. If the model is built objectively and honestly, it is *not* designed a priori to generate the global pattern. The model should be developed solely based on observations and experimental data concerning the subunits of the system and their interactions. Ideally the researcher would not even need to know the characteristics of the global pattern until the end of the study. In many situations, this is not the case. As pointed out in Chapter 7, it may be temping to vary the parameters of a model until the desired global pattern is produced. Once this is done, it is further tempting to make ad hoc explanations for a particular choice of parameter or assumption in the model. Such an approach is strongly discouraged. It may make for an "attractive" model, but it hardly increases understanding of the system under study.

One goal in studying a system that we suspect may be self-organized is objective model building. A second goal is to verify the model. A match that may be found between the global pattern and the output of the model is not particularly strong evidence of the model's validity. The next step is to manipulate

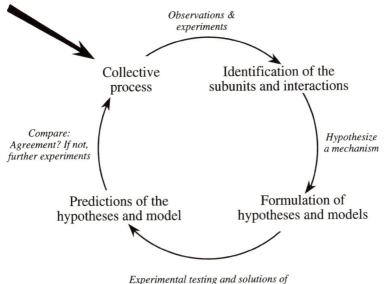

Observations &
experiments

Collective
process

Identification of the
subunits and interactions

Compare:
Agreement? If not,
further experiments

Hypothesize
a mechanism

Predictions of the
hypotheses and model

Formulation of
hypotheses and models

Experimental testing and solutions of
the model/simulations

Figure 6.2 A typical cycle of studies (shown above) involved in the study of how patterns and structures arise through self-organization.

the system in some way and observe whether behavior predicted by the model matches observed behavior of the system after the manipulation.

Ideally, one designs a manipulation that can generate different predictions for each of several alternative hypotheses of how the system works. If the manipulation generates responses predicted by the self-organization hypothesis but not by alternative hypotheses, this provides strong support for self-organization. A goal of this chapter, and of the book, in general, is to stimulate and provide useful guidelines for research on self-organized systems.

A fixed sequence of discussions is followed for each biological system discussed in Part II of the book: (1) description of the collective process; (2) identification of the lower-level subunits and a description of their interactions; (3) formulation of a model hypothesis; (4) generation of predictions of the model; and (5) comparison of the predictions of the model with the collective process. Possible alternative models are discussed at the same time along with experiments that have attempted to distinguish between self-organization and other mechanisms of pattern- and process-formation.

The Purposes of Models

As discussed in the next chapter, some biologists question the value of models. To support our approach, therefore, we suggest three useful purposes for constructing models: to predict what will happen in the physical world; suggest and direct further experimentation and observations; and foster a greater understanding of the system under study.

Most of these uses are relevant to the study of biological self-organization. Regarding the first point, if a model we formulate corresponds well with reality it will have predictive power. If it accurately predicts the outcome of a novel situation that is verified experimentally, it increases our confidence in the validity of the model. This interplay between model and empirical work also helps to improve our understanding of the biological system. A model serves as a precisely formulated, quantitative hypothesis about how a system works and makes specific predictions that often can be tested experimentally, allowing one to falsify the hypotheses upon which the model was based. Once one has achieved high confidence in the model, one can begin to use the model to perform thought experiments or collect data that would be impossible or difficult in real life, thereby further increasing our understanding of the biology of the study system. Ultimately, the model can enable us to examine crucial features of the system that are not open to experimental analysis, since a model allows us easily to modify the type and magnitude of the interactions within the system. This also allows us to determine which features are most important for generating particular system-level properties.

In the formulation of a model, the investigator is forced to be explicit about each detail of the mechanics of a system. One consequence of this is that in the process of model building, the investigator often realizes that no information is available about a particular part of the system. This stimulates further analysis to acquire the missing information. Thus building a model can help one appreciate the biological richness of one's subject. For example, in formulating a model of ant foraging, one quickly realizes that the model can be set up with ants laying pheromone trails either as they leave the nest, as they return to the nest, or both. Also, trails may be laid down in greater or lesser amounts depending upon whether the foragers have found food. Model building makes clear the need for information about such details.

Model building can be considered an art in its own right. We feel, however, that modeling for modeling's sake often results in a diminution of interest in the subject being modeled. For the biologist who is generally interested in using models as a means of increasing his understanding of a particular living system, model building for its own sake may seem a frivolous endeavor.

Specific Models of Self-Organized Systems

In preparation for Parts II and III, a brief introduction is offered here to techniques for formulating models of biological systems and to the types of models one can use. We will first present a simple example of group-collective activity and model it by two different approaches: first with differential equations, and then by Monte Carlo simulation. Next, we describe a more complex example and show how it can be modeled as a cellular automaton. This exercise will illustrate the advantages and disadvantages of each type of model.

Differential Equation Models

Consider the model illustrated in Figure 6.3 of ants foraging without recruitment. A food source (a feeder with sugar syrup) is placed outside a small colony of ants. The nest has a total of a hundred foraging ants that pass between each compartment (nest, food, etc.) at an average rate (r) equal to the reciprocal of the time spent in each compartment. The total time in each compartment includes the travel time to the next compartment. One can observe these ants and gather data about the behavior of individual foragers. Assume that many ants were observed to behave as follows: (1) an ant spends an average time $(1/r_N)$ of five minutes inside the nest before leaving to forage; (2) an ant that left the nest to find food generally finds it within one minute $(1/r_D)$, or, if it is unsuccessful spends an average thirty minutes $(1/r_S)$ searching before giving up and returning to the nest; (3) once an ant has found the feeder, it takes an average two minutes $(1/r_F)$ to load up and return to the nest. For simplicity, assume that ants do not improve their searching or foraging abilities over time (no learning), so that these parameters are constant during the course of the observations. We would like to know the behavior of the system, in other words a description of the time course of the distribution of the ants. This is a hypothetical simplified situation, but it introduces the more realistic and complicated models presented in Chapters 12 and 13.

The number of ants in the nest, at the feeder, and still searching for food at any given time is determined by solutions of the following simultaneous

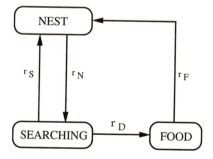

Figure 6.3 In a compartment model of foraging, ants leave the nest to search for food; successful ants find the food and load up before returning home, while unsuccessful ants return directly home. The parameters, r_N, r_S, r_D and r_F are rate constants for movement between compartments N, S, and F.

differential equations:

$$\frac{dN}{dt} = r_F F + r_S S - r_N N,$$

$$\frac{dS}{dt} = r_N N - r_S S - r_D S,$$

$$\frac{dF}{dt} = r_D S - r_F F,$$

where

N = number of ants in the nest,

F = number of ants at the feeder,

S = number of ants searching for food,

r_N = rate at which ants leave the nest,

r_F = rate at which ants leave the food source,

r_S = rate at which ants return from unsuccessful searching,

r_D = rate at which ants discover food during searching.

The first equation, for example, states that the change in the number of ants in the nest per unit time is equal to the rate at which ants return from the food source ($r_F F$), plus the rate at which ants return from unsuccessful searching ($r_S F$), minus the rate at which ants leave the nest ($r_N N$). In addition to these equations, we need to know the total number of foraging ants and the initial conditions of the system. We assume, for simplicity, that a hundred ants are foraging, and all start in the nest (at time $t = 0$, $N_0 = 100$). This set of equations may be solved using a variety of analytical or numerical techniques for integrating differential equations. A computer program to solve this system of equations may be written (see, for example, Press et al. 1986), or one may use available software for modeling differential equations. With any of these methods, the solutions of the equations for F, S, and N are as shown in Figure 6.4. The distribution of ants among the compartments reaches an equilibrium after about five minutes, with 63.3 ants in the nest, 24.5 at the food source, and 12.2 ants searching. Note that in our differential equation model it is possible to have fractional numbers of ants in a compartment. In such a simple differential equation model it is easy to solve for the stationary distribution of ants in each compartment without recourse to the computer. One can simply set all the derivatives—dN/dt, dF/dt and dS/dt—equal to zero (indicating there is no change in the numbers of ants in the compartment over time) and algebraically solve the system of equations. Recalling that the total number of foraging ants, $N_0 = N + F + S = 100$, we can derive the following equations for the number

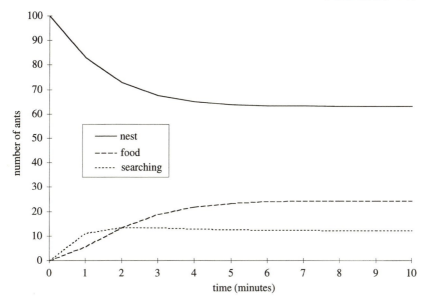

Figure 6.4 Numerical solutions of the differential-equation model shown in Figure 6.3 indicate the number of ants in each compartment (nest, searching, food) over time.

of ants in each compartment after the system has reached equilibrium:

$$S = \frac{r_N N}{r_S + r_D},$$

$$F = \frac{r_D S}{r_F} = \frac{r_D r_N N}{r_F (r_S + r_D)},$$

$$N = \frac{N_0}{\left(1 + \dfrac{r_N}{r_S + r_D} + \dfrac{r_D r_N}{r_F (r_S + r_D)}\right)}.$$

This illustrates one of the advantages of employing differential equation models—it is sometimes possible to derive explicit solutions.

Monte Carlo Simulations

Differential-equation models do not treat each ant as an individual behaving in a probabilistic (stochastic) manner, but assume the ants are a population of essentially divisible individuals behaving deterministically. This is reasonable when the number of ants is large and a continuous model provides an adequate description of the global behavior of the system. Simulation models, in contrast, often use a discrete approach that models the behavior of each individual in the system. Such simulations can yield the kind of group-level pattern that might emerge from an actual experiment. The ant-foraging system can be implemented as a Monte Carlo simulation that incorporates the randomness one observes in biological units such as social insects.

The Monte Carlo simulation presented here keeps track of each individual. The actions of each individual are implemented in a probabilitistic manner based on behavioral rules of thumb derived from experimental observations or reasonable assumptions. Figure 6.5a shows the results of a Monte Carlo simulation of the ant-foraging model. A pattern similar to the results of the differential equation model is evident except that the number of ants in each compartment continually fluctuates around the mean value obtained from the differential equation model, as expected from the probabilistic nature of the simulation. Nonetheless, after many runs of the simulation the average results of the Monte Carlo model are the same as those of the differential equation model. Figure 6.5b, for example, shows that the average of ten simulations begins to approach the values of the differential equation model[1] seen in Figure 6.4.

Cellular Automaton Models

A cellular automaton (CA) simulates groups of biological components by arranging the components as points in a lattice or grid and allowing the components to interact according to simple rules. The CA simulation is usually performed in a two-dimensional lattice. Subunits change their states in a series of discrete time steps based on their proximity to neighboring subunits, their own state, and the states of their neighbors. (See Box 6.1. Ermentrout and Edelstein-Keshet [1993] offer an excellent review of CA models in biology.)

An example will make the process clear. Consider a simple model of the propagation of a contagious disease through a community of individuals.[2] Here the subunits are individuals living in particular locations. The interaction between individuals is contagion—the likelihood that one individual will infect another with the disease. In this simplified model, an individual can be in one of three states: not infected, infected, or recovered from the infection. For simplicity, all noninfected individuals are equally susceptible to infection, all infected individuals are equally contagious, and all recovered individuals are no longer susceptible or contagious. We assume that individuals do not move around but remain fixed in their locations, so that they only infect their nearest neighbors. We thus have a system of discrete individuals, located on a discrete space. The interaction—infection—is a function of many factors including the virulence of the pathogen, susceptibility of individuals to the infection, duration of the illness, and the density and pattern of distribution of individuals. The question is: What is the temporal and spatial pattern of disease-propagation in the population?

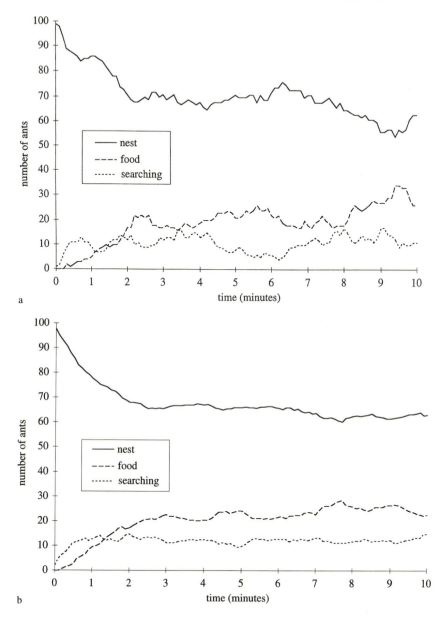

Figure 6.5 Results of a Monte Carlo simulation of the ant-foraging model outlined in Figure 6.3 are shown in (a), while (b) shows results of the same Monte Carlo simulation of ant foraging averaged over ten iterations. The results after ten iterations begin to converge on the values for the solution to the differential equation model in Figure 6.3.

This model of disease transmission has been simplified to just three parameters: the initial density of individuals in the community, the probability that an infected individual will infect his neighbor, and the duration of the illness. As discussed in Chapter 3, any of the parameters can be tuned to affect the pattern of disease transmission. Certain combinations of these three parameters will result in one behavior of the system, namely an infection that dies out without spreading through the population. Other combinations, such as a high density of individuals (crowding), high infectivity, and long disease duration, result in an epidemic that spreads through the community (Figure 6.6a, b, c, d). The parameter space of the model's behavior can be explored by systematically varying the three parameters and plotting the results in three dimensions. The graph will show two-dimensional surfaces that demarcate transitions (bifurcations) between three-dimensional regions that exhibit different behaviors.

We can imagine that the goal of a pathogen is to increase its fitness by assuring efficient spread throughout the population. Natural selection favors variants of the pathogen that efficiently reproduce and propagate, and thus, acts on such a system by selecting pathogens with characteristics most favorable for propagation within a particular host. The process of selection is equivalent to the process of parameter-tuning. Over time, natural selection tunes the behavioral and physiological parameters of the pathogen, changing the characteristics of the host-pathogen system.

Similarly, from the host's point of view natural selection tunes the host response to minimize detrimental effects on host fitness. The point is that by describing infectious processes as dynamical systems modeled with a CA in no way contradicts or minimizes the role of natural selection in the process. Rather, the dynamic system approach emphasizes that natural selection can modify the interactions among system components.

As it stands, this simplified contagion model is a stylized caricature of disease transmission, but there is no reason why it cannot be made more quantitative and realistic. We could specify a particular spatial scale for the lattice, a realistic time step, and allow each individual to move about in a certain pattern and at a specified speed, corresponding to experimentally obtained data. We could obtain a more quantitative measure of the probability of disease transmission from individual to individual based on actual epidemiological data, and include individual variability in these measures. We could also add additional states to the model, for example an immune state, representing individuals who already had the infection or were immunized, and specify a duration of immunity. In this manner, it might be possible to use the model to determine the proportion of randomly chosen individuals who should be immunized to prevent an epidemic or the average number of infected individuals

Figure 6.6 This cellular automaton (CA) model of the spread of infection through a community has white circles representing uninfected individuals, grey circles infected individuals, and black circles post-infection recovered individuals.

that would result in an epidemic. CA models have become increasingly popular for modeling various natural phenomena and are being implemented with greater mathematical sophistication to yield useful quantitative predictions as well as graphically striking qualitative results.

Benefits and Drawbacks of Different Models

There are several important differences between the three modeling approaches. Differential-equation models are the most common modeling approach. By solving a system of differential equations, we obtain a precise quantitative description of the behavior of a system over time. We can analyze the system, characterizing equilibrium conditions and bifurcation phenomena. The disadvantage of differential equations is that it may be difficult to formulate a set that accurately describes the biological system. Furthermore differential-equation models are most appropriate for continuous rather than discrete processes, so that some assumptions must be made when using such equations to model discrete processes. In addition, especially for a nonmathematician, differential-equation models can be so abstract that it is difficult to intuitively understand how the system works. Yet another problem for most biological systems is that the equations rarely have an analytical solution and so computers must be used to obtain numerical solutions.

The most important advantage of Monte Carlo and CA simulations is that they are often easier to implement than differential-equation models. CA models in particular provide striking visual results where the patterns of activity often resemble the patterns observed experimentally. This is clearly seen in the CA model of disease presented above.

In the Monte Carlo simulation of ant foraging, the only way to determine the equilibrium distribution of the ants is to run the simulation many times and compute the average results of all the runs. Systems with large numbers of subunits, moreover, can require a long period of real-time simulation. Nonetheless, it is often easier to incorporate a variety of biological nuances into the model. In CA models, as the grid size becomes large one reaches a point where the simulation becomes extremely slow, since at each clock tick the program must make calculations for each cell in the entire array based on the state of each neighboring cell. Finally, just as a certain mathematical expertise is needed to solve differential equations, a certain expertise in computer programming is needed to design simulations. A variety of software programs are available for implementing systems of differential equations that actually make it easier to implement that type of model of ant foraging than the Monte Carlo simulation.

Throughout this book we will present models based on differential equations, Monte Carlo simulations, and CA. This diversity reflects the fact that for a particular case one type of model often is easier to implement (or is more useful) than the others. Unfortunately, in most cases only one or another model has been developed to explain a particular pattern or superstructure. In a few cases, however (for example, Chapter 16), different models are available for comparison.

Box 6.1 Cellular Automaton Models—Details and Examples

In a cellular automaton (CA) model, the subunits (cells) of the system are discretely arrayed on a grid or lattice, often in two dimensions (Figure 6.7). Each cell is characterized by a location and its condition (state). A cell interacts with a specified set of neighbors, based upon a set of simple rules. The rules take into account the state of the cell itself, and the state of its neighbors. A typical rule specifies the transition of the cell from one state to another as the system evolves in a series of discrete time steps called generations. One of the best known CA models is Life[3] (Figure 6.7), invented by John Conway and popularized by Martin Gardner (1970, 1971). Although this model epitomizes a complex system, there is nothing complicated about the individual components of the system. A few simple rules govern component interactions using only local information. The system may consist of numerous components but they are all similar.

Although this is merely a stylized model of the dynamical evolution of a population of organisms and without real biological relevance, its simplicity makes it a good didactic example. It also has been used to explore a number of concepts in self-organization such as self-organized criticality (Bak et al. 1989). The neighborhood of each cell in the model consists of its eight immediate neighbors. Each cell can be in one of two states (alive or dead—on or off), and the following four rules determine the time evolution of the system:

1. A live cell surrounded by two or three live cells at time, t, will survive (remain alive) at time $t + 1$.

2. A live cell with no live neighbors or only one neighbor at time, t, will be dead at time $t + 1$. (It dies of loneliness.)

3. A live cell with four or more live neighbors at time, t, will be dead at time $t + 1$. (It dies of overcrowding.)

4. A dead cell surrounded by three live cells at time, t, will be alive at time $t + 1$. (It will be born.)

One of the fascinations with this particular CA model is the diversity of patterns and structures that can evolve over time with such a simple set of rules—an example of emergence.

Although the Life CA serves merely as a metaphor for population dynamics in an ecological system, more realistic models of biological,

Box 6.1 continued

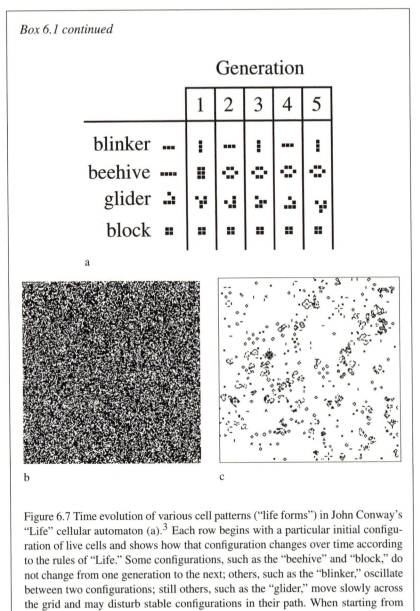

Figure 6.7 Time evolution of various cell patterns ("life forms") in John Conway's "Life" cellular automaton (a).[3] Each row begins with a particular initial configuration of live cells and shows how that configuration changes over time according to the rules of "Life." Some configurations, such as the "beehive" and "block," do not change from one generation to the next; others, such as the "blinker," oscillate between two configurations; still others, such as the "glider," move slowly across the grid and may disturb stable configurations in their path. When starting from a dense random array of cells (b), far more complicated configurations occur (c) and the pattern may not settle down for hundreds or thousands of generations. The underlying grid, in which each cell has eight neighbors, is not shown in these diagrams.

Box 6.1 continued

chemical and physical phenomena have been presented. Examples include the chemical system of the Belousov-Zhabotinsky reaction (Gerhardt et al. 1990; Markus and Hess 1990); wind-induced ripple patterns of sand dunes (Forrest and Haff 1992); stripes of stones in alpine and polar regions (Werner and Hallet 1993); interspecific competition among grasses (Silvertown et al. 1992); vertebrate skin patterns (Young 1984; Murray 1981, 1988); the formation of ocular dominance stripes in the brain (Swindale 1980; Miller et al. 1989); mollusc-shell pigment patterns (Lindsay 1977; Ermentrout et al. 1986); and the growth of bacteria and fungi (Ermentrout and Edelstein-Keshet 1993). The models have also been extended to include systems in which the cells can move across the lattice, as in the case of fibroblast aggregation (Edelstein-Keshet and Ermentrout 1990; Ermentrout and Edelstein-Keshet 1993) and the formation of stable foraging trails in ants (Ermentrout and Edelstein-Keshet 1993).

Various CA simulations (including many of the examples mentioned in the previous paragraph) can be downloaded at our website:

Figure 6.8 (starting next page). Some examples are shown of biological-pattern formation and the simulation of similar patterns by means of cellular automaton models. Although a striking resemblance may be found between the biological pattern and its simulation, the actual mechanisms of pattern formation remains to be confirmed experimentally in most of these systems. (a) Vermiculated rabbit-fish (*Siganus vermiculatus*) skin pattern and its simulation[4] by a two-dimensional cellular automaton. Simulation of the rabbit-fish (*Siganus vermiculatus*) skin pattern (a)i–(a)iii shows the emergence of the stable pattern from an initial, random distribution of pigmented and unpigmented cells (i, time = 0; ii, time = 1, iii, time = 10, in arbitrary units). Simulation of the zebra (*Equus burchelli*) (b) skin pattern[5] started from a random array of cells [i, ii, and iii as in (a)]. The porphyry olive (*Oliva porphyria*) shell pattern is simulated by a one-dimensional cellular automaton, (c) and (c)i, and the lichen-growth pattern, (d) and (d)i, is simulated by diffusion-limited aggregation. (Sander 1986, 1987)

Box 6.1 continued

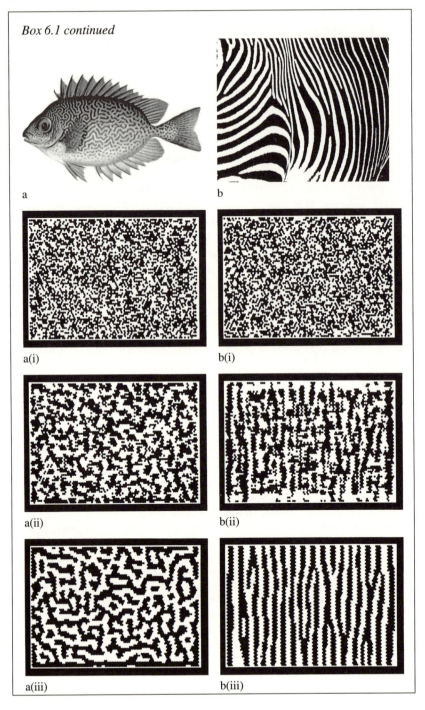

a

b

a(i)

b(i)

a(ii)

b(ii)

a(iii)

b(iii)

Box 6.1 continued

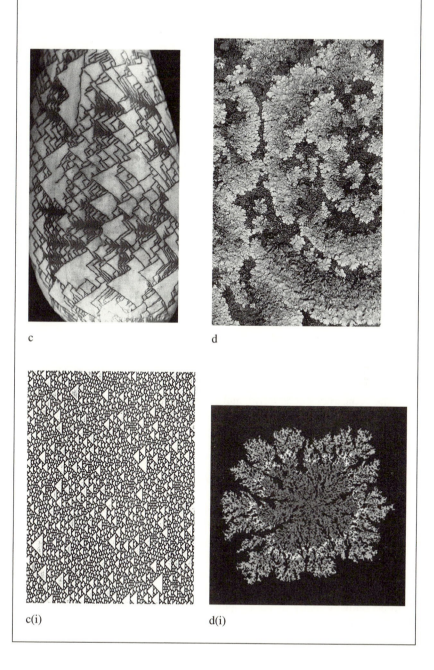

c

d

c(i)

d(i)

Box 6.1 continued

http://beelab.cas.psu.edu. Links to other sites containing similar simulations are also provided. In particular, we recommend visiting the StarLogo website described in Box 6.2.

Box 6.2 StarLogo—A Programming Language for Simulating Decentralized Processes

StarLogo is a programmable modeling environment developed by the Epistemology and Learning Group at the MIT Media Lab for understanding decentralized systems. As such, it is an excellent didactic tool for studying the self-organizing processes presented in this book. In the words of the developers of StarLogo, the programming language was designed for

> exploring the workings of decentralized systems — systems that are organized without an organizer, and coordinated without a coordinator.... In decentralized systems, orderly patterns can arise without centralized control. Increasingly, researchers are choosing decentralized models for the organizations and technologies that they construct in the world, and for the theories that they construct about the world. But many people continue to resist these ideas, assuming centralized control where none exists—for example, assuming (incorrectly) that bird flocks have leaders. StarLogo is designed to help students (as well as researchers) develop new ways of thinking about and understanding decentralized systems.

StarLogo allows one to simulate and gain insight into many real-life phenomena, such as bird flocks, traffic jams, ant colonies, and market economies. The program can be downloaded from the MIT StarLogo website at http://www.media.mit.edu/starlogo. Included at the website are a variety of demonstration simulations. In addition, we have set up a StarLogo website at http://beelab.cas.psu.edu with programs written to accompany many of the chapters in this book. Simulations of Conway's Life, epidemics, animal coat patterns, slime mold aggregations

Box 6.2 continued

(Chapter 8), beetle aggregations (Chapter 9), synchronized firefly flashing (Chapter 10), bird flocks (Chapter 11), honey-bee cluster thermoregulation (Chapter 15), and honey-bee comb pattern development (Chapter 16) can be downloaded from our site.

7

Misconceptions about Self-Organization

Perception uses rules of thumb, short-cuts, and clever
sleight-of-hand tricks that are acquired by trial and error
through millions of years of natural selection. This is a
familiar strategy of biology.
— Francis Crick (*The Astonishing Hypothesis:
The Scientific Search for the Soul*)

The study of self-organization in animal societies is a relatively new approach to understanding the mechanisms of social behavior. With any new approach, some misunderstandings are apt to arise. In this section we address issues that may trouble colleagues unfamiliar with the methodology presented in this book. On the other hand, some of these misconceptions may appear obvious to readers familiar with this field.

Misconception #1: Self-organization and natural selection are alternative explanations of evolution. Over the past few years, the excitement over developments in the study of self-organization has led to a number of statements which suggest that self-organization is an alternative to or, at least, a missing element in the explanation of evolution. For example, Waldrop (1990, p. 1543) states: "Complex dynamical systems can sometimes go spontaneously from randomness to order; is this a driving force in evolution?... Have we missed something about evolution—some key principle that has shaped the development of life in ways quite different from natural selection, genetic drift, and all the other mechanisms biologists have evoked over the years?... Yes! And the missing element...is spontaneous self-organization: the tendency of complex dynamical systems to fall into an ordered state without any selection pressure whatsoever."

Stuart Kauffman (1993, p. 10 and 78) urges us to reformulate Darwinian theory to include self-organization: "Since Darwin we have come to view selection as the overwhelming, even the sole, source of order in organisms. ... We [now] may have begun to understand evolution as the marriage of selection and self-organization."

Perhaps, therefore, the most important misconception to dispel is the notion that mechanisms of pattern formation based on self-organization somehow minimize the importance of natural selection. We wish to emphasize that this is not our view. Instead, we believe that natural selection is intimately linked to self-organization, since it molds the rules of interaction among the components of a living system. We suspect that the statements by Waldrop and Kauffman reflect the paradox presented in Chapter 3 that biological systems display far more complexity than can be accounted for by direct genetic coding. One of the revelations of self-organization studies is that the richness of structures observed in nature does not require a comparable richness in the genome but can arise from the repeated application of simple rules by large numbers of subunits. There is no contradiction or competition between self-organization and natural selection. Instead, it is a *cooperative "marriage"* in which self-organization allows tremendous economy in the amount of information that natural selection needs to encode in the genome. In this way, the study of self-organization in biological systems promotes orthodox evolutionary explanation, not heresy.

> It is a capital mistake to theorize before one has data.
> Insensibly one begins to twist facts to suit theories, instead
> of theories to suit facts.
> —Sherlock Holmes, *A Scandal in Bohemia*
> (Quoted in Winfree, 1980)

Misconception #2: The models that support hypotheses based on self-organization are often generated by tweaking the parameters of the model until the outcome of the model matches the observed data. If this is the case, they do not help the biologist understand how animal societies actually work. As already discussed in Chapter 6, the ideal model of pattern formation at the group level is a bottom-up model based on detailed empirical studies of the behavior of individuals within a group. Some aspects of the behavior inevitably will remain unknown even after extensive study, so in formulating a model for a particular group activity one must include hypotheses regarding parameters that cannot be measured experimentally and some estimates of their values. It should not be necessary, nor is it honest, to adjust parameters until the results of the model generate the desired global outcome.

Part of the utility of a model comes in creating the model, since the creative process helps pinpoint the gaps in our understanding, while the hypotheses formed to deal with these gaps become the foci of further empirical studies. Even before these studies have been accomplished, the simulations and reality checks with the model enable us to assess the plausibility of the proposed hypotheses for the unknown behavior of the group's members.

The ideal model not only faithfully generates what is observed in the natural situation, but also accurately predicts the outcome of perturbations of the system. If our model can provide a match between theoretical predictions and the observed behavior of the system after it has been experimentally manipulated in novel ways, then we have much stronger evidence for the validity of the model. At present, one of the greatest weaknesses of the self-organization approach in biology (e.g., morphogenesis or ecological oscillations) is that only a few examples exist of such experimental tests. (In Chapters 14 and 16, some experimental tests of the models are described.)

> The obvious point is that, even when we have the
> correct premises, it may be very difficult to discover what
> they imply.
>
> —Herbert Simon (1981)

Misconception #3: Self-organization models merely predict what one's intuition could tell us without the aid of a mathematical model. As mentioned in the previous chapter, the human mind can be extremely poor at intuitively predicting what happens when large numbers of individuals interact. Thus models really do help us explore the implications of behavior rules that we discover for single individuals but in nature are executed by tens, hundreds, or thousands of individuals as an ensemble. To cite an example of this from the later chapters, consider the development of foraging trails in ant colonies (see Chapter 13). The mathematical model of this pattern-formation process does make one admittedly trivial prediction, namely that when an ant colony is presented with two food sources differing in sugar concentration it will preferentially exploit the sweeter one. But the model also makes several nonintuitive quantitative predictions, including the minimum number of ants needed to develop a trail, the time it takes for a trail to form, and the pheromone volatility required to build a permanent trail.

Often models also make it easier to compare different systems. In the case of the foraging models of ants and bees presented in Chapters 12 and 13, we will see that although both foraging models involve similar mechanisms of recruitment and positive feedback, subtle differences in the interactions among individuals lead to striking differences at the collective level. These differences are clearly seen when one compares the structures of the two models.

> We should make things as simple as possible, but not
> simpler.
>
> —Albert Einstein (Quoted in Francis Crick, 1994)

Misconception #4: The self-organization approach fosters a falsely simplified view of nature. Living systems are immensely complex, so any explanation of a biological phenomenon involves some simplification in order for the human mind to begin to comprehend it. Nonetheless, we believe that often one can explain much—though certainly not all—of a given biological phenomenon in terms of a small set of surprisingly simple mechanisms. Our personal challenge, therefore, is to see whether particular instances of adaptive, group-level pattern formation can be explained largely or fully in terms of a small set of relatively simple behavioral rules for members of the group. These rules are often implemented in the form of a mathematical model or simulation. The primary goal of such models is not to include in minute detail every aspect of the system's biology, but rather to capture its essence. Therefore, a model based upon self-organization will often appear to be an overly simplified caricature of a biological process. Note that this approach is not a search for simplicity for simplicity's sake, but an honest attempt to determine the essential individual-level rules underlying the collective complexity. Whether the emerging results accurately represent real biological processes will be determined by the results of empirical studies.

> Here 'emergence' does not mean mysteries popping out of
> the undergrowth; it means that with a sufficient
> understanding of interactive processes, we should come to
> understand why a complex whole has properties its parts
> lack on their own, and how the parts are modified by the
> context in which they lie.
> —Richard L. Gregory (1994)

Misconception # 5: Emergent properties are a mystical notion without scientific basis. Self-organization studies have shown that the implementation of simple behavioral rules by large numbers of components in a system can yield unexpected structures and events not present at the level of the individual components. Because the global (collective) properties of the system often defy intuitive understanding of their origins, those properties may seem to appear mysteriously. There is nothing mystical or unscientific about their emergence, however.

Part I I

Case Studies

Figure 8.1 A schematic diagram of the life cycle of the cellular slime mold, *Dictyostelium discoideum*, shows that during starvation the cells aggregate to form a slug that develops into a fruiting body. Spores form at the top of the fruiting body and are released to continue the cycle. In the background, spiral waves of *Dictyostelium* amoebae move in response to cyclic adenosine monophosphate (cAMP). (Illustration © Bill Ristine 1998)

8

Pattern Formation in Slime Molds and Bacteria

Just at this time, I discovered an old and at the time little known paper by Alan Turing (1952). Turing demonstrated that a hypothetical system of interacting chemicals, reacting and diffusing through space, could generate a regular spatial structure which, he speculated, would provide a basis for subsequent morphogenetic development.

What was appealing about this view was that it offered a way out of the infinite regress into which thinking about the development of biological structure so often falls. That is, it did not presuppose the existence of prior pattern, or difference, out of which the observed structure could form. Instead, it offered a mechanism for self-organization in which structure could emerge spontaneously from homogeneity . . . . I am suggesting that the story of pacemakers in slime mold aggregation provides an unusually simple instance of the predisposition to kinds of explanation that posit a single central governor; that such explanations appear both more natural and conceptually simpler than global, interactive accounts; and that we need to ask why this is so.
    —E.F. Keller, *Reflections on Gender and Science*

Pattern Formation in Unicellular Organisms

The concept of pattern-formation as a result of self-organization is common in such disciplines as chemistry and physics. For instance, in Chapter 1 we discussed the patterns formed by the Belousov-Zhabotinsky reaction, the ripples on a sand dune, and Bénard convection. In each of these examples, no imaginitive leap is needed to assume that each individual in the pattern is unaware of the global pattern but is reacting only to local information, and that no overall leader organized the pattern-formation. Although local heterogeneities in the environment may well modify the final pattern, the pattern-forming processes themselves are classic examples of self-organization. More difficulties arise, however, when the interacting individuals are complex entities in their own right. It might be clear that individual sand grains are unaware of the overall ripple pattern on a sand dune, but it is much less clear whether individual honey bees are aware, for example, of the overall pattern of honey and pollen deposition on the comb. Before addressing patterns generated by higher organisms, it is useful first to consider patterns generated by organisms midway in complexity between, say, the molecules of the Belousov-Zhabotinsky reaction and individual honey bees. The classic example of pattern formation in such a system is that of the slime mold, *Dictyostelium discoideum*.

Various species of unicellular organisms—bacteria, myxobacteria, myxomycetes and cellular (amoebic) slime molds—exhibit a remarkable interplay between single-cellular and multicellular behavior. One of the best studied examples is that of the cellular slime mold, *D. discoideum* (Bonner 1967, 1983; Goldbeter 1996) (Figure 8.1). In the presence of a plentiful supply of bacteria (which the amoebae eat), each amoeba acts independently of its neighbors, growing and dividing with a doubling time of around three hours. This is called the growth phase of the life cycle. When *Dictyostelium* amoeba are starved, however, they enter the developmental phase and begin to aggregate, forming complex spatial patterns.

In the initial stages of aggregation multiple concentric circles and spiral patterns of amoeba are formed (Figure 8.2a). Individual amoebae move in a pulsatile manner toward the spiral cores or the circle centers and form clumps. In some strains the amoebae form streaming patterns as they move towards the centers of the pattern (Figure 8.2b,c). Aggregation leads to the formation of a multicellular organism called a slug, comprising about 10,000 to 100,000 cells that can move about on the substrate for some time. Eventually, the slug develops into a fruiting body, a spherical stalk with a cap on top that contains spores. Under appropriate conditions the spores can be released and germinate, thus completing the life cycle (Figure 8.1). Although all amoebae initially are identical, by the time the fruiting body has formed the cells have differentiated into a number of different cell types. Thus, *D. discoideum* provides a particularly interesting model system for studying not only intercellular interactions and the formation of spatial patterns but also cellular differentiation and morphogenesis.

Similar behaviors are observed in many bacterial species, particularly the myxobacteria (Kaiser 1983; Zusman 1984; Dworkin and Kaiser 1985). Like cellular slime molds, myxobacteria are predators that feed on other microorganisms, and travel in large clusters that have been likened to a wolf pack. Inside the cluster the movement of each individual bacterium exhibits little order, but the overall movement of the swarm is highly coordinated. Individual cells rarely leave the cluster permanently but may leave and return. In clusters of *Myxococcus xanthus*, periodic waves of movement, called ripples, are often seen. Ripples occur with a period of around 15 minutes, and appear to emanate from localized areas in a swarm. On starvation, the growth phase ends and

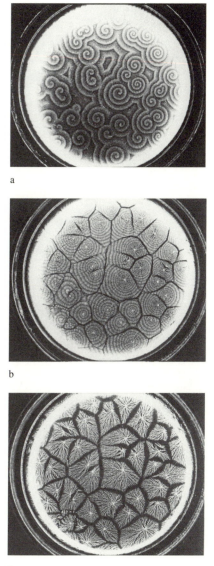

a

b

c

Figure 8.2 Aggregation of *Dictyostelium discoideum* on an agar plate (5 cm in diameter) reveals the formation of spiral waves of cAMP that induce (a) cell movement, (b) the onset of cell streaming, and (c) well-developed stream morphology. The photos were taken approximately 30 min apart. Using dark-field photography, the position of the cAMP waves in (a) and (b) can be inferred from the differential light scattering responses of elongated (moving) and rounded (stationary) cells. In (b), each spiral wave defines an approximate domain from which it has begun recruiting cells. After another 30 min (c) each domain is clearly separated from its neighbors. Inside each domain the cells have begun to form streams as they move toward the center. (Photographs courtesy of P.C. Newell)

the bacteria begin aggregating to form fruiting bodies in which the cells have differentiated into a number of different cell types.

Many bacterial species, although not exhibiting the same kind of swarming behavior, show extraordinary spatial patterns when grown on agar plates under special conditions, as shown in Figure 8.3 (Budrene and Berg 1991,

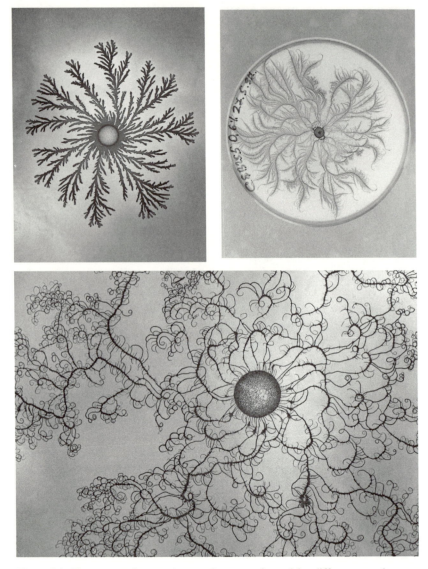

Figure 8.3 Three examples are shown of patterns formed by different morphotypes of *Bacillus subtilis* bacterial cultures grown on agar plates under various conditions. (Photos courtesy of Eshel Ben-Jacob)

1995; Ben-Jacob et al. 1994, 1995; Shapiro 1988, 1992). When a culture of *E. coli* is grown on agar from a single point, for example, the culture initially advances in a swarm ring that moves out from the point of inoculation in a con-

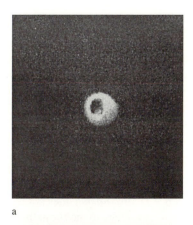

a

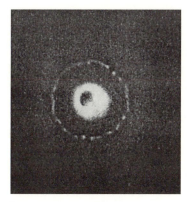

b

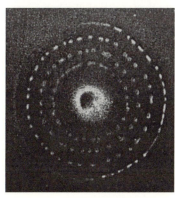

c

Figure 8.4 Evolution of an aggregation pattern of *Salmonella typhimurium* grown on agar: time after inoculation is (a) 38 h, (b) 40 h, and (c) 46 h. (Reproduced from Woodward et al. 1995, figure 2; used with permission)

centric manner. When the substrate concentration (succinate) is low, no other spatial patterns appear. As the succinate concentration is increased, however, the swarm ring begins to leave concentric rings of high cell density in its wake. The rings break into separate clumps. Thus the overall pattern is one of separate cell aggregations arranged in concentric rings. As the succinate concentration is increased still further, the cell aggregations begin to arrange themselves in more complicated spatial patterns, including clockwise and counterclockwise spirals and hexagonal lattices.

In a closely related species, *Salmonella typhimurium*, once the colony has spread across the culture, pattern formation begins at the center and moves radially outward, forming concentric rings of high cell density that spontaneously destabilize into concentric rings of separate aggregations (Figure 8.4). In contrast to the waves in *Dictyostelium* or the ripples in *Myxococcus*, the concentric rings are temporally stable.

Adaptive Significance of Multicellular Aggregation

Why bacteria and cellular slime molds should form such spatial patterns and aggregations is not always clear. There is a definite advantage for myxobacteria to feed in large groups, as they depend on enzymes, either secreted extracellularly or membrane-bound, to kill or immobilize their prey. The greater the concentrations of such enzymes, the greater the ability of the bacteria to digest resistant material, and to use a variety of food sources. Presumably, bacterial aggregation and the formation of a fruiting body confer advantages in the dispersal of the species. When food in one area is depleted, cellular aggregation allows them to reproduce in a way that increases the chances that progeny will be carried to places where there is more food. Since *D. discoideum* does not feed cooperatively, clearly there must be a significant advantage to the formation of multicellular fruiting bodies independently of other multicellular behavior.

One particularly intriguing possibility is that the spiral nature of the waves during the initial aggregation phase may itself have adaptive significance. As we shall see, these waves are probably the result of spiral waves of cyclic adenosine 3′, 5′-monophosphate (cAMP) in an excitable medium. A common property of such excitable media is that spiral waves often rotate at a higher frequency than other periodic waves. Once formed, spiral waves tend to predominate by pushing any other wave activity out to the boundaries. In *D. discoideum* this serves the purpose of increasing the size of the region from which cells are recruited to form the slug. Since larger slugs presumably confer benefits in terms of larger fruiting bodies and better dispersal, there may very well be selective pressure for the formation of spirals as opposed to simple periodic, concentric waves.

Alternatives to Self-Organization

When one considers how patterns arise in slime molds, no plausible alternative to self-organization seems possible. Each amoeba obviously responds to information in its local environment and each cell, in turn, has a limited range of relatively simple responses. It is not reasonable to suppose that an individual cell is at all aware of the overall pattern on the agar dish or in the colony. Hence, this process of pattern formation typifies self-organized activity in which local rules can lead to global patterning. Such observables as the speed and period of the spiral waves (discussed in detail below) are not generated by a master amoeba directing the movements of each individual, nor do they arise from each amoeba following a preset recipe or template. They result entirely from the diffusion and reaction of chemicals and the interaction of each amoeba with the local concentration of each chemical. Although the fine details are still under investigation and dispute, the overall mechanism is clear.

It is important to note that this does not rule out the possibility that spatial or cellular heterogeneity may significantly influence the overall pattern. For instance, in the 1960s it was widely believed that aggregation in cellular slime molds resulted from the emission of periodic signals by specialized cells that act as a central founder or pacemaker (Keller 1985). Although pacemakers may exist in the system, they are now believed to be those cells that, by chance, are the first to undergo development triggered by starvation (Lauzeral et al. 1997), and hence are a result of cellular heterogeneity.

Biological Basis of Aggregation in *D. discoideum*

Many models have been advanced to explain mechanisms of pattern formation and aggregation in bacteria (Keller and Segel 1971; Keller and Odell 1975; Lapidus and Schiller 1976; Lauffenburger et al. 1984; Agladze et al. 1993; Ben-Jacob et al. 1994; Woodward et al. 1995). Since an even greater amount of work has been done on cellular slime molds, in particular *D. discoideum*, we shall concentrate on this organism.

Although there is still disagreement about the exact mechanisms underlying the spatial patterning and aggregation seen in this slime mold, it is generally agreed that cAMP plays a vital role in the intercellular communication that makes such patterns possible. Starved *Dictyostelium* amoebae respond by producing cAMP and releasing it into the extracellular environment. What is particularly interesting about this secretion is that it can occur in two qualitatively different ways, depending on the exact experimental conditions. These two modes of secretion are usually called *oscillatory release* and *relay*. Gerich and Hess (1974) showed that starved amoebae in continuously stirred cell suspensions released a chemoattractant (now known to be cAMP) in an oscillatory fashion, with a period of around five to ten minutes. Since this period is identical with the period of pulsatile movement in the aggregation phase of the amoebae, it seems likely that such periodic secretion is one of the fundamental mechanisms underlying cell aggregation. The second type of dynamic behavior, the so-called relay, was first seen by Roos et al. (1975) and Shaffer (1975). If a cAMP pulse of sufficient amplitude is added to the medium in the continuously stirred cell suspension, the cells respond by secreting a much larger cAMP pulse.

Both secretory modes are the result of positive feedback at the level of the cAMP receptor. This is shown by the observation that increased concentration of extracellular cAMP causes increased production of intracellular cAMP, which may then leave the cell. In this manner, cAMP is produced by a process of positive feedback where cAMP stimulates its own production. If this were the only mechanism operating, the concentration of cAMP would quickly grow beyond reasonable bounds, and so there must be some mechanism to limit its production. Although it was initially thought that depletion of ATP was the

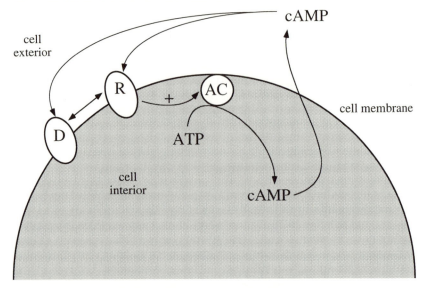

Figure 8.5 Schematic diagram of the mechanisms underlying control of cAMP secretion in *Dictyostelium* indicating that binding of cAMP to its receptor (R) leads to the activation of adenylate cyclase (AC) and subsequent production of further cAMP that is transported out of the cell. This positive feedback loop is indicated by a "+". In the negative feedback loop extracellular cAMP inactivates the cAMP receptor, converting it to the inactive form (D), thus leading to a decrease in further cAMP production.

limiting step, subsequent experimental results showed this was not the case, necessitating the development of more complex models. One class of models (Goldbeter and Segel 1977, 1980; Martiel and Goldbeter 1987; Tyson et al. 1989; Tyson and Murray 1989; Höfer et al. 1995) is based on the assumption that high concentrations of cAMP desensitize the cAMP receptor. Thus, cAMP has both a positive and a negative feedback effect on the production of cAMP. An alternative class of models (De Young et al. 1988; Monk and Othmer 1989, 1990; Tang and Othmer 1994) assumes that, when cAMP binds to its receptor, it opens calcium channels in the cell membrane thus increasing intracellular calcium concentration that inhibits the cAMP production. As it appears to be supported better by current experimental evidence, we shall focus on the first model in which receptor desensitization is responsible for limiting the production of cAMP (Figure 8.5).

A principal goal of the modeling will be to explain how this sequential positive and negative feedback in the production of cAMP can lead to the formation of concentric circles and spiral waves of cAMP concentration. As we shall see, a single model is able to account for both modes of cAMP secretion—oscillatory release and relay—merely by slight variations in some of the parameter values. Furthermore, the model shows how coupling of individual cells by the diffusion of cAMP can lead to the formation of regular spatial patterns as observed experimentally. Thus, the model provides a unified explanation for what may initially appear to be diverse phenomena.

The waves of cAMP concentration are only part of the story, however. Not only does each cell detect increased concentrations of cAMP and respond by producing cAMP itself, it also moves in the direction of increasing cAMP as shown experimentlly by Tomchik and Devroetes (1981). How exactly the cell does this is not known, but the result is clear. Periodic waves of cAMP moving through the culture cause corresponding periodic movement of amoebae (Tomchik and Devroetes 1987), which results in the clumping of cells at the core of the spirals or at the center of the concentric circles. However, the amoebae move at a speed of about 0.03 mm/min, which is approximately one-tenth the speed of cAMP waves. Thus, the problem of spatial patterning in this organism can be studied in two stages. First, models are used to study how cAMP molecules can be organized into spiral waves and concentric circles, while the background amoeba-density remains practically constant. Since the amoebae move so much slower than the cAMP waves, this is a good approximation. Second, the models for how cAMP waves are generated are coupled with models for the movement of the amoebae in response to cAMP gradients, so that interactions between these two processes can be studied.

Modeling Pattern Formation in *D. discoideum*

Our first goal will be to understand the mechanisms underlying oscillatory release and relay. Once we understand how these behaviors arise, we can study the consequences of coupling individual cells through diffusion of cAMP. This overall procedure is very common in modeling studies. In general, wave phenomena cannot be well understood until we have a good understanding of the behavior of the model in the absence of diffusion.

Oscillations and Relay

Martiel and Goldbeter (1987) proposed a model for the cAMP receptor and showed how the dynamics of the receptor could result in oscillatory secretion of cAMP. The cAMP receptor in their model can exist in four different states, as illustrated in Figure 8.6: R, the receptor active with cAMP unbound; D, inactive with cAMP unbound; RP, active with cAMP bound; and DP, inactive with cAMP bound. When the receptor is in the form RP it can combine with and activate adenylate cyclase, an enzyme that catalyzes the formation of cAMP from ATP, and increases the rate of intracellular cAMP production. The original model consisted of simultaneous equations describing movement of the receptor between these states, as well as the rates of production and degradation of extracellular and intracellular cAMP. However, Martiel and Goldbeter

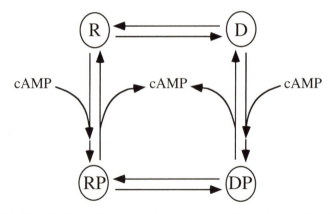

Figure 8.6 The cAMP receptor assumed to exist in four states: R, the active state with no cAMP bound; D, the inactive state with no cAMP bound; RP and DP, the active and inactive states, respectively, with cAMP bound. State RP combines with and activates AC.

showed that the equations could be simplified to a system with only three variables that represent changes in cAMP and the fraction of active receptors. It is only necessary to use the fact that the conversions from R to D and RP to DP are the rate-limiting steps, i.e., that the other transitions among receptor states are much faster than these two, and that there is approximately a constant supply of ATP (the precursor of cAMP), as shown experimentally by Roos et al. (1977). The variables of the final model are ρ, γ, and β, where

ρ = total fraction of receptors in active form,
β = nondimensional intracellular concentration of cAMP,
γ = nondimensional extracellular concentration of cAMP.

It is assumed that extracellular cAMP is degraded by a phosphodiesterase at a rate equal to $k_e\gamma$ and secreted by the cell at a rate $(k_t/h)\beta$, and diffuses with diffusion coefficient d. The constant parameter h is the ratio of the extracellular to the intracellular volume. Thus, the equation for the conservation of extracellular cAMP is, in English and in mathematical notation,

rate of change of extracellular cAMP = secretion by cells − degradation,

$$\frac{\partial \gamma}{\partial t} = \frac{k_t}{h}\beta - k_e\gamma.$$

Similarly, the equation for the rate of change of intracellular cAMP is

rate of change of intracellular cAMP = synthesis by cells − secretion − degradation,

$$\frac{\partial \beta}{\partial t} = \Phi(\rho, \gamma) - k_t \beta - k_i \beta.$$

In these equations k_e, k_t, and k_i are constant parameters, Φ is the rate of cAMP synthesis in terms of the proportion of receptors in the active form and of the extracellular concentration of cAMP. To avoid excessive complication, we don't give its exact form here. We merely note that, as expected from the discussion above, it is an increasing function of ρ and γ. Hence, as the extracellular concentration of cAMP increases (as γ increases), the rate of intracellular cAMP production also increases. This is the positive feedback step. However, increased intracellular cAMP concentration leads to increased extracellular cAMP, which inactivates the receptor. As the fraction of active receptors decreases (as ρ decreases), the rate of cAMP production also decreases. This is the negative feedback step.

The final equation in this model describes the rate at which receptors are converted to the active form:

rate of change of active form of receptor = resensitization of receptor

— desensitization of receptor,

$$\frac{\partial \rho}{\partial t} = f_2(\gamma)(1 - \rho) - f_1(\gamma)\rho.$$

As with Φ, we don't give the exact forms of the functions f_1 and f_2. We just note that f_1 is an increasing function of γ and f_2 is a decreasing function of γ. Hence, as γ increases the receptors become desensitized, but as γ decreases, they recover their sensitivity. Note that the rate at which the receptors become resensitized is proportional to the fraction of receptors that are not in the active state $(1 - \rho)$.

It is useful to describe in detail what happens when the extracellular concentration of cAMP (γ) is suddenly increased (Figure 8.7). Ignoring diffusion for the time being, an increase in γ has a number of effects: first, it increases Φ, and thus β also increases as the rate of cAMP synthesis increases. An increase in β then causes an increase in γ via the secretion term $k_t\beta$, thus closing the positive feedback loop. Simultaneously with the increase in Φ, an increase in γ causes an increase in f_1 and a decrease in f_2. Thus, receptor-inactivation rate is increased and the receptor-activation rate is decreased, leading to a decrease in ρ. Because Φ is an increasing function of ρ, a decrease in ρ leads to a decrease in Φ and hence a corresponding decrease in γ that closes the negative feedback loop.

Although the foregoing effects can be seen intuitively—indeed, a mathematical model is unnecessary for general descriptions of the feedback loops— the consequences of these interactions are not at all intuitively clear. Here the value of the mathematical model becomes apparent. Under what circumstances

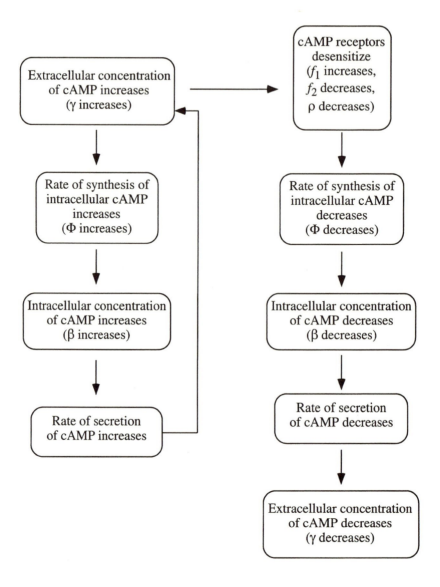

Positive feedback loop   Negative feedback loop

Figure 8.7 Positive and negative feedback loops are involved in the control of cAMP secretion.

will the feedback loops generate oscillations in the concentrations of cAMP? Is it possible for oscillations to exist under any conditions at all? If oscillations do exist, what is their period and how do they compare to experimental results? The only way these questions can be answered is by studying the mathematical model. Each isolated part of the system—the rate of intracellular cAMP degradation for instance—can be measured experimentally, but experiment alone cannot determine what happens when all the parts are allowed to interact.

It turns out that for parameter values in the physiological range, the Martiel-Goldbeter model exhibits sustained oscillations in the extracellular cAMP concentration. The oscillations have a period of about 10 min, in good agreement with experimental observations, and the intracellular concentration of cAMP reaches a maximum of about 25 μM.

For parameters similar to those associated with sustained oscillations, the Martiel-Goldbeter model also exhibits relay, which, the modeling literature usually refers to as excitability. When the model is unperturbed, the cAMP concentration reaches a steady state and remains there indefinitely. In mathematical terminology this is called a stable steady state. If this state is slightly perturbed by, say, a small increase in the extracellular cAMP concentration, the perturbation dies away and the model returns to the original steady state. If the perturbation is greater than some threshold, however, the model exhibits a large transient change in cAMP concentration before returning to the original steady state. In other words, the model predicts that if a large enough cAMP dose is added to the slime mold population, the cells initially produce a large amount of cAMP before becoming quiescent again. This is the result of positive feedback in cAMP production. In fact, excitability is closely connected to the oscillations discussed above. For some parameters initial cAMP production triggers sustained oscillations, while for other, similar parameters no oscillations occur and the cAMP pulse eventually disappears.

Coupling by Diffusion of Extracellular cAMP

The difficulties inherent in trying to understand complex systems through intuition alone are magnified when extracellular cAMP diffusion is included in the model. In Box 8.1 we present a brief derivation of how diffusion is included in the model equations. However, a detailed understanding of diffusion (and, later, chemotaxis) is not necessary to understand the results of the model.

In the presence of extracellular cAMP diffusion the range of possible behaviors is greatly expanded, as wave solutions of various types can exist, such as expanding concentric circles and spirals. Nevertheless, some general observations can be made.

Excitability and oscillations are at the heart of wave phenomena, and this is true of this particular model. Consider, for example, a row of cells in one

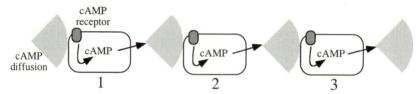

Figure 8.8 A wave of cAMP is propagated along a line of excitable cells.

dimension, each of which is excitable in the sense discussed above. Suppose that the cell on the left side of the row (which we shall call cell 1, as shown in Figure 8.8) is stimulated by the addition of extracellular cAMP. Because cell 1 is excitable, it responds to the stimulus by producing a large amount of cAMP and releasing it into the extracellular environment. This cAMP is free to diffuse to the neighboring cells, in this case cell 2, where it can act as a stimulus to cell 2. If the diffusion-dependent stimulus is great enough it can stimulate a response in cell 2, which, in turn, will release still more cAMP into the extracellular environment. Repetition of this process can lead to a wave of cAMP production and secretion that travels along the line of cells. One might think that the cAMP released from cell 2 can initiate a response in cell 1 as well as cell 3, thus producing a wave in each direction; this is not so, however, because cell 1 has just responded to a stimulus of cAMP and most of its receptors are inactivated for a period and the cell is refractory. For this reason wave propagation is in one direction only.

In this way, local excitability can be organized into traveling waves with a speed and period determined by the parameters of the model. However, it is important to note that an excitable system cannot, by itself, generate a series of waves organized in concentric rings. Such waves may be obtained when the parameters are such that each cell is in the oscillatory regime, i.e., each cell is independently capable of oscillatory cAMP secretion. Expanding concentric waves of cAMP can then be generated, as can spirals.

As we have seen, the model can generate either oscillations or excitability, merely by small changes in parameter values. Thus a single mechanism suffices to explain the wide range of observed wave behaviors.

Rarely is it possible to determine the wave speed or shape analytically. Recourse must usually be made to numerical solutions. A typical model spiral-wave of cAMP concentration is shown in Figure 8.9 (Höfer et al. 1995). The period and speed of the model spirals are in excellent agreement with those observed experimentally. In general, spiral waves are formed when a traveling wave front meets an obstacle and is broken. The broken end of the wave can then curl up to form a spiral core, which then rotates indefinitely. It is thought that this is one way in which spirals of electrical activity are generated in ventricular cardiac tissue. If the wave of electrical activity moving across the

Figure 8.9 A model spiral wave. (Reproduced from Höfer et al. 1995, figure 2a; used with permission)

ventricle meets an area of dead tissue, the wave front can be broken and spirals of electrical activity can form, resulting in cardiac arrhythmias. Another way of generating a spiral wave that may be more relevant for *Dictyostelium*, has been described by Pálsson and Cox (1996) and Lauzeral et al. (1997). In their model, which is based on the Goldbeter-Martial model presented here, individual cells can spontaneously fire off pulses of cAMP in a random manner. If one of these pulses occurs behind a preexisting wave front the wave generated by the pulse cannot travel in all directions (as it cannot propagate into the refractory region behind the preexisting wave). Thus, the second wave will curl up to form a pair of spirals rotating in opposite directions. In simulations, one of the spirals in this pair is often pushed out by the other and eliminated entirely, leaving a single spiral, although exactly how this happens is not completely understood.

A Model for Cell Movement in Response to cAMP

We now consider the next level of complexity in modeling *Dictyostelium* aggregation—the movement of the cells in response to the cAMP waves described above. Recall that the model discussed above assumed that the cell density was uniform and the cAMP waves were superimposed on this uniform density. Although this is an accurate approximation for small time intervals, before cells have had a chance to move very far, for longer time intervals cell movement will begin to affect the cAMP waves. For instance, if cells begin to clump in one area, the production and concentration of cAMP in that area will be greater than in areas with less clumping. The resulting gradient in cAMP concentration will affect cell movement, and so cell behavior is determined by a coupling in which cAMP affects the cells, and vice versa.

To model cell movement we make three additional assumptions. The first is that each cell can detect the cAMP gradient in its vicinity and move in the direction of increasing cAMP concentration. Exactly how cells accomplish this is unspecified; only the end result is modeled. The second assumption is that a prolonged cAMP stimulus decreases the cells' ability to detect cAMP gradients (the analogue of the receptor desensitization that played an important role in the model for cAMP waves). Although the desensitization mechanism is unknown, a plausible model is that it is due to the same kind of receptor desensitization that occurs in the model for cAMP waves. Finally, we assume that cell-cell adhesion comes into play once the cells are close enough to one another, and thus once a clump forms it cannot quickly disperse.

Before the model equations are presented, let us consider briefly the evidence suggesting that the behavior of each cell is more complicated than just a simple movement up the cAMP gradient. It has been observed experimentally that a wave of cAMP is, to a good approximation, symmetrical; the shape of the back of the wave is the same as the shape of the wavefront and is merely reflected around the middle of the wave. Suppose that a given cell responds only to the cAMP gradient and that when the cAMP wavefront reaches the cell, the cell moves in the direction of increasing cAMP, toward the peak of the wave, in the direction opposite to the wave motion. After the wave-peak has passed the cell however, the cell will still move toward the wave peak and thus will move in the same direction as the wave. If the wave is symmetrical about its peak, the result will be a net movement of the cell in the direction of the wave. This is because as the cell moves through the front of the wave, it is moving in a direction opposite to the wave and so reaches the wave peak quickly; but when the cell is moving through the back of the wave it is moving in the same direction as the wave and thus takes longer to reach the wave-peak. The cell therefore spends a longer time traveling in the same direction as the wave—thus its net motion is in the direction of wave propagation (Figure 8.10a). However, this is not observed experimentally. In fact, cells appear to move towards the wave peak when the wave front first reaches them (i.e., they move up the cAMP gradient) but do not seem to move after the wave peak has passed, and this results in a net movement in the opposite direction to the wave. One way to explain this is to suppose that cells respond to the initial cAMP gradient, but by the time the wave peak has passed their cAMP-detection mechanism has been desensitized and they can no longer detect the cAMP gradient (Figure 8.10b). This assumption is the basis of the model presented here.

In the new model, the equation for the rate of change of the active form of the receptor is the same as in the previous model. Extracellular cAMP is synthesized at a rate that depends on local cell density, the fraction of active receptors, and the extracellular cAMP concentration. Further, it is degraded at a rate that depends on the local cell density and on the concentration of extracellular phosphodiesterase. As before, the rate constant for phosphodiesterase activity is k_e. Finally, cAMP diffuses with diffusion coefficient d. Putting these

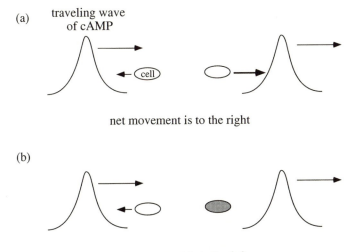

Figure 8.10 A cell may have two responses to a traveling wave of cAMP: (a) the cell responds continually to the cAMP gradient and moves in the direction of the traveling cAMP wave (as discussed in the text); (b) the cell responds to the initial cAMP gradient at the wavefront, but then becomes desensitized (shown shaded). Unable to respond to the cAMP gradient after the wave has passed, the cell moves opposite to the wave.

terms together gives

rate of change of extracellular cAMP = secretion by cells
 − degradation + diffusion,

$$\frac{\partial \gamma}{\partial t} = \theta(n)\Phi(\rho, \gamma) - [\theta(n) + k_e]\gamma + d\nabla^2\gamma.$$

In this equation, n denotes the cell density and is a function of the spatial variables x and y, θ is an increasing function of n that incorporates the increasing rates of secretion and degradation of cAMP as the cell density increases. As before, $\Phi(\rho, \gamma)$ is the rate of cAMP synthesis.

Additional equations are needed for the diffusion of extracellular cAMP and the movement of cells by random diffusion and by chemotaxis (see Box 8.1). The equations are based on the assumptions that cells diffuse much more slowly when they are densely packed (they tend to clump together), and that cells move towards regions of higher cAMP concentration.

Although the rules of behavior underlying the model are simple, the resulting equations are complex and it is not possible to obtain analytic solutions. The equations can be solved numerically, however, and typical solutions are shown in Figure 8.11 (Höfer et al. 1995). After five minutes the distribution of cells corresponds to the distribution of cAMP waves. A spiral wave of cAMP

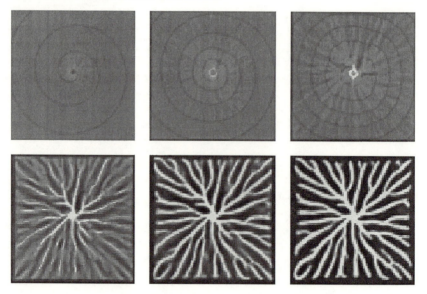

Figure 8.11 Solutions of the model equation exhibit cell streaming during the aggregation phase. The cell densities are shown at 5, 25, 35, 65, 75, and 80 min. (Reproduced from Höfer et al. 1995, figure 2b; used with permission)

has formed and at 25 minutes a corresponding spiral pattern of cells has formed a small clump at the core of the spiral. After 35 minutes the clump at the spiral core is better defined. At 65 minutes the spiral has begun to lose its form but is still clearly a recognizable spiral. In addition to forming a clump at the spiral core, the cells start to form streams of cells that move toward the core. At 80 minutes the streaming is well developed and cells are clearly moving along well-defined paths into the spiral core.

This complex behavior is the result of coupling between cells and cAMP concentrations. When there is no coupling (as in the model for cAMP waves), regular spirals of cAMP form and keep their form over a long time period; but when cells can move in response to the cAMP wave they form patterns that tend to disrupt the wave and allow formation of the more complex streaming patterns.

It is important to realize that the streaming patterns and the clump at the spiral core are a spontaneous result of simple rules governing the behavior of each cell and do not result from any predisposition toward streaming in the underlying medium. As each cAMP wave passes a cell, the cell reacts by moving in the direction opposite to the wave (as described above). Intuitively, one can imagine that regular waves of cAMP moving through a homogeneously distributed cell population can lead to uniform movement toward the spiral core. In this scenario each cell would move directly toward the core and no cell-streams would form. Although it is possible for the cells to aggregate this way, it is not a stable condition. Inevitably, random perturbations cause small inhomogeneities to appear even in an initially homogeneoue cell-distribution. Because cells tend to stick to one another, with a stickiness that increases with higher cell density (this comes from the term $\mu(n)$ in Box 8.1), cell-clumping is autocatalytic, or a positive feedback process. When an initial clump forms due to the random nature of cell movement, the likelihood increases that other cells will stick to the clump until a dense stream of cells forms, all moving towards the spiral focus.

Box 8.1 Derivation of the Model Equations

Here we present a brief derivation of the model equations when diffusion and chemotaxis are included.

Equation for diffusion of extracellular cAMP

In general, the rate of change of extracellular cAMP is described by the equation,

$$\frac{\partial \gamma}{\partial t} = -\nabla \cdot \mathbf{J},$$

where the vector, $\mathbf{J}$, is the flux of cAMP and $\nabla = (\partial/\partial x, \partial/\partial y)$. This means that the rate of change of extracellular cAMP at any particular region is just the flux of cAMP into that region, minus the flux of cAMP out of that region. For a small region, this flux difference is the negative of the gradient of the flux, i.e., $-\nabla \cdot \mathbf{J}$. The challenge is to determine the most appropriate form for the flux that best models the way in which cAMP moves. Simple diffusion is modeled by *Fick's Law*, which states that the flux is proportional to the gradient of the cell density, in which case we would have $\mathbf{J}_{\text{diffusion}} = -d\nabla\gamma$, where d is the diffusion coefficient. This leads to the well-known diffusion equation, $\partial\gamma/\partial t = d\nabla^2\gamma$. Thus the complete equation for the extracellular cAMP, including both reaction and diffusion is

rate of change of extracellular cAMP = secretion by cells
    − degradation + diffusion,

$$\frac{\partial \gamma}{\partial t} = \frac{k_t}{h}\beta - k_e\gamma + d\nabla^2\gamma.$$

Box 8.1 continued

Equation for cell movement

The equation for the movement of the cells is somewhat more complicated as it must account for two processes, diffusion and chemotaxis, i.e., movement in response to chemical gradients. As for cAMP, the rate of change of cell density is described by the equation $\partial n/\partial t = -\nabla \cdot \mathbf{J}$, where the vector $\mathbf{J}$ is the flux of cells. Since cell movement is due to diffusion and chemotaxis, the expression for $\mathbf{J}$ will involve two terms. Simple diffusion is again modeled by Fick's Law, and thus $\mathbf{J}_{\text{diffusion}} = -\mu \nabla n$, for a variable diffusion coefficient, μ. Unlike the constant diffusion coefficient of cAMP, μ is variable since cell diffusion varies with cell density. When cell densities are low, the cells diffuse at some background rate. When cell densities are high, however, the cells diffuse much more slowly as they clump together. One way to model this activity is to let μ be a function of the cell density. When n is small, we may assume that $\mu = \mu_1 + \mu_2$, but as n increases, μ decreases to μ_1. How exactly this function is described mathematically is not particularly important. What matters is the overall shape. A function with the required properties is

$$\mu(n) = \mu_1 + \frac{\mu_2 N^4}{N^4 + n^4}.$$

When $n = 0$, $\mu = \mu_1 + \mu_2$, but μ turns sharply down at the threshold value $n = N$. This means that cell-clumping has little effect until the threshold cell density is reached, at which point the cells quickly become very sticky. With the variable diffusion coefficient, our model for diffusion becomes

$$\frac{\partial n}{\partial t} = \nabla \cdot [\mu(n)\nabla n].$$

It remains to model the movement of cells due to chemotaxis. The chemotactic flux of cells consists of three terms. First, the flux is proportional to the gradient of the cAMP concentration, or ∇u. Secondly, it is proportional to the cell density, because a given cAMP gradient in a region of high cell density will cause a greater total flux than in a region of low cell density, as in the former case there are more cells available to move. Finally, it is proportional to a term, $\chi(\rho)$, which describes the desensitization of the cells to a maintained cAMP gradient. Recall that ρ

Box 8.1 continued

is the fraction of receptors in the active state. As ρ increases it is reasonable to suppose that the cells respond better to cAMP gradients, and so $\chi(\rho)$ should be an increasing function of ρ. As for our choice of $\mu(n)$, the actual equation used to describe the shape of χ is unimportant. All that matters is the generic behavior. We use the increasing function,

$$\chi(\rho) = \frac{\chi_0 \rho^m}{A^m + \rho^m},$$

for some constants A, m, and χ_0. Note that $\chi(0) = 0$, and that χ rises to a maximum of χ_0 as ρ increases. A sketch of χ is given in Figure 8.12. Combining the chemotactic terms gives $\mathbf{J}_{\text{chemotaxis}} = \chi(\rho) n \nabla u$.

From the expressions for the diffusive and chemotactic cell fluxes we get the cell density equation,

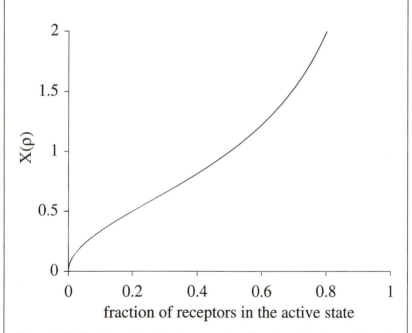

Figure 8.12 The shape of the function $\chi(\rho)$ describes desensitization of the cells to a maintained cAMP gradient.

Box 8.1 continued

rate of change of cell density = diffusive flux + chemotactic flux,

$$\frac{\partial n}{\partial t} = \nabla \cdot [\mu(n)\nabla n] - \nabla \cdot [\chi(\rho)n\nabla u].$$

Note that the sign of the chemotactic flux is the opposite of the sign of the diffusive flux. This is because cells diffuse *away* from regions of high density, but are attracted *toward* regions of high cAMP concentration. This completes the specification of the model.

Box 8.2 Simulating Spiral Waves of cAMP in *Dictyostelium*

With it's ability to simulate large numbers of subunits (called turtles), as well as characteristics of the substrate (called patches), StarLogo is a useful programming environment for simulating a system of slime-mold cells that interact with one another and secrete cAMP into their local environment. The StarLogo program can be downloaded at http://www.media.mit.edu/starlogo.

The following are the basic portions of the code used in the simulation. A complete version of the simulation can be downloaded at our website: http://beelab.cas.psu.edu. We include the program code here in part to show how economically StarLogo allows one to simulate this system. In addition, the comments in the program code provide an explanation of how the program works and its relationship to the biology and models in the chapter. These comments are preceded by a semicolon and are ignored by StarLogo.

```
_____ Declarations _____

patches-own [chemical        ;amount of cAMP in patch
             refractory]      ;remaining time that patch
                              will be refractory


_____ Initialization of the environment _____

TO SETUP
ca                            ;clears display, patches, and
                              turtles
setchemical 0                 ;resets chemical to zero
```

Box 8.2 continued

```
ifelse ((random 100) < density)
  [setpc white                       ;colors "density" percent of
                                      patches white
   setrefractory 0]                  ;sets the refractory period
                                      of the cell to 0
  [setpc black                       ;colors other patches grey
   setrefractory -1]                 ;sets "refractory" to -1, i.e.
                                      empty & never receptive
repeat number                        ;gives a random patch 300
                                      units of chemical,
[osetchemical-at random screen-size random screen-size 300];
"number" of times
```

———— Main program ————————————————————————

```
TO GO
repeat 1 [diffuse chemical 0.5]     ;each patch shares 50% of its
                                     chemical with its 8 neighbors

if refractory >=0
  [ifelse refractory = 0
    [ifelse chemical > threshold
      [setrefractory period          ;receptive patches that detect
                                      a threshold level of
       setpc red                     ;chemical become refractory,
                                      turn red, and emit 100
       setchemical chemical + 100]   units of chemical
      [setpc white]]                 ;receptive patches with
                                      chemical concentrations less
                                      than the threshold are
                                      colored white
  [setrefractory refractory - 1      ;refractory patches decrement
                                      "refractory", decrement
   setpc gray                        ;chemical, and are colored
                                      grey
                                     ;remove 1/period of the cAMP
                                      each time step
                                     ;(& don't allow negative
                                      values)
setchemical max 0 chemical - int (100/period + 1)]]
end
```

What does the simulation do?

The program simulates waves of motion and chemical relaying in the cellular slime mold *Dictyostelium discoideum*. When *Dictyostelium*

Box 8.2 continued

amoebae are starved on an agar surface they begin to aggregate, and form complex spatial patterns. Aggregation leads to the formation of a multicellular organism called a slug consisting of about 10,000 to 100,000 cells that can move about on the substrate for some time. Eventually, the slug develops into a fruiting body—a spherical stalk with a cap on top that contains spores. Under the appropriate conditions the spores can be released and germinate, thus completing the cycle.

The amoebae coordinate their movement by secreting cyclic adenosine monophosphate (cAMP) and moving against the resulting cAMP gradient. The simulation ignores cell motion because cells move several times slower than cAMP waves propogate. Accordingly, the rules governing the cells' behavior include:

- Cells that sense a concentration of cAMP above the relay threshold (which is believed to be higher than the movement threshold), emit 100 units of cAMP and enter a refractory state for a specified number of time steps.
- Cells in the refractory state are insensitive to cAMP, thereby disabling cAMP secretion, and gradually break down the cAMP in their locality by means of the enzyme phosphodiesterase.
- With each time step, the patches on which the amoebae reside share 50% of their cAMP content with the eight neighboring patches on the rectilinear array.

How the program works

The program creates a random distribution of slime mold cells (amoebae) in the environment. Empty sites on the substrate are shown as black cells on the display. Grey cells indicate refractory amoebae, white cells receptive amoebae, and red cells amoebae relaying cAMP.

The initial conditions include a few randomly selected amoebae considered to have become starved, prompting them to release a pulse of 300 units of cAMP into the environment. The user can select the initial density of amoebae on the substrate and the number of cells that begin to release cAMP at the start of the simulation. A threshold can also be set to specify the amount of cAMP needed in a patch to stimulate a cell to synthesize and release more cAMP. Finally, the user can also specify the duration of the amoebae's refractory period.

Box 8.2 continued

What the program demonstrates:

Changes in the threshold, density, and refractory period parameters affect the extent of wave propogation. In a homogeneous environment with a uniform distribution of amoebae covering the substrate, concentric waves of cAMP develop. Spirals of cAMP form only when there are sufficient "defects" (empty spaces) in the substrate that break the wave fronts and allow them to curl into spirals.

Figure 8.13 A "snapshot" from the StarLogo simulation, stationary slime mold, showing concentric waves and spirals of cAMP.

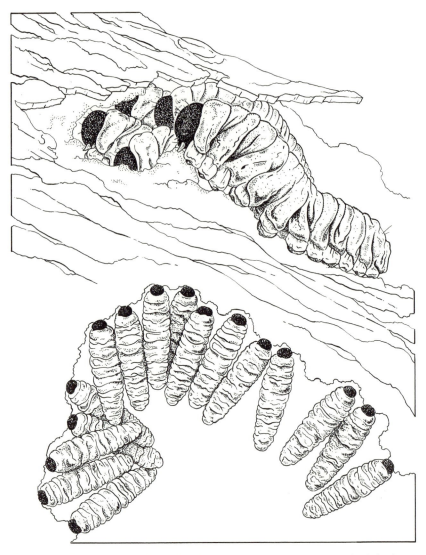

Figure 9.1 Cluster of scolytid bark beetle larvae, *Dendroctonus micans*, in their chamber beneath the tree bark. (Illustration © Bill Ristine 1998)

9

Feeding Aggregations of Bark Beetles

Birds of a feather flock together.
  —George Wither, *Abuses*

Introduction to Clustering Processes

Animals are not distributed homogeneously in the environment (Pielou 1977) but tend to be aggregated. These heterogeneous aggregations often reflect an underlying heterogeneity in the environment. Individuals often use differences in illumination, humidity, or temperature as cues to guide their aggregation behavior. Cues are templates for the aggregation process in such cases. Many studies of animal behavior have focused primarily on the individual's response to these environmental cues (Classic reviews of the subject have been published by Fraenkel and Gunn (1961) and Allee (1931)). Environmental cues do not always explain aggregation, however.

Aggregation also may result from social interactions involving attractions among the members of the group (Figure 9.1, Figure 9.2a–f and Table 9.1). This is especially obvious in uniform environments with no apparent inhomogeneities. In such situations, the clustering of individuals can only be explained by communication and interactions among group members. To the extent that aggregation involves interactions among members of the group, one should consider the possibility that self-organization plays a role in the process. This was the theme of the previous chapter dealing with aggregation and pattern formation among slime molds.

During aggregation individuals generally do not rely solely on predetermined environmental cues or on interindividual communication. Most processes involve the interplay between individual responses to the environmental template *plus* interactions among the members of the group. That interplay is addressed in this chapter. Regardless of the mechanisms used, the term *aggregation* or *cluster* is used here to describe any assemblage of organisms that it is at a higher density than in surrounding regions.

We are interested in the mechanisms and kinetics of clustering in animal groups. Our goal here, as in the other chapters, is to explore the link between the two levels of empirical analysis: the individual level involving the physiology and behavior of a single organism (how it decides to move, stop, or orient,

a

50 cm

b

Figure 9.2 Different examples of clustering in which self-organization plays a role, at least in part, in the aggregation of individuals: (a) crèche of young penguins clustering for warmth (illustration courtesy of Stéphane Porta); (b) cluster of webs of the social territorial spider, *Philoponella republicana*; although the webs are clustered, each spider builds its own web (illustration courtesy *La Recherche*);

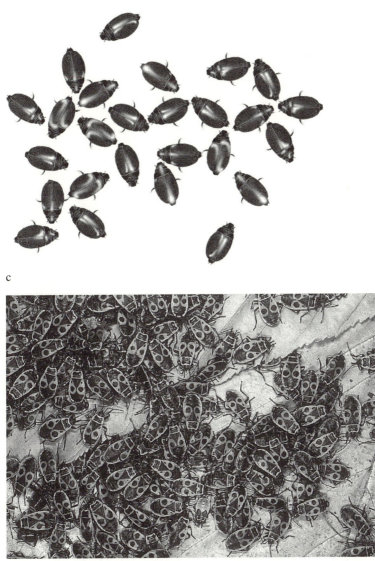

c

d

figure 9.2 continued
(c) cluster of whirligig beetles (*Gyrinus* sp.) swimming on the surface of a pond
(illustration courtesy Anne-Catherine Mailleux); (d) firebug (*Pyrrhocoris apterus*)
overwintering aggregation (photo © Scott Camazine); (e) flamingo (*Phoenicopterus
ruber roseus*) nests in the estuaries of the Rhône River in Camargue, southern France;
the regularly distributed, aggregated nesting sites suggest the interplay of positive
feedback (gregarious behavior) and negative feedback (territorial aggression), and

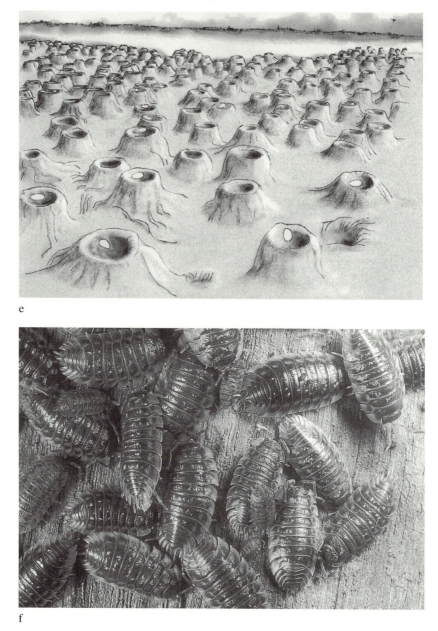

e

f

figure 9.2 continued
recalls the polygonal nest sites of fish (Figure 2.1) (based on a photo by Bob Krist/Black
Star/Rapho in Adam, 1988) (illustration courtesy of Anne-Catherine Mailleux); (f) a
sow bug aggregation in a desirably humid region (photo © Scott Camazine).

Table 9.1 Examples of Cluster-Forming Invertebrate Animals

Organisms	Scientific Name	Signal	References
Honeycomb worm	*Phragmatopoma californica*	Chemical	Morse 1993
Serpulid	*Spirorbis borealis*	Chemical	Newell 1972
Brittle stars	*Ophiothrix fragilis*	Mechanical	Broom 1975
Cirripedes	various species	Chemical	Crisp and Meadows 1962
Sphecid wasps	*Steniolia obliqua*	Unknown	Linsley 1962
Jack pine sawfly	*Neodiprion pratti banksianae*	Chemical	Ghent 1960
Flies and mayflies	various species	Visual	reviewed in Höglund and Alatalo 1995
Cockroach	*Blattella germanica*	Chemical	Rivault and Cloarec 1998
Pyrrhocoride	*Dysdercus cingulatus*	Chemical	Farine and Lobreau 1984
Flour Beetles	*Tribolium confusum*	Chemical	Naylor 1959
Whirligig beetle	*Gyrinus* spp.	Mechanical	Tucker 1969
Social spiders	*Agelena consociata*	Mechanical	Krafft 1975
Social caterpillars	various species	Chemical and mechanical	Fitzgerald 1995

for example) and the collective level of analysis involving the spatial distribution and global behavior of the group. Our interest in self-organizing systems focuses on the crucial role of interindividual attraction in creating biological patterns and collective behavior. In this chapter we investigate the clustering behavior of *Dendroctonus micans*, a scolytid bark beetle that exhibits gregarious behavior during its larval stage. We believe it provides a good example for the analysis of clustering and, in particular, the formation of self-organized clusters (Figure 9.1).

Adaptive Significance of Collective Feeding by Bark Beetle Larvae

D. micans is an important pest of spruce, in the Palearctic region (Grégoire 1988). It attacks living trees, which usually remain alive during the pest's entire life cycle. The beetle's eggs are laid in batches in intracortical chambers and, on hatching, the young larvae aggregate and feed side by side on the living host phloem. The feeding groups may split around an obstacle, such as a branch, and merge afterwards. Separate groups also may fuse if nearby brood clusters meet. The feeding larvae induce a strong host reaction in the tissues surrounding the larval chambers, in the form of copious resin production. The secretion of large quantities of resin can entrap and kill the larvae. When larvae feed cooperatively, they are able to overwhelm and outmaneuver the reaction of the host tree (Grégoire 1988). As in the case of myxobacteria described in the previous chapter, communal feeding appears to be adaptive. The key questions are: How do larvae feeding in a favorable place attract other individuals? How do local heterogeneities of the feeding substrate affect the geometry of the advancing larval horde?

Individual Behavior of the Larvae

Two main aspects of larval behavior are believed to contribute to the collective feeding process—larval production and emission of pheromone, and larval response to the pheromone gradient. The larvae use monoterpenes in the tree resin, especially α-pinene, which is oxidized to produce the aggregation pheromones, trans- and cis-verbenol, verbenone, and myrtenol. These pheromones are emitted when the insect eats host tissue or even after the larvae have been exposed to α-pinene vapors for several hours (Grégoire et al. 1982). *Dendroctonus* larvae are remarkably tolerant to α-pinene, even though the compound is toxic to many other insects. The tolerance may be achieved, in part, by the larva's ability to convert the α-pinene to less harmful derivatives, such as verbenols, verbenone, and myrtenol.

As the aggregation pheromones diffuse in air, larvae respond to the gradient of concentration by moving toward the zone of highest concentration. It is believed that larvae also respond to physical contact with other larvae by attempting to remain in contact with neighbors (thigmotaxis). However, the magnitude of this thigmotactic response is thought to be less important than the larval response to pheromone (Grégoire et al. 1982).

Collective Behavior under Laboratory Conditions

To document more carefully the range of collective behaviors exhibited by *Dendroctonus* aggregations, experiments were made with fourth or fifth instar larvae collected from logs and kept in the laboratory (Deneubourg et al. 1990a). These controlled experiments were good approximations to natural conditions and provide an excellent system for the analysis of clustering kinetics.

The experimental arena consisted of two square glass plates (30 × 30 cm) sandwiched together, but separated by a 3-mm space to allow free movement of the larvae. The floor plate was covered with a circular piece of filter paper, 24 cm in diameter. The larvae were released on the filter paper, either homogeneously arrayed or with a specified number of larvae in a nucleus located away from the center of the filter paper.

In the first group of experiments, larvae were evenly distributed in the arena at densities between 0.025 larvae/cm^2 to 0.33 larvae/cm^2. In each experiment, a single cluster formed at or near the center of the arena. Figure 9.3 shows this phenomenon starting from an initial density of 0.17 larvae/cm^2. Within 5 min, approximately 50 percent of the larvae had clustered at the center of the arena, and by 20 min 90 percent had clustered. At lower densities, the larvae clustered more slowly and the clusters were proportionally smaller (Figure 9.4). At a low density of 0.04 larvae/cm^2, a loose cluster appeared very slowly, gathering only 25 percent of the population.

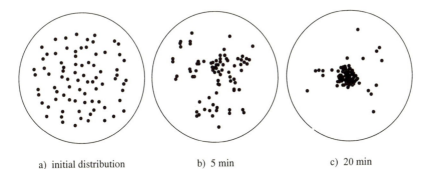

a) initial distribution b) 5 min c) 20 min

Figure 9.3 Starting from a random array at a high density (0.17 larva/cm^2), *Dendroctonus* larvae soon formed a tight cluster in an arena containing 80 larvae. Clustering is shown for the initial distribution, (a), (time = 0 min), (b) 5 min, and (c) 20 min. (Data were obtained from a photograph, figure 5 in Deneubourg et al. 1990a)

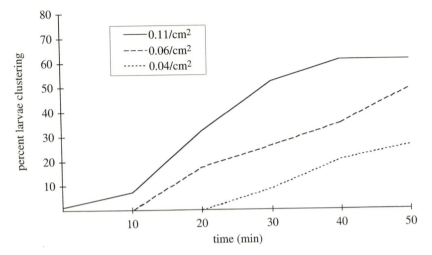

Figure 9.4 Percentage of the total population clustered at the center of the arena for three different densities lower than that in Figure 9.3. The larvae were homogeneously distributed at the beginning of each experiment.

In a second set of experiments, the larvae were evenly distributed, except for a nucleus of larvae added in a peripheral position. The following observations were made:

1. When the initial nucleus was large (20 or 30 larvae comprising 25 percent or more of the total 80 larvae), it grew and eventually "captured" 80–90 percent of the larvae (Figure 9.5). No central nucleus formed. The larger the initial cluster, the faster it reached its final size. In 20 minutes the 30-larvae nucleus achieved an average twofold increase, whereas the 20-larvae nucleus had a negligible increase in size. The 20-larvae nucleus required 33 min to double in size.

2. When the initial nucleus was small (10 larvae in a total population of 80), an initial cluster failed to develop, but a central cluster formed and grew, comparable in size to the cluster obtained with homogeneous initial conditions (Figure 9.5).

3. For intermediate values of the initial nucleus (15 larvae in a total population of 80), three different patterns developed:

   a. The initial cluster grew in some cases and became dominant (Figure 9.6a), as in (1) above;
   b. the initial cluster disappeared in some cases (Figure 9.6b) but a new cluster developed, as in (2) above;

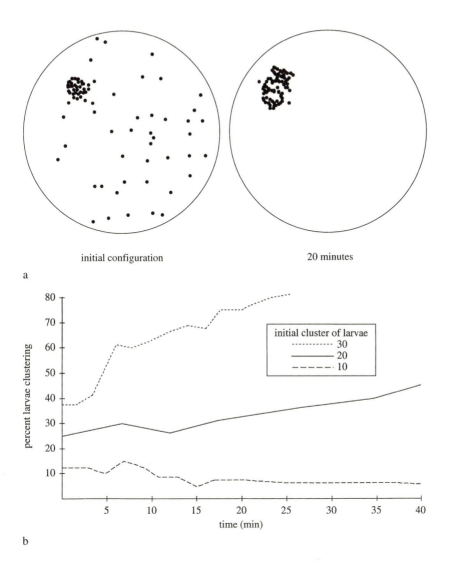

Figure 9.5 (a) An initial 30-larvae eccentric cluster and its spatial distribution after 20 minutes. Total population was 80 larvae (density: 0.17 larvae/cm^2). (Data were obtained from a photograph, Figure 7, in Deneubourg et al. 1990a.) In (b), an initial eccentric cluster of 30 or 20 larvae grew among homogeneously distributed larvae. Also shown in (b) is the decline of an initial 10-larvae eccentric cluster. Total population was 80 larvae (density: 0.17 larvae/cm^2).

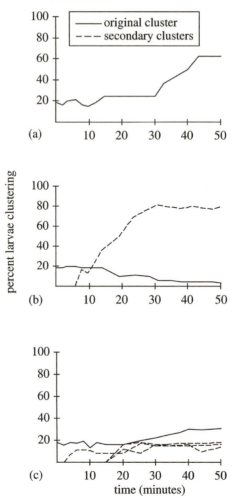

Figure 9.6 Decline or growth is shown for an initial 15-larvae eccentric nucleus among homogeneously distributed larvae and a total population of 80 larvae (density: 0.17 larvae/cm^2). In (a) the initial cluster grew and became dominant; in (b) the initial cluster disappeared and a new cluster developed. In (c), the initial cluster grew and other, secondary clusters appeared and coexisted with the initial cluster.

    c. the initial cluster grew in some cases but other, secondary clusters also appeared; the final state of this pattern was characterized by the coexistence of two, three, or four clusters (Figure 9.6c). This was the only experimental setup in which multiple clusters of larvae appeared.

In each pattern associated with 15 larvae, 80–90 percent of the total population was aggregated by the end of the experiment (approximately 60 min), similar to the proportion that aggregated in the first set of experiments.

A Model of *D. micans* Aggregation

A model is described in this section which demonstrates that a simple self-organization mechanism can explain the experimentally observed patterns of aggregation in *D. micans*. The model is based on the suggestion by Grégoire et al. (1982) that cluster formation results from competition between two opposing forces: the random movement of larvae, and the attraction of each larva to pheromones produced by other larvae as they feed. The specific assumptions of the model are as follows:

1. Each larva emits pheromone continuously and at a constant rate, α, the same for each individual.
2. Pheromones diffuse in the brood chamber, forming a gradient of pheromone concentration.
3. Larvae move along the pheromone-concentration gradient toward areas of higher concentration.
4. In the absence of a sufficient (threshold) pheromone gradient, larvae move randomly.
5. Other behaviors of the larvae, such as thigmotaxis, are ignored. This, in part, tests the hypothesis that responses to pheromone are sufficient to account for larval aggregation. Furthermore, although thigmotaxis undoubtedly plays a role in larval aggregation, it cannot be an important factor in the initial phase of aggregation when the larvae are more widely separated and not in contact with one another.

We develop the model in two dimensions, similar to natural conditions in the larval gallery, but with several simplifications: the arena is completely homogeneous, without obstacles, and open to free pheromone diffusion at its edge.

X and C are the larval density and the pheromone concentration, respectively, at a given time t. The temporal development of larval density is given by the following equation, which is similar in form to the equation for the diffusive and chemotactic movement of slime-mold cells as described in Chapter 8:

$$\frac{\partial X}{\partial t} = d\nabla^2 X + \gamma\nabla(X\nabla C). \tag{9.1}$$

This partial differential equation expresses the idea that larvae are attracted to the pheromone gradient (∇C) but move randomly in the absence of detectable pheromone. The first term on the right side represents the random walk of the larvae. The second term represents the attraction of the animals to the pheromone gradient. d is proportional to the speed of the random walk, estimated to be 1–2 cm²/min based upon experimental observations of the time required for a larva to leave a region 1 cm², and γ is a constant that specifies the larval attraction to the gradient. The numerical value of γ, approximately 2–3 cm⁴/min, has been estimated by comparisons between the model and experiments. Equation 9.1 does not account for a threshold in perception of the gradient. This simplification induces a higher sensitivity in the equations. In our open experimental arena, the steep gradients were probably higher than the threshold.

The pheromone concentration is governed by the equation:

$$\frac{\partial C}{\partial t} = \alpha X + d_c \nabla^2 C. \tag{9.2}$$

The first term specifies the production of the pheromone, where α is the amount of pheromone emitted per animal per unit time. The second term specifies diffusion of pheromones produced by the larvae. d_c is the diffusion constant. Its numerical value has been estimated by comparison with other organic cyclic compounds such as toluene and benzene (Wilson et al. 1969) and has been maintained constant in all the simulations. By dividing Equations 9.1 and 9.2 by the coefficient α, C can be measured in terms of the individual emission rate. With this change of units, α is no longer present in the model.

The boundary condition for X is zero flux, meaning that the larvae do not leave the arena. C is assumed to have zero concentration outside the arena. These conditions correspond to those in the experimental setup where the pheromones freely diffuse at the edges of the arena.

Simulations of a square arena were performed using the finite difference method to approximate the partial differential equations. Since larvae gathering in a cluster enhance the attractiveness of that cluster, this system behaves in an autocatalytic (positive feedback) manner. The higher the larval density level in a region, the stronger the gradient and the greater the tendency to move toward the crowded region. The final stationary distribution of the larvae results from the interplay between this positive feedback, which brings the larvae together, an individual's random walk, which opposes aggregation, and, given the experimental conditions, the physical constraint of overcrowding. In the model, however, density was not constrained to prevent overcrowding or excessive superposition of individuals. Still, the results of the simulation do not exhibit unrealistic densities of individuals since larval diffusion is sufficient to reduce overcrowding.

Simulation Results

The model was used to simulate the effect on aggregation patterns of different initial larval densities and of other initial conditions, including homogeneous larval distributions with or without an initial nucleus of larvae. The simulation results were compared to actual experimental results as a test of the model. (A version of the model written in StarLogo is available at our website: http://beelab.cas.psu.edu.)

In the first set of simulations (corresponding to the first set of experiments), clusters always formed at the center of the arena. Figure 9.7, which should be compared to the experimental results shown in Figure 9.4, shows the temporal evolution of larval density in the center of the arena for three initial densities. It appears that density determines both the rate of aggregation and its efficiency defined as the number of larvae that are part of the final cluster. At high density (0.11 larva/cm^2), a cluster formed and rapidly grew at the center of the system, gathering 80 percent of the population within 25 min. At lower density (0.06 larva/cm^2), 60 percent of the population aggregated before 50 min. At still lower density (0.04 larva/cm^2), a small inhomogeneity (ten percent of the total population) formed after 50 min. These results are similar to the experimental data. Note that although our models include the escape of pheromones at the periphery of the arena, and thus causes a pheromone-concentration gradient, this gradient is not responsible for the clustering phenomenon but does determine the *location* of the cluster within the arena.

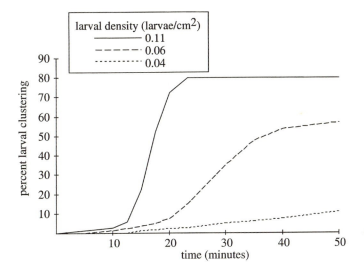

Figure 9.7 Simulation of the percentage of the aggregated population that appears in the center of an arena at three densities. $d_c = 4.2$ cm^2/min; $d = 1.2$ cm^2/min; $\gamma = 2.4$ cm^4/min. The simulations were made on a square arena 22×22 cm. Equations were solved numerically by the finite difference method with the area divided into 121 squares.

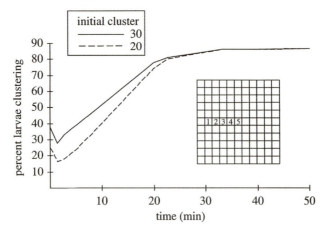

Figure 9.8 The simulation shown are for larvae clustered at square 1. In addition to a homogenous background of larvae, an initial cluster of either 20 or 30 larvae was placed at square 1. The values of d_c, d, and γ, are as in Figure 6, and the total population was 80 larvae (density: 0.17 larvae/cm^2).

In the second series of simulations (corresponding to the second set of experiments), larvae were uniformly distributed, but an additional nucleus of larvae was added in a peripheral position. The temporal evolution of larval clustering was measured. Figure 9.8 shows the growth of nuclei of 20 and 30 larvae, comparable to experimental results shown in Figure 9.5. Aggregation occurs very rapidly and results in the capture of nearly 90 percent of the total population. During the first few minutes, the population in the initial nucleus decreases slightly, but the larvae remain scattered over a very small area around the initial site. If the initial nucleus contained only five larvae, it was unable to persist and a new stable cluster gradually developed in the center of the arena (Figure 9.9) (Compare with Figure 9.5). For initial values of the nucleus between five and twenty larvae, the final cluster formed midway between the position of the initial nucleus and the center of the arena, being closer to the center of the setup for smaller values of the initial nucleus. In the latter two cases, as in the first one, about 90 percent of the larvae end up in the final cluster.

Discussion of the Model

The simulations suggest that aggregation can be explained as a response by individual larvae to pheromone emission by neighboring larvae. This simple model agrees well with experimental results and accurately simulates the role

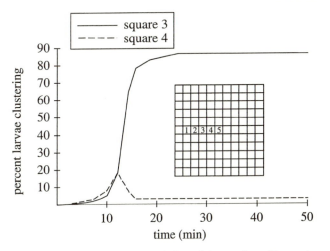

Figure 9.9 Simulation of the temporal evolution of the number of larvae in squares 3 and 4, with an initial cluster of 5 larvae introduced in square 1. The values of d_c, d, and γ, are as in Figure 6, with a total population of 80 larvae (density: 0.17 larvae/cm^2).

of initial density in determining the final pattern as well as the aggregation rate of the larvae and size of the resulting cluster, and the effects of initially homogeneous or heterogeneous distributions of larvae.

The experiments with preformed nuclei demonstrate that, with certain critical configurations, such as an initial nucleus of fifteen larvae, the system may exhibit several different patterns (Figure 9.6). Which one of these patterns occurs is governed by random events linked to the individual behavior of the larvae and to small, uncontrollable differences in the initial conditions of each experiment. These variations are also present in the other experimental situations (homogeneous initial conditions or large or small initial nuclei) but not sufficiently strong to influence the outcome.

In simulations and experiments with identical density, arena size, and pheromone diffusion, the initial conditions (the presence or absence of a preformed nucleus of larvae) determined the final pattern of larval distribution. This is characteristic of systems exhibiting multistationary states. The final pattern that develops in self-organizing systems can be extremely sensitive both to the system's parameter values and to initial conditions, determining which of several outcomes emerge. This concept of multistationarity will be seen later in other examples presented in this book.

During the aggregation of *Dendroctonus*, thigmotaxis may complement pheromone chemotaxis. Once a larva contacts another individual, there may be a tendency to maintain contact. Such behavior could reinforce the responses to pheromone, as well as dampen the effect of a larva's random motion. It may explain the discrepancy between the model and experimental results when initial conditions include larval nuclei of intermediate size (15 larvae). In the range of the parameter values used for the simulation, the model does not exhibit stable multicluster solutions as found during the experiments (Figure 9.6). The multicluster solutions of the model are only transient states. Thigmotaxis, which has not been taken into account in the model, may have a stabilizing effect on the small clusters. However, the good agreement between the model and experimental data support the validity of the hypotheses and the secondary role of the thigmotaxis.

Summary

This example of clustering was chosen because the aggregation of *Dendroctonus* larvae is simple to analyze experimentally and theoretically and thus has a didactic value. The other examples in this section—fish schooling (Chapter 11) and synchronized flashing of fireflies (Chapter 10)—also exemplify aggregation but involve more complicated processes. A school of fish is a dynamic, ever-changing cluster in three-dimensional space. In addition, fish probably use more than one sensory modality to modulate their behavior. The synchronization of firefly flashing is an aggregation of signals in time that also serves to gather together individuals in three-dimensional space. Because of their relative complexity, the mechanisms involved in both of these examples are more difficult to understand and analyze than the *Dendroctonus* example.

An Alternative to Self-Organized Pattern Formation

An alternative explanation for the clustering of organisms is that it is a response to an external cue that acts as a template. Let us consider this explanation to see how it differs from self-organized aggregation.

Figure 9.10 illustrates clustering that results from a response by individuals to an external stimulus. The stable fly, *Stomoxys calcitrans*, aggregates in regions where temperature is between 24° and 32° C. Clustering occurs when the flies move to a zone of preferred temperature (Fraenkel and Gunn 1961). In effect the template is a temperature gradient independent of time to which the flies respond. This aggregation process exhibits different global properties than self-organization clustering. At least two different mechanisms, klinokinesis and orthokinesis, may explain this type of aggregation (Fraenkel and Gunn 1961).

In klinokinesis, an individual's probability of *turning* increases as the individual finds itself closer to the zone of preferred temperature. As a result, an individual in an unfavorable region tends to maintain a more or less straight path that gets it out of that area. But when an individual reaches the preferred temperature zone its path becomes more convoluted and it tends to remain in the same area. The human body louse, *Pediculus humanus corporis*, shows such klinotactic behavior (Figure 9.11).

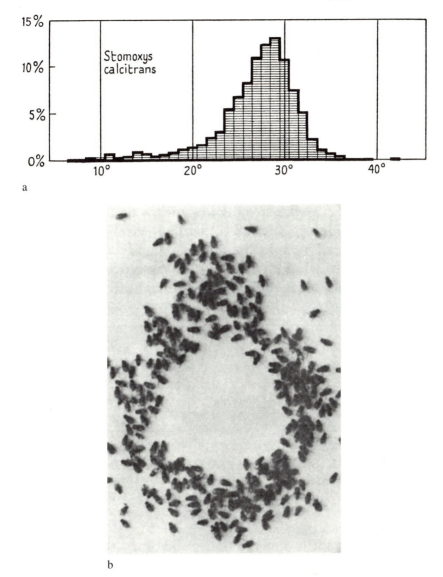

a

b

Figure 9.10 The spatial distribution and clustering of stable flies, *Stomoxys calcitrans*, in a thermal gradient (a) indicates animals are clustered around their prefered temperature, approximately 29° C, and nearly all individuals are found between 24° and 34° C. The temperature preference of blowflies, *Phormia terranovae*, is illustrated in (b). A light bulb behind a vertical sheet of opaque paper created the temperature gradient. The flies avoided both the hot central region and the colder periphery, clustering in a middle zone of about 30° C (from Fraenkel and Gunn 1961, figures 93 and 94). (Used with permission of Dover Publications)

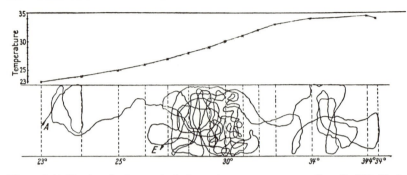

Figure 9.11 Tracks of a human body louse in a temperature gradient (in °C). Notice the convoluted path just above 30° C typical of klinotactic behavior (from Fraenkel and Gunn 1961, figure 9.7). (Used with permission of Dover Publications)

In orthokinesis, in contrast, an individual varies its *speed* depending upon whether it is in a preferred zone. The woodlouse, *Porcellio scaber*, shows this type of response to humidity. An individual in a region of low humidity, moves constantly. As the humidity increases toward 100 percent, fewer animals move and cluster in regions of higher humidity (Fraenkel and Gunn 1961).

It is important at this point to ask how self-organized aggregation differs from aggregation based on a template. In particular we would like to know what kinds of tests can be used to distinguish between these two different mechanisms of aggregation. The most important difference is that in self-organized systems the signals that individuals respond to are dynamic and change over time due to the behavior of the individuals involved. When clustering is based on an environmental template, the stimulus is generally a fixed feature of the environment. As a result, the response of the individuals is independent of the density of individuals. In the woodlouse example, the individuals always aggregate in the region of highest humidity and do so regardless of the initial density of individuals.

To test for the presence of an environmental template, one would first examine the environment in search of cues that individuals might use to guide their aggregation. On finding such a cue (perhaps a temperature gradient), one could gently disrupt the aggregation, remove the environmental cue suspected of leading to aggregation, and see if reaggregation occurs. If the individuals fail to aggregate, one experimentally restores the cue and determines if the individuals are able to form their original cluster. Such results would provide good support for the hypothesis that aggregation is based upon environmental cues rather than interindividual cues.

In self-organizing systems, the signal that ultimately causes aggregation is provided by other individuals. The interactions among individuals can often lead to a social amplification of the signal. As more individuals aggregate they provide an even stronger impetus for aggregation. This positive feedback often results in more interesting and more complex behaviors of the system. For example, different patterns may result from minor differences in the past history of the system, especially its initial conditions. In contrast, systems that use external cues reach the same final state regardless of the initial conditions.

The global patterns of self-organizing systems are sensitive to the density or number of individuals involved, which affects whether a pattern will emerge at all and the rate of emergence. Moreover, such self-organizing systems are able to produce clusters from a homogeneous distribution of animals in a homogeneous environment.

Given these differences, can a model based upon a template explain the results obtained in the experiments with *D. micans*? Are the larvae merely responding to a pre-established external gradient from the center to the edge of the boundary? This cannot be the case because the clustering process of the beetle larvae strongly depends upon the initial density and configuration (homogeneous or clustered) of larvae. In a model based upon an external cue, these initial conditions will not effect the final configuration of the larvae.

Another alternative hypothesis is aggregation based on signals from a leader. If one could identify an individual in the group that is the leader who decides where to aggregate, this would provide strong evidence against a self-organized process involving multiple local interactions among group members. In testing this hypothesis, one would have to identify the individual or individuals in the group who act as leaders and experimentally determine whether the group could aggregate in the absence of the leader. Some difficulties might be encountered. It might not be possible to identify a leader, even if one existed. Perhaps the individual who acted as a leader during the aggregation process plays no special role (and would be difficult to identify) after the individuals have clustered. And, even if such a leader could be identified and removed, it is possible that another individual would take its place. Despite these technical difficulties, thorough examination and experimentation to exclude these alternative hypotheses should be undertaken before one confidently states that a particular instance of aggregation is based on a self-organizing mechanism.

Unfortunately, a current weakness of the self-organization approach is that these types of experiments and tests are seldom performed. In most cases presented in this book, we were forced to rely on plausibility arguments. In some cases, the researchers have sought alternative mechanisms that are not self-organized. They may have searched for a leader or template that determines the pattern-formation process and found none. Instead, they may have found that individuals appear to be using local interactions among group members to guide their behavior. Although one would certainly like stronger evidence for a process based upon self-organization, on the grounds of plausibility, the self-organization hypothesis has often been accepted. This is clearly a weakness of the self-organization approach, but we hope that by acknowledging this weakness efforts will be made to address it.

Interactions between External Cues and Self-Organization

In many clustering processes, both external cues and interindividual communications are involved. It is desirable therefore to determine the relative contributions of external cues and self-organization in pattern formation. One can attempt experimentally to eliminate all external cues and observe whether clustering still occurs. This was the rationale behind the design of the *Dendroctonus* experiments. But even here it is possible that certain external cues could affect the outcome of the pattern by acting as a preformed template for individuals to follow. For example, in the experimental situation the edge of the arena is open to the air and can serve as a sink for the pheromone. Thus, a small pheromone gradient occurs at the boundaries of the arena and might favor the formation of a cluster at the center.

In natural conditions the larval environment is obviously more heterogeneous, making it more difficult to rule out the role of external cues. For example, a zone of soft wood may become a center for clustering if larvae eat more easily in this region and so emit more pheromone. In this experimental setup, the initially weak gradient is amplified by clustered individuals. In many foraging situations an environmental heterogeneity is modulated or amplified by the individuals that find food. Individuals arriving at the site may have been attracted by the sight or odor of food. Here, the initial aggregation occurs solely by means of an external cue. However, the discovery of a food source can be followed by the release of a signal that attracts other individuals to the food. This is the case for the recruitment systems of social insects (See Chapters 12, 13, and 14). In other cases individuals may not have evolved a special signal (such as pheromone) to recruit nestmates or conspecific individuals, but other individuals may be attracted to the site by visual or auditory cues from individuals feeding at the site. In either case, discovery of the food source by an animal increases the probability that others discover and utilize this source. For example, the jack pine sawfly, *Neodiprion pratti banksianae*, also feeds in a group and its behavior is similarly believed to be a feeding adaptation (Ghent 1960). Instar larvae at first have difficulty cutting into the tough cuticle of the pine needles, but as soon as a few larvae make feeding incisions in the needles other larvae aggregate at the site through olfactory cues from odors of the pine foliage and saliva of the feeding insects.

Similar interplays between external cues and self-organized clustering appear in the life cycle of marine invertebrates. Early in life these organisms are planktonic, floating in the sea. This is followed by a settlement phase. The choice of a suitable microhabitat for settling is of fundamental importance since a mistake in settling cannot be corrected and may lead to death. The cues used for settlement can be classified in two groups: cues provided by the environment and by the conspecifics. For example, an environmental cue for barnacles and mussels is the texture of the substrate (Crisp and Barnes 1954; De Block and Geelen 1958). In the case of the cirripedes (Crisp and Meadows 1962) or the serpulid *Spirorbis borealis* (Newell 1972), the conspecific signal is in the form of a chemical produced by previously settled adults.

Another aspect of the interplay between individuals and the environment is a situation in which the environment interferes with interindividual communication or decreases the range of signals between individuals. The classical experiments of Allee (1931) on clustering of the brittle star illustrates this point. These echinoderms are known to produce very large clusters. The clustering mechanism is based on short range interactions approximately the length of the starfish arms. Allee placed the brittle stars in an aquarium without intervening obstacles, such as seaweed or stones. The animals quickly clustered. However, clustering was strongly reduced or inhibited when the animals were placed in an aquarium containing seaweed. There is no need to suppose that the behavioral rules used by the starfish differ as a function of the amount of seaweed in the environment. The brittle stars apparently are showing similar thigmotactic behavior towards both conspecifics and seaweed, clinging to either. As a result, individuals in contact with seaweed move less frequently and fail to form clusters.

The interplay between individuals and the environmental substrate can be complex. Although seaweed can inhibit cluster formation in one spatial scale of interaction, at another scale—when patches of seaweed are widely and inhomogeneously distributed—they can act as nuclei to stimulate clustering.

Although the cases discussed here are rather simple, they provide an opportunity to show how diverse responses can occur with the same set of basic rules and how spatial clustering patterns are affected by the interplay between environmental cues, interindividual communications, population density, and initial conditions.

The role of aggregation in the social life of animals and the relative simplicity of its experimental investigation provides an important source of material for the study of self-organization and its alternatives. The authors believe that self-organized aggregation occurs in many situations despite the current lack of strongly supportive experimental data. This belief is based on many examples (Table 9.1) in which attractions among individuals are described as playing important roles.

We expect that as an increasing number of cases are carefully analyzed we will see that—despite the different modes of communication involved, the diverse functions that aggregations serve and the range of environmental conditions under which it occurs—self-organized aggregation has common features regarding the density of individuals, perturbations, initial conditions, and types of patterns that arise.

Figure 10.1 Artist's depiction of an aggregation of fireflies flashing synchronously in a tree in southeast Asia. (Illustration © Bill Ristine 1998)

10

Synchronized Flashing among Fireflies

The Glowworms ... represent another shew, which settle on
some Trees, like a fiery cloud, with this surprising
circumstance, that a whole swarm of these Insects, having
taken possession of one Tree, and spread themselves over
its branches, sometimes hide their Light all at once, and a
moment after make it appear again with the utmost
regularity and exactness, as if they were in perpetual
Systole and Diastole.

> —Perhaps the earliest documented description of
> synchronous flashing, from the Dutch physician
> Engelbert Kaempfer's (1727) account of a trip
> along the banks of the Chao Phraya (Meinam)
> River in Thailand in 1680. (Quoted in J. Buck,
> *Quarterly Review of Biology*)

Synchronous Rhythmic Flashing

Fireflies are a familiar sight on warm summer evenings. Flashing their abdominal lanterns as they fly over lawns and meadows, one generally discerns no
pattern to the flickering display. Among the common North American genera
Photinus and *Photuris*, the males rove about singly, searching for females that
rest on low vegetation, and are thus called roving fireflies. Courtship involves a
Morse-like code of alternating signals between the sexes. The male modulates
the duration and intensity of his flash, as well as the rate and number of flashes
in a species-specific pattern. This allows a female to recognize a conspecific
male, to which she responds with a flash after a species-specific time delay. At
about 24 °C, for example, the female *Photinus ignitus*, an eastern U.S. species,
answers the male after three seconds (Lloyd 1981). Among the roving fireflies,
successful mating often requires a period of continuous species-specific communication involving alternating, accurately timed flashes that guide a flying
male to the sedentary female. (For reviews see Lloyd 1971, 1979, 1981.)

Fireflies are beetles (family Lampyridae), and occur as more than 2000
species worldwide. Not surprisingly different species have evolved a great variety of luminescent mating signals. However, one particular form of luminescent behavior, seen from India east to the Philippines and New Guinea,
has particularly fascinated explorers and naturalists for hundreds of years. In

this vast region, enormous aggregations of fireflies gather in trees and flash in near-perfect synchrony (Figure 10.1). None of the authors has had the opportunity to see these fascinating displays in the wild, but films of synchronously flashing fireflies in Malaysia can be seen in the documentary nature program, "Talking to Strangers," in the "BBC Trials of Life" series of videos by David Attenborough, available from Time-Life Videos. The hypnotic excitement of witnessing such displays was described seventy years ago by Howard (1929) as "a strange sight... there in the darkness was a tree just filled with lightning bugs. The strange thing was that they all flashed at the same time. One second everything would be dark, the next second the whole tree would be aglow with a beautiful light!"

Adamson (1961) made similar observations:

It is then too that one sees the great belt of light, some ten feet wide, formed by thousands upon thousands of fireflies whose green phosphorescence bridges the shoulder-high grass. The fluorescent band composed of these tiny organisms lights up and goes out with a precision that is perfectly synchronized, one is left wondering what means of communication they possess which enables them to coordinate their shining as though controlled by a mechanical device.

In the same vein, Smith (1935) wrote:

Imagine a tree thirty-five to forty feet high thickly covered with small ovate leaves, apparently with a firefly on every leaf and all the fireflies flashing in perfect unison at the rate of about three times in two seconds, the tree being in complete darkness between the flashes. Imagine a dozen such trees standing close together along the river's edge with synchronously flashing fireflies on every leaf. Imagine a tenth of a mile of river front with an unbroken line of [mangrove] trees with fireflies on every leaf flashing in synchronism, the insects on the trees at the ends of the line acting in perfect unison with those in between. Then, if one's imagination is sufficiently vivid, he may form some conception of this amazing spectacle.

Recent descriptions of the phenomenon provide more specific details. In the firefly *Pteroptyx cribellata*, for example (Buck and Buck 1976, 1978; Lloyd 1973a), before sunset large numbers gather in certain swarm trees, having apparently arrived on previous evenings. Thousands of individuals begin flashing soon after sunset, while synchrony builds up slowly through the night. One can see a great variety of light emissions at these congregations. Most prominent are the males' rhythmic synchronous flashes. Perched males flash synchronously. Airborne males also synchronize with others as they move in a slow hovering flight up to several meters from the foliage, attracted to other females. Males that land on foliage can be seen approaching females. The males group in small clusters around a perched female.

Unsynchronized flashing also occurs. Females emit arrhythmic long-duration glows that are dimmer than the male's signal, males respond rapidly to the females with glows and twinkles, and other males emit continuous glows as they chase females in flight. These flashing displays continue through the night and successful encounters result in copulations.

As Buck and Buck (1976) remark, neither the beauty of such a spectacle nor its mesmerizing effect can entirely account for the fascination of these displays. During the past 300 years, not only have there been many dozen descriptions, but also lively debate over how fireflies accomplish their rhythmic flashing. Some early authors felt that the phenomenon was merely an illusion and adamantly denied its existence (see review by Buck 1938). Much of the fascination as well as disbelief during the early 1900s arose because people had difficulty imagining any mechanism to explain *how* fireflies coordinate their synchronous flashing. There was also the mystery of *why* fireflies flash in unison. Most of this chapter is devoted to answering the "how" question; however, we will first briefly explore the possible adaptive significance of this activity.

Adaptive Significance of Synchronous Rhythmic Flashing

The reproductive significance of luminescence is well established for the roving fireflies and the consensus among entomologists is that collective synchronized flashing also is related to reproductive behavior (Figure 10.2). However, the significance of synchrony for mating is still not entirely clear. Even the most recent review of the subject (Buck 1988) presents a bewildering array of potential explanations, some of which are presented in the next section, but none of which is entirely convincing. Furthermore, the early functional interpretations of synchronous flashing suffered from erroneous group selectionist views and other misunderstandings concerning evolutionary theory. Some of that confusion persists in current discussions (see Buck and Buck 1978, 1980; Lloyd 1971, 1973a, 1973b; Otte 1980 for detailed discussions). Finally, we must emphasize that fireflies employ a variety of mating systems (Buck and Buck 1978; Lloyd 1966, 1973a; Otte and Smiley 1977), and so a single functional explanation may be inadequate to explain all instances of synchronous flashing.

Rather than review the entire literature, at this point we will outline what appears to be the most likely explanation for synchronized flashing in *Pteroptyx malaccae*, the Thai species, that Smith (1935) described. As in other species, synchronized flashing is performed entirely by males that maintain positions spaced on individual leaves or similar territories in trees for long periods. Females fly to these trees to mate with males. The females emit irregularly timed, longer-duration flashes that are dimmer than the male's. Mating pairs are commonly found in the trees (Buck and Buck 1978). The mating system does not

Figure 10.2 Thousands of fireflies flashing in synchrony as shown in a time exposure photo of a nocturnal mating display. (From Strogatz and Stewart 1993)

involve a repetitive male-female dialog as in the roving species, but rather a massed congregation of males that females visit for mating. In certain verte-brates and invertebrates, such male-mating assemblies are termed *leks* (Brad-bury and Gibson 1983). According to Gibson and Bradbury (1986), "Lek sys-tems are commonly defined by four criteria: absence of parental care by the male, clusters of displaying males, location of mating aggregations away from resources required by females, and apparent freedom of females to choose mates."

The mating system of *Pteroptyx* may involve such a lek system (Buck and Buck 1976; Otte and Smiley 1977; Lloyd 1979), and it has been argued that synchrony among the males could serve at least three purposes in this context. First, synchrony could accentuate the males rhythm (Otte and Smiley 1977). If, as in *Photinus* (Lloyd 1966), females select males based upon a species-specific flash repetition rate, then the chaotic flashing of dense aggregations of males may thwart the female's ability to assess the males' flashing pattern. By mutually synchronizing their rhythm, each male benefits by allowing its flashing rate to be evaluated by a potential mate.

Second, a similar argument based on enhanced detection applies if mating requires that males visually detect females (Otte and Smiley 1977). If a portion of the mating sequence requires that males detect the flash of nearby females, synchronization could serve as a noise-reduction mechanism allowing males better to find females in the dark interflash periods.

Third, synchronization could be a signal enhancement mechanism (Otte and Smiley 1977) enabling small groups of synchronizing males to attract larger numbers of females. According to (Lloyd 1979), "Ability to keep the identifying species-specific rhythm with conspecific males is critical to a male's success because it permits him to enhance the attractiveness of his sublek in the swarm as it competes with others for incoming females...."

Note that these explanations operate within a rather localized environment. In this regard, Lloyd (1973a, p. 991) warns that the huge *mass* displays encompassing areas of one or many trees may not have adaptive significance. He states "Although mass synchronous flashing of fireflies is conspicuous and associated with mating behavior, it probably is of little reproductive significance. This phenomenon may merely be the gross consequence of individual males synchronizing with their neighbors as they compete in small clusters for females on an extremely localized level." In other words, synchronization among small groups of males in a local lek is likely to be adaptive, but *mass* synchrony may be an epiphenomenon that arises as local synchronies coalesce at higher male densities.

In summary, it seems reasonable to assume that among *Pteroptyx*, at least at the level of small clusters of males, synchronous flashing could serve adaptive functions related to mating. Similar arguments have been suggested to explain the synchrony occasionally seen in other species such as roving *Photinus* fireflies (Otte and Smiley 1977).

Whether or not we accept the thesis that synchronized flashing has adaptive value, the question we will address in this chapter is: What is the mechanism of synchronization? We will see that through the natural selection of certain behavior and physiology, some firefly species evolved a simple, yet elegant self-organized mechanism for synchronous flashing.

Some Early Hypotheses of the Mechanism

It is instructive to see how an adequate explanation of synchronous flashing emerged gradually over many decades. Many mechanisms were offered before 1938 to explain synchronous flashing (reviewed by Buck, 1938). Hypotheses based on leadership and templates were seriously considered and, for many years, were difficult to refute. Part of the problem of formulating a satisfactory explanation was the lack of experimental data to support or refute specific hypotheses. The absence of an adequate conceptual framework also hindered the development of good hypotheses. As a result, a variety of rather naive explanations were initially suggested. We will briefly review these early hypotheses, and see how they were supplanted only recently by a satisfactory explanation.

Anthropomorphic and Other Inadequate Explanations

Several early writers refused to believe that synchronization occurred. Laurent (1917) was so skeptical of the phenomenon that he attributed synchronization to the rapid rhythmic twitching of the observer's eyelids! Craig (1917) reasoned that whenever a "large number of fireflies are flashing at slightly different rates there must be a great amount of accidental synchronism." Craig (1916) also contended that, "Viewing any large assortment of instances without statistical methods, one can see in them whatever one is predisposed to see; and we are always predisposed to perceive a rhythm—this is a well-known psychological fact."

Explanations denying the very existence of synchronization of course were abandoned as soon as recording photometers and high-speed movie cameras offered incontestable evidence that synchronized flashing was more than a quirk of human perception. However, such evidence was unavailable until 1965, when several expeditions to Borneo, Thailand, New Guinea, and Southeast Asia gathered data to verify centuries-worth of anecdotal observations by naturalists and explorers (Buck and Buck 1976).

Other early writers believed in the existence of synchronization, but invoked explanations that were clearly naive and anthropomorphic in their reliance on an animal's consciousness of its own rhythm. For example, Wheeler (1917) described pelicans in flight with their wings apparently beating in synchrony. He attributed this to a "fine sense of rhythm on the part of each individual." He went on to describe the synchronous flashing of fireflies in similar terms and stated, "In fireflies the initiation of the simultaneous flashes must be due to optic stimuli, as it is in people endeavoring to keep in step with one another, but the continuation of the established rhythm would seem to depend on a kind of 'Einfühlung'," which is German for "empathy." (In the collo-quial sense, the verb "einfühlen" means to be on someone's wavelength.) Many years later, Buck and Buck (1968) observed that the mechanism was still be-lieved to be analogous to certain human activities: "Among 'sense of rhythm' behaviors in man the rhythmic synchronous handclapping that may break out spontaneously at sporting events, though not involving a preexisting rhythm in the participants, has some resemblances to firefly synchrony." The mechanism was termed *anticipatory synchrony* because the human participants seemed to be using information from the preceding cycle or cycles "to predict the proper time to [clap] in order to attain synchrony," rather than to be acting in response to immediate cues from the handclapping of others (Buck and Buck 1968).

No Environmental Influences Organize Synchronous Flashing

Nature provides many sources of rhythmic activity, such as seasonal tem-perature changes, daily tides, and diurnal light-dark cycles. Therefore, it is not surprising that someone would suggest an environmental trigger to explain the rhythmic flashing of fireflies. This is a hypothesis based on a kind of template, one of the commonly assumed alternatives to a self-organization mechanism. However, it is difficult to imagine an environmental stimulus of sufficient reg-ularity and proper frequency to impose the observed synchrony. Buck (1938) cites a discussion in the *Transactions of the Entomological Society of Amer-ica*, in 1865, in which the far-fetched suggestion was made that puffs of wind stimulate the fireflies alternately to expose and conceal their lights.

Eventually it was realized that *repetitive*, external rhythmic stimuli could not explain the phenomenon, so it was suggested that perhaps a *single* stimu-lus was responsible. It was argued that the synchrony might arise from a single triggering stimulus in combination with the fireflies' regular flashing rhythm. If the rhythm were regular enough and if all the fireflies were to begin flashing at the same time, then synchrony might result automatically. This may ap-pear to make sense, but cannot be the mechanism. Termed *inertial synchrony*, this slightly more sophisticated, but equally inadequate, explanation was con-vincingly discounted by Buck and Buck (1968). First, even if an abrupt event, such as a particular level of ambient light, a clap of thunder, or a bolt of light-ning initiated flashing, the individuals would soon fall out of rhythm unless the rate of flashing were exceedingly regular, far more regular than that observed. Second, we know that aggregations of synchronously flashing fireflies are dy-namic, with individuals arriving and leaving through the night (Lloyd 1973a), making it unlikely that all the participants could have been subjected to a com-mon initiating stimulus. Finally, detailed observations by numerous observers have failed to reveal any evidence of a single triggering stimulus.

Leaders Control Synchronous Flashing

Having discounted the possibility that synchronization is imposed by external stimuli, the only other logical explanation seemed to be that the organizing influence came from an individual within the group of fireflies itself—a hypothesis based upon a leader. In the early part of this century, Blair (1915) read a paper before the South London Entomological Society and Natural History Society. He believed that the "probable explanation of the phenomenon is that each flash exhausts the battery, as it were, and a period of recuperation is required before another flash can be emitted. It is then conceivable that the flash of a leader might act as a stimulus to the discharge of their flashes by the other members of the group, and so bring about the flashing concert by the whole company."

Two valid objections make it unlikely that any leader coordinates the rhythmic flashing. First, it is inconceivable that a single leader could act as the pacemaker for the extensive mass congregations of the thousands of fireflies observed in species such as *Pteroptyx*. How could a single leader be visible to all the members of the congregation? The second objection is that even if one invokes a mechanism with several leaders relaying information through the system, experimental data readily rules out any sort of follow-the-leader mechanism. A critical finding was that among the Thai species *Pteroptyx malaccae*, the time interval beginning with the first flash of a member in the group and ending with the last flash in the group was extremely short, only about 30 ms. The minimum delay for flash generation, even by direct neural stimulation near the light organ, was experimentally measured at 55–80 ms. This proved that during group synchronization, fireflies cannot be using the sight of a neighboring flash to initiate their own flash in a follow-the-leader type of mechanism. Their reaction time is simply too slow (Hanson et al. 1971) (also see Box 10.2).

Of all these early hypotheses of synchronization, only one came close to what we now believe to be the correct mechanism. Richmond (1930), lacking any experimental data, but with surprising insight, theorized: "Suppose that in each insect there is an equipment that functions thus: when the normal time to flash is nearly attained, incident light on the insect hastens the occurrence of the event. In other words, if one of the insects is almost ready to flash and sees other insects flash, then it flashes sooner than otherwise. On the foregoing hypothesis, it follows that there may be a tendency for the insects to fall in step and flash synchronously."

As discussed later, Richmond captured the essence of a self-organization system of interacting oscillators optically coupled to one another. We don't know whether Richmond understood the full implications of his description, but it is clear that Buck (1938) did not, because he viewed the theory merely as a "simplified form of the leader theory," and he remarked that, "Richmond's theory... like all the other theories, appears to be completely inadequate to explain displays of synchronism of the magnitude observed in Siam...."

Although Richmond suggested that his mechanism could be extended to a system involving many flashing units, we do not know whether he simply envisioned the most rapidly flashing individual acting as a leader (the interpretation that Buck seems to have taken), or whether he truly conceived of a collective self-organizing mechanism for coupling large numbers of fireflies. Whatever the case, Richmond's insight, tucked away in the back pages of *Science*, appears to have been largely forgotten. However, Hanson et al. (1971) do cite Richmond's article in the introduction of their own *Science* article describing a mechanism for synchronization based on the photic resetting of a flash-timing oscillator in the brain.

Let us now examine some experiments that served as a prelude to the formulation of a mechanism for synchronized flashing.

Neurophysiology of Individual Flashing

Experimental work on the mechanism of flashing began with neurophysiological studies of the central nervous system of fireflies (Case and Buck 1963). Hanson et al. (1971), Bagnoli et al. (1976), and Case and Strause (1978) reviewed the evidence suggesting that rhythmic flashing of male fireflies is controlled by a neural timing mechanism in the brain that oscillates at a constant frequency. Case and Buck (1963) and Buonomici and Magni (1967) found that each flash is triggered by nerve impulses in the brain that travel down the ventral nerve cord and lanternal nerves to the firefly's lantern. Experiments using ablation and local electrical excitation further supported the role of the brain as the central timer (Bagnoli et al. 1976).

Once the oscillator was identified, experiments revealed details of how it worked. Appropriate photic input could enhance or inhibit flashing (Buck 1937; Magni 1967). With systematic studies of the effect of exogenous light signals on the flash rhythm, a mechanism for synchronous flashing emerged. It was also discovered that different species use different synchronization mechanisms (Hanson 1978).

To avoid confusion, let us initially restrict our discussion to the synchronization mechanism of *Pteroptyx cribellata*, an extensively studied species from Papua New Guinea (Buck et al. 1981; Hanson 1978; Hanson et al. 1971). A set of experiments by Buck et al. (1981) revealed the effect of artificial pulses of light on the male rhythm. Individual fireflies were restrained and prevented from seeing their own flashes. In a darkened room, a fiberoptic system guided pulses of white light of 40 ms duration to the firefly's eye. The firefly flashed in response to the stimulus and each flash was recorded for analysis. In the absence of any stimulation, one male was found to flash regularly at a rate of nearly one flash per second (more precisely, its free run period was 965 ± 90 ms). After measuring this spontaneous periodicity, single 40 ms pulses were interjected randomly within its spontaneous flashing cycle every

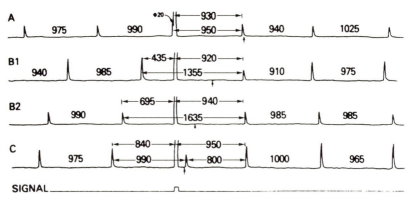

Figure 10.3 Responses of a firefly to single, artificial 40 ms pulses of light. See text for a discussion. (From Buck et al. 1981)

10 s or so. The firefly was tested with 21 pulses over a period of about 250 cycles. Three different responses to the artificial photic signals were observed, depending upon when the signal was imposed relative to the last preceding free run flash (Figure 10.3):

1. In response (A), one of the 21 signals happened to occur almost simultaneously with the firefly's own spontaneous flash (see response A). Since the first post-signal flash occurred 930 ms later—almost exactly when expected had there been no signal—this signal appeared to have no effect on the firefly's normal rhythm.

2. Each of the 17 signals that occurred between 110 and 840 ms after the firefly's own spontaneous flash inhibited the next expected free run flash responses (B_1 and B_2). Instead of occurring when expected, the flash was delayed. It occurred approximately 1 s after the signal pulse. In response B_1, the flash occurred 920 ms, and in B_2 940 ms, after the signal pulse. Viewed another way, the signal inhibited the expected flash, delaying it by approximately the length of time into the cycle that the signal occurred. These relationships should be clear from careful examination of responses B_1 and B_2.

3. Three signals fell late in the firefly's flash cycle, 840 ms or more after the previous flash but prior to the next expected flash. These late signals did not inhibit the next expected flash, which occurred when expected. In response C, the flash arrived 990 ms after the previous flash. However, the next flash arrived *early*, after 800 ms in response C.

In all cases, flashes occurring after the affected flash were followed by a series of flashes of normal free run duration. In other words, the signal pulse did not alter the firefly's normal period, but only affected the timing of the next flash (in B_1 and B_2) or the timing of the flash after the next flash (in C).

At first this behavior must have seemed confusing, as it may seem confusing to the reader at first inspection. A perplexing feature was that identical light signals resulted in two very different effects on the flashing rhythm. The type B effects caused the expected flash to arrive late, and the type C effects caused the following flash to arrive early.

Nonetheless, based upon the results of this one set of simple experiments the authors were able to hypothesize a model mechanism that was consistent with all the data and explained synchronized flashing. Let us examine the proposed model.

A Model Based on Coupled Oscillators

A breakthrough in our understanding of the mechanism of synchronous flashing actually came a decade earlier than the experiments just described when Hanson et al. (1971) presented results of preliminary experiments with *Pteroptyx cribellata*. The second sentence of their brief two-sentence abstract in *Science* succinctly states their hypothesis: "Since the interval between the pacer signal and the firefly's flash of the next cycle approximates the firefly's normal free-run period, it is suggested that the pacer signal resets the flash-timing oscillator in the brain, thus providing a mechanism for synchronization." This idea was suggested even earlier by Winfree (1967), who presented a theoretical paper on the behavior of populations of coupled oscillators. Referring to the "astonishing but persistent" reports of synchronizing fireflies in southeast Asia, he stated, "We will see ... that innate individual rhythmicity with phase-dependent sensitivity to mutual influences can give rise to ... striking community synchronization."

To understand the implications of Winfree's observation consider an individual firefly flashing at its normal free run period (a, in Figure 10.4). The pacemaker resetting model assumes that some property of the oscillatory center in the firefly brain—called "excitation"—gradually changes over time. Excitation of the brain's pacemaker rises from its baseline level to a threshold triggering level that elicits a flash. It is assumed that once a flash is triggered, the excitatory state spontaneously falls back to the baseline level and restarts the cycle of rising excitation. Although this excitation has never been directly measured in the firefly brain, there is a useful electrical analog of an oscillator that behaves similar to the firefly oscillator (Strogatz and Stewart 1993).

An oscillator can be modeled as circuit consisting of a resistor in parallel with a capacitor (Figure 10.5). A constant input current supplied by a battery charges the capacitor by increasing the voltage across the capacitor plates. Once the threshold voltage is reached, the capacitor discharges, and the cycle repeats.

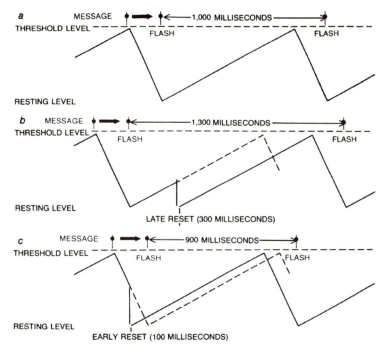

Figure 10.4 Model of a resettable pacemaker within a firefly. See text for discussion. (From Buck and Buck 1976)

Similarly, each time the oscillatory center in the brain reaches the triggering level, a volley of signals passes down the nerve cord to the firefly's lantern. The rising excitation (charging) phase of the cycle takes about 800 ms. The time it takes from the signal leaving the brain to the onset of flashing in the firefly's lantern is about 200 ms. During this same 200 ms the brain's oscillator is also reset (discharged) from its threshold level back to zero. As a result of these neurophysiological processes, the normal interflash period is about 1 s.

When no external stimuli are applied, the flashing remains regular and rhythmic. Now suppose the firefly receives a photic signal during the charging portion of its cycle (b, in Figure 10.4). A signal of sufficient intensity abruptly resets the excitation back to its zero baseline. If that happens, the flash does not occur at the expected time but is delayed until the excitation has again built up from zero to the triggering level. The flash is delayed by a period equal to the interval from the baseline level to the onset of the light signal.

Finally, suppose the firefly receives a signal shortly after the threshold level for triggering the flash has already been reached (c, in Figure 10.4). The model assumes that the impulse to trigger the next flash has already left the brain and is traveling down the ventral nerve cord. At this point, any light signal to

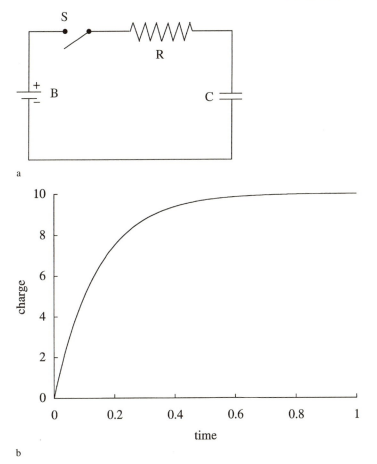

Figure 10.5 An oscillator modeled as a circuit (a) with a resistor (R) in parallel with a capacitor (C). The charge on the capacitor depends on the battery voltage, the resistance, and the capacitance, as shown in (b).

the brain cannot affect the timing of the flash cycle in progress and the flash occurs as expected. However, the light signal does affect the oscillatory center in the brain. Instead of taking the normal 200 ms to reset (discharge) back to the zero baseline, the light signal causes an immediate reset. As a result, the flash following the next flash occurs earlier by an amount of time equal to the time remaining for the normal reset.

Based on this single process of photic resetting of the neural oscillator in the brain, self-organizing synchronized flashing occurs. Each firefly acts as an intrinsic oscillator flashing at its own characteristic frequency. But in addition, each firefly interacts with its neighbors. Each firefly is coupled to its neighbors through light perceived from other flashing fireflies; the sight of a neighbor's flash shifts the individual's rhythm.

A Critical View of the Model

This model for synchronization seems consistent with experimental data showing how a single male firefly's flashing frequency rapidly becomes entrained to an artificial pacing flash. Data also show that a group of fireflies exhibits remarkable flash synchrony, so much so that recordings of light emission from a single firefly and a group of fireflies is practically indistinguishable (see figure on page 78 of Buck and Buck 1976). However, a crucial aspect of the model is missing. Although we now can understand how an artificial pacing signal can entrain a single firefly, we have not been shown that this mechanism is sufficient to explain what happens when a *group* of fireflies, all flashing out of phase, is brought together. The collective situation is far more complicated. Many flashes are emitted concurrently and there are reciprocal effects of one firefly on another. We must also consider the differing intensities of the flashes, a function of light intensity falling off inversely with the square of the distance from the source. In nature, during the incoherent initial stages of the process, each firefly sees a barrage of conflicting light emissions. Furthermore, each firefly has its own slightly different intrinsic rhythm which is somewhat variable from flash to flash. The problem is one of extrapolating the results of experiments in which a single artificial pacer flashing in a precise rhythm entrains a single firefly to the more general situation of a population of real fireflies in the field.

The problem of synchronization among a population of oscillators requires a more rigorous theoretical and mathematical approach. The problem has received considerable attention partly because of its intrinsic mathematical interest and partly because of the importance and ubiquity of such processes in biology. Some of the recent work (Mirollo and Strogatz 1990; Strogatz et al. 1992; Strogatz and Stewart 1993) is inspired by the medically important subject of the origin of synchronicity in the heart's natural pacemaker, a cluster of about 10,000 cells termed the *sinoatrial node* (Peskin 1975; Jalife 1984; Michaels et al. 1987). Mirollo and Strogatz (1990) have analyzed mathematically a population of oscillators interacting through a mechanism similar to that found in *Photinus pyralis* (See below: Flash Synchronization in Other Firefly Species.) Their model, however, makes a number of critical simplifying assumptions in the interest of mathematical tractability. They assume that all oscillators in the population are *identical*, that each oscillator is sensitive to incoming light impulses throughout its charging cycle, and that the increase in excitation is concave downward as in the electrical analog (Figure 10.5), rather than linear as assumed by Buck and Buck (1976) and Buck et al. (1981) (see Figure 10.4).

Under these assumptions, a population will become synchronized under almost *all* initial conditions. The system synchronizes rather slowly at first, but then builds up more rapidly (Mirollo and Strogatz 1990). Although the assumption of *identical* oscillators makes the problem more tractable mathematically, we would really like to know the properties of a system of realistic firefly oscillators, with similar but not identical intrinsic flash frequencies.

Strogatz and Stewart (1993) discuss this more complicated situation in a recent article in *Scientific American*. (For a more technical presentation, see Strogatz et al. 1992 and Ermentrout 1991.) Their main conclusion is as follows: "The behavior of communities of oscillators whose members have differing frequencies depends on the strength of the coupling among them. If their interactions are too weak, the oscillators will be unable to achieve synchrony. The result is incoherence, a cacophony of oscillations." (Strogatz and Stewart 1993, p.107). As the variation in frequencies of individual oscillators falls below a critical threshold, a portion of the system suddenly synchronizes. The combined signal of this synchronization cluster stands out above the background noise of random flashes and "captures" additional oscillators, further amplifying its collective signal. This infectious positive feedback results in an epidemic of synchrony.

Strogatz and Stewart's conclusions provide an excellent commentary on the relationship between self-organization and natural selection. (See Misconception #1 in Chapter 7.) If flash synchronization has adaptive value, as we assume it does, then natural selection should result in fireflies with the necessary strength of coupling among individual fireflies to achieve flash synchrony. By acting on the physiological "rules" of interaction among individual fireflies, evolution has determined the final structure of the self-organized process.

Flash Synchronization in Other Firefly Species

Based on comparative studies (Hanson 1978) (see also the review by Buck 1988) it appears that several different mechanisms for self-organized flash synchrony have evolved. Studies suggest that *Photinus pyralis*, for example, has a different pacemaker resetting mechanism than *Pteroptyx cribellata*. Instead of causing an immediate resetting of the pacemaker to its baseline level, a light signal appears to advance the flash by raising excitation to the threshold triggering level. A number of other species also appear to utilize flash-advance synchronization rather than the phase-delay synchronization seen in *Pteroptyx cribellata* (Buck 1988). In a detailed study by (Hanson 1978), *Luciola pupilla* (unlike *Pteroptyx cribellata*) required many cycles of an imposed light signal to become synchronized. This slow entrainment suggests that an external light signal does not advance the pacemaker of *Luciola* immediately to threshold but only raises the excitation level of the pacemaker partially towards a triggering level. Regardless of the details of each mechanism, the important feature is that fireflies are optically coupled through their ability to reset their neighbor's pacemaker.

Even though an enormous amount of sophisticated neural circuitry and neurophysiology undoubtedly underly the firefly's central nervous system oscillator and lantern, this self-organizing mechanism is conceptually simple and can explain all the observed features of flash synchronization. With the pieces of this fascinating mystery laid out in front of us, and with the advantage of hindsight, perhaps the synchronization mechanism now appears rather trivial. This is often the case with mechanisms based on self-organization.

Even with the mystery largely solved, many unknown details still remain. For example, the actual timer in the brain has never been isolated. Bagnoli et al. (1976), on the basis of their surgical ablation and electrical stimulation studies conclude that "(a) the photomotor neurons of the firefly's brain are located in the deep protocerebral neuropile; (b) their rhythmic activity is the result of the interaction with an oscillator located in the optic lobes, possibly in the lobula." Isolating the firefly flash-control center and characterizing the presumed neural network responsible for its oscillatory properties will prove to be a formidable task. A comparable oscillatory system that has been well studied is the rhythmic bursting neuron of the sea mollusk, *Aplysia* (Pinsker 1977a, b). But as Buck et al. (1981) remark, "the firefly brain is not much larger than the largest *Aplysia* neuron." The difficulties of doing studies at the single-cell level within such a tiny brain make it unlikely that the system will be analyzed in detail, at least within the foreseeable future.

Flash Synchronization as a Self-Organizing Process

In what ways is the mechanism of flash synchronization a self-organizing process? A key feature of the system is that the pattern emerges as a result of multiple interactions among the fireflies. Synchronization is not imposed by any influence outside the system, such as a leader, a supervisor or external physical cue. Synchronization arises from within, based on local interactions among fireflies that follow the simple rule: A neighbor's light emission shifts the timing of one's own light emission (Figure 10.6). The rule does not directly code for synchrony, as with a conductor's baton beating time in a predetermined cadence, yet this simple phase-shifting rule is sufficient to coordinate the rhythm of the group.

As discussed previously, the ability of a local group of synchronized individuals to capture additional oscillators is a form of positive feedback, a common feature of self-organizing systems. Negative feedback is also present, in the form of a physiological constraint that keeps positive feedback from self-perpetuating out of control: A firefly can be stimulated to flash only within a fixed range of frequencies. Although its normal period is about 1000 ms, it can be paced only at a rhythm of between 800 and 1600 ms. Within a certain range

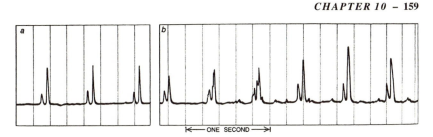

Figure 10.6 Solo and chorus flashing of *Pteroptyx malaccae* males are compared: a single firefly emits a two peaked flash every 500 ms in a pattern virtually duplicated by synchronized fireflies. (From Buck and Buck 1976)

the system is refractory, providing negative feedback that brakes and shapes the positive feedback, helping to create a precise temporal pattern.

How might this system have evolved? One possibility is that natural selection favored the basic oscillatory flash apparatus of male fireflies as a mating adaptation because it was inherently a self-organizing mechanism for group synchronization based on individual pacemaker reactions to light signals from local neighbors. A collective pattern emerges using an optical coupling mechanism to provide networks of local interactions among fireflies.

Box 10.1 Demonstration of Synchronization in Humans

Buck and Buck (1976) describe two simple demonstrations of synchronized activity in humans that can be performed in the classroom. The first demonstration shows that synchronized finger-tapping in human subjects does not occur as a result of individuals reacting to one another in a follow-the-leader manner. The second demonstration suggests that the actual mechanism involves individuals listening to the tapping rhythm and adjusting their own frequency to the collective rhythm.

Demonstration 1: The person conducting the demonstration instructs participants to close their eyes and hold a coin in their fingers. Participants wait for the instructor to tap his or her coin on the tabletop. Participants then tap their coins as quickly as possible. This is a test of reaction time performed as a group. As the graph in Figure 10.7a shows, very few participants can respond in less than 150 ms.

Demonstration 2: The instructor gives the participants these simple instructions: "Close your eyes and start tapping your coin on the tabletop in a comfortable, regular rhythm trying to synchronize with your

Box 10.1 continued

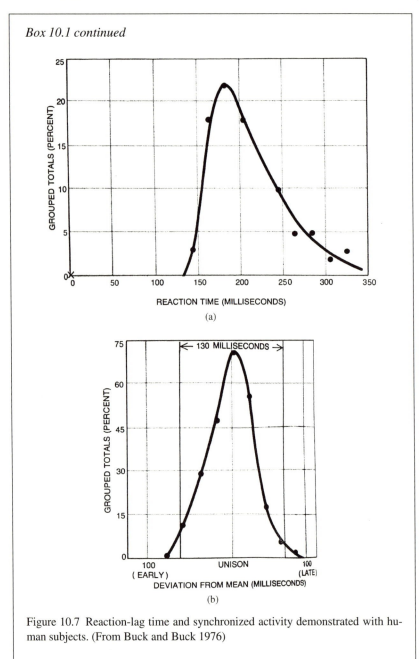

Figure 10.7 Reaction-lag time and synchronized activity demonstrated with human subjects. (From Buck and Buck 1976)

Box 10.1 continued

neighbors." As Buck and Buck point out, with these instructions an audience of several hundred people will synchronize within a few cycles, usually at a frequency of 2 to 3 taps/s. More remarkably, the time interval between the first taps and the last taps in a volley are generally less than 130 ms (Figure 10.7b), shorter than the reaction time just demonstrated!

What is the mechanism for synchrony in demonstration 2? As the reaction time data confirm, the participants cannot be waiting to hear a neighbor's tap before initiating their own tap. The explanation offered by Buck and Buck (1976) is that each person must have "measured the passage of time up to the instant when it was necessary to initiate the neural message that caused his fingers to move in synchrony with the fingers of all the other participants." Thus, based on the subject's perception of the rhythm of the taps, the subject anticipates the time to the next tap and factors in the motor delay between the brain and finger. This is the "anticipatory time-measuring" or "sense of rhythm" explanation earlier suggested by Buck and Buck (1968) for the mechanism of rhythmic synchronous flashing of fireflies. We should now be better able to appreciate the difference between the mechanism of synchrony used by humans in finger tapping and the mechanism of firefly flash synchronization. The cognitive processes that humans use to synchronize finger tapping are not those used by fireflies. Among fireflies, the perception of a neighbor's photic signal automatically resets an oscillator in the firefly's brain and triggers a flash. Among humans, there is no oscillator for an arbitrary tapping rhythm. Humans do indeed have a sense of rhythm, they can measure time intervals and anticipate when to initiate a motor message to tap.

With a means for recording the sounds of subjects as they tap, and a device for displaying the recording, it is possible to generate graphs similar to those in Figure 10.7. This can be done with most personal computers using a microphone and the appropriate sound analysis software.

Box 10.2 Temporal Patterns in Other Organisms

The natural world abounds in examples of collective temporal patterns other than synchronized rhythmic activity in insects. The mechanisms for many of these examples are unknown, but we expect that many eventually will be shown to involve self-organizing mechanisms.

Box 10.2 continued

Three categories of synchronization and rhythmicity may be distinguished (Table 10.1). The simplest category (I in Table 10.1) involves a burst of synchronized activity based partly on mutual, positive feedback interactions. For example, many sea birds that nest in large colonies stimulate each other to breed earlier and within a shorter period of time compared to members of small colonies. Termed the *Fraser Darling effect* (Darling 1938) in herring gulls, for example, this results in higher survival of the chicks since the percentage mortality (due largely to predation) of offspring per unit time is relatively constant. The adaptive significance of such a collective phenomenon is obvious. It is the temporal equivalent of Hamilton's so-called selfish herd geometry (Hamilton 1971). In this situation a mother bird gains a fitness advantage by hiding her offspring from predation in a large crowd of conspecifics within a narrow window of time. The density-dependence of the effect indicates that it does not result solely from an external environmental influence, such as the change in day length, but suggests the additional role of a self-organized mechanism based on behavioral and physiological interactions among group members.

Another, similar example is the synchronous release of spermatozoa by marine sponges (Reiswig 1970). In this case, the spread of activity to nearby colonies when they come into contact with sperm-laden water suggests some form of communication, possibly chemical, among individuals rather than the sole influence of an environmental trigger. Many other examples in this category are known, such as synchrony of vocalizations in frog choruses or bird leks. Only a few examples are listed in the table.

A second category of synchronized rhythms (II in Table 10.1) involves activities for which individuals show no intrinsic rhythmicity but are ·rhythmic as a group. An example is *Leptothorax* ants, in which colonies exhibit synchronous rhythmic bursts of activity approximately every 20 min (Franks and Bryant 1987; Franks et al. 1990; Cole and Trampus 1991, 1998; Goss and Deneubourg 1988). Another example is pancreatic beta-cells that secrete insulin. These cells are electrically coupled by their gap junctions. In sufficiently large clusters, the cells exhibit regular electrical bursting activity whereas isolated cells show disorganized spiking activity (Sherman et al. 1988; Sherman and Rinzel 1991).

The third category (III in Table 10.1) involves rhythmic individual activity synchronized at the group level. Synchronization of flashing among fireflies is just one example. Others include the synchronized choruses of certain crickets, katydids, and cicadas; the synchronization of human fe-

Box 10.2 continued

Table 10.1 Synchronized Rhythmic Activity (see Box 10.2 for an explanation of categories I, II, and III) among Groups of Organisms

Organism	Scientific Name	Process and Category	Coupling Mode	References
Hydrozoans	(Many species)	Coordinated movements of zooids (I)	Mechanical, electrical	Mackie 1973
Tropical marine sponges	*Verongia archeri, Geodia* sp. *Neofibularia nolitangere*	Synchronized sperm release (I)	Unknown	Reiswig 1970
Red abalones	*Haliotis rufescens*	Synchronized sperm and egg release (I)	Chemical (prostaglandins)	Morse 1993
Fiddler crab	*Uca annulipes*	Synchronized claw waving (I)	Visual	Gordon 1958; Backwell et al. 1998
Harvestmen	Family Phalangidae	Synchronized group movements[a] (I)	? Mechanical (tactile)	Newman 1917; Wheeler 1917
Spiders	*Anelosimus eximius* (Theridiidae)	Synchronized prey capture activity (II)	? Web vibrations	Krafft and Pasquet 1991
Ants	*Leptothorax* spp.	Synchronized activity cycles (II)	? Mechanical (tactile)	Franks and Bryant 1987; Franks et al. 1990; Cole 1991
	Eciton burchelli (army ants)	Periodic foraging activity (II)	?	Schneirla 1949, 1956; Gotwald 1995
	Campanotus spp. and others	Synchronized alarm drumming (I)	Substrate vibrations	Fuchs 1976a,b; Hölldobler and Wilson 1990
	Messor pergandei (harvester ants)	Periodic foraging activity (II)	Pheromone	Rissing and Wheeler 1976; Goss and Deneubourg 1989
Honey bees	*Apis mellifera*	Synchronized respiration (I)	Unknown	Moritz and Southwick 1992
Saharan silver ant	*Cataglyphis bombycina*	Synchronized foraging activity (I)	Unknown	Wehner et al. 1992

Box 10.2 continued

Table 10.1 continued

Organism	Scientific Name	Process and Category	Coupling Mode	References
Hornet	Vespa orientalis	Synchronized vibrations (III)	Mechanical	Barenholz-Paniry et al. 1988
Migratory locust	Schistocerca gregaria and others	Mass migrations (I)	Visual, ? chemical, tactile	Uvarov 1928
Fall webworm larvae	Hyphantria cunea	Synchronized group movements[a] (I)	? Mechanical (tactile)	McDermott 1916; Peairs 1917
Aphids	None given	Synchronized body movements while feeding[a] (I)	?	Tanner 1930
Termites	Unidentified Indian species	Synchronized chewing[a] (I)	?	Connor 1933
Firefly	Pteroptyx spp., Photinus spp. Luciola spp., others	Synchronized flashing (III)	Photic (optical)	Buck 1988 and references therein
Snowy tree cricket	Oecanthus fultoni	Synchronized chirping (III)	Acoustic	Walker 1969
Katydids	Mecopoda sp., Neoconocephalus spiza	Synchronized chirping (III)	Acoustic	Sismondo 1990; Greenfield and Roizen 1993
Periodical cicada	Magicicada spp.	Synchronized chirping (III)	Acoustic	Alexander 1967, 1975
Springtails (collembola)	Hypogastrura spp.	Synchronized moulting (III)	? Olfactory (pheromonal)	Leinaas 1983
Herring gulls, & other colonial sea birds	Larus spp. and others	Synchronized breeding (I)	Unknown	Darling 1938
Human females	Homo sapiens	Synchronized menstrual cycles (III)	? Olfactory (pheromonal)	McClintock 1971; Russell et al. 1980

[a]Observations not well documented in the literature.

Box 10.2 continued

male menstrual cycles; and the synchronized molting of springtails. All probably involve interactions among coupled oscillators. Many other well-documented examples of this type of mutual synchronization in biological systems not listed in the table include those in which individual units are single cells, such as sinoatrial-node pacemaker cells (Jalife 1984; Michaels et al. 1987); *Saccharomyces* yeast cells undergoing glycolytic metabolism (Winfree 1980); the unicellular marine plant *Gonyaulax* with its circadian rhythm of bioluminescence (Winfree 1980, 1987); and the intensively studied example of oscillations in chemotactic cyclic AMP signals in the slime mold, *Dictyostelium*, described in Chapter 8.

Table 10.1 also lists examples of coordinated activity among groups of organisms that are not sufficiently documented to know whether the process is rhythmic and synchronized. These are the examples involving harvestmen, webworm larvae, aphids, ants, termites, and probably fall into the simplest category noted at the outset.

A type of mutual synchronization seen in honey bees (Southwick and Moritz 1987), deer mice (Crowley and Bovet 1980), and red wolf/coyote hybrids (Roper and Ryon 1977) was omitted from the table because the activities are circadian rhythms probably triggered largely by diurnal cues. They are of interest here, however, since experiments have shown them to be modulated by social interactions among group members.

Figure 11.1 A school of fish showing the regular alignment of a large number of individuals and suggesting the ability of the school to move as a synchronous, cohesive structure. (Illustration © Bill Ristine 1998)

11

Fish Schooling

"and the thousands of fishes moved as a huge beast,
piercing the water. They appeared united, inexorably bound
to a common fate. How comes this unity?"

—Anonymous, 17th century

One of nature's most awe-inspiring phenomena is the sight of hundreds, thousands, or even millions of animals moving together as a coordinated unit (Figure 11.1). Noisy flocks of a thousand starlings burst into flight at the approach of a barking dog and maneuver gracefully around tall city buildings. At Carlesbad Cavern millions of Mexican free-tailed bats stream from the cave each evening to begin their nightly feeding foray. A herd of several hundred thousand wildebeest moves fluidly along the Serengeti plains of Africa. In the sea schools of millions of Atlantic cod swim together in tight formation as they migrate along the cold, deep waters off the Newfoundland coast. These impressive collective phenomena, and others listed in Figure 11.2 and Table 11.1, seem to suggest an intelligence that far transcends the abilities of each group member. In this chapter we examine the remarkable ability of fish to coordinate their movements in a group. We have chosen the example of fish schools primarily because it has been well studied and sufficient quantitative data are available to permit the formulation of reasonable hypotheses and models of the schooling process.

The Behavior of Schools

A school of fish maneuvers gracefully, with all its members moving in parallel in the same direction. When a school suddenly changes direction, all its members rapidly respond, moving cohesively, almost in unison, as flawlessly as if they were parts of a single organism.

These behaviors suggest that a school possesses special group-level properties. One such property is the rapid transfer of information throughout the school that enables the entire group to execute swift, evasive maneuvers at the approach of predators (Partridge 1982). This action is the so-called Trafalgar Effect (Treherne and Foster 1981), by analogy to the rapid transfer of battle-flag signals along a chain of ships in Admiral Nelson's fleet at Trafalgar. Such

a

Figure 11.2 Additional examples of groups of animals exhibiting collective motion that may be based on self-organizing mechanisms: (a) locust plague in Kenya, Africa; (b) flock of starlings (photo © John Bova/Photo Researchers); (c) Mexican free-tailed bats leaving a cave (photo © Merlin Tuttle/Photo Researchers).

evasive maneuvers occur in a group of individuals as a wave of reaction propagating many times faster than the approach of the predator. Potts (1984) hypothesized that this group coordination is based on individuals observing the wave of reaction by neighboring fish and timing their own execution to coincide with its arrival, in the same way that a human chorus line performs a wave of high kicks.

b

c

Figure 11.2 Continued.

One evasive maneuver is flash expansion, in which the school rapidly expands and bursts radially (Figure 11.3). Another evasive tactic is the fountain effect, in which a school of small, slow-moving prey outmaneuvers a predator by splitting into two groups, each of which moves in opposite directions and regroups behind the predator (Figure 11.4).

Coordinated maneuvers are employed not only by prey to avoid their predators but also by predators to trap their prey more efficiently. For example, schools of giant bluefin tuna may coordinate their movements to capture other

Table 11.1 Examples of collective motion of organisms with selected references on their biology and modeling.

Organism	Scientific Name	Collective Process	Selected References
Fish	Many species	Coordinated movements of fish in a school	Shaw 1962; Partridge et al. 1980; Partridge 1982; Pitcher et al. 1976; Rose 1993
Passerine bird flocks	Many species	Coordinated movements of birds in a flock	Higdon and Corrsin 1978, May 1979
Canada geese	*Branta canadensis*	Chevron-shaped flight formation	Lissaman and Shollenberger 1970; Gould and Heppner 1974, May 1979
Wildebeest (and other mammals)	*Connochaetes taurinus* and other species	Coordinated movements of animals in a herd	Sinclair and Norton-Griffiths 1979
Spiny lobster	*Panulirus argus*	Mass migration of individuals in single file	Bill and Herrnkind 1976; Herrnkind and McLean 1971; Kanciruk and Herrnkind 1978
Processionary moth	*Cnethocampa (Bombyx) processionea*	Nighttime columns of feeding caterpillars	Frost 1959

Table 11.1 *continued*

Organism	Scientific Name	Collective Process	Selected References
Flies in a swarm	many species of Diptera and Ephemeroptera	Coordinated motion of individuals in the swarm	Okubo 1980, 1986
Locusts	*Schistocerca gregaria*	Coordinated motion of individuals in the swarm	Uvarov 1928
Fibroblast cell cultures	—	Oriented movements of cells in cultures	Edelstein-Keshet and Ermentrout 1990a, 1990b, 1990c
Slime molds	*Dictyostelium discoideum, Polysphondylium violaceum* and other species	Cohesive movements of individuals in the aggregation	Bonner 1983; Keller and Segel 1970; Steinbock et al. 1991; Schaap 1986
Myxobacteria	*Myxococcus xanthus, Chondromyces apiculatus* and other species	Cohesive movements of individuals in the swarm	Shimkets and Kaiser 1982; Stevens 1990; Zusman 1984; Shimkets 1990; Lauffenburger et al. 1984; Pfistner 1990

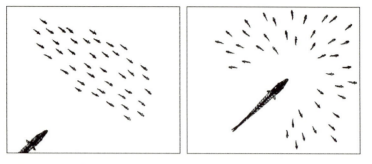

Figure 11.3 A school of fish displaying the evasive maneuver, termed *flash expansion*. (From Partridge 1982, used with permission)

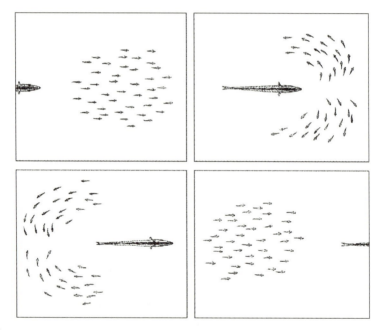

Figure 11.4 A school of fish displaying the evasive maneuver, termed the *fountain effect*. (From Partridge 1982, used with permission)

schooling fish (Partridge 1982). The tuna swim in a parabolic formation, driving a school of prey between the outstretched arms of the parabola and then surrounding the prey.

To capture prey efficiently or avoid a predator effectively, these collective motions require that individuals coordinate their movements with those of other members of the group. During these maneuvers, individuals seldom collide with their neighbors even in the frenzy of an attack.

These descriptions of fish schools are based largely on observations in nature and have been supplemented by more detailed studies of the position and internal, three-dimensional configuration of fish in a school. The studies were conducted by filming individually marked fish swimming in a 10-m-diameter annular tank, 1.2 m deep (Partridge 1981; 1982; Partridge et al. 1980). Schools of twenty to thirty fish were trained to swim in view of a video camera mounted on a gantry rotating above the school. A bright light attached to the moving gantry was aimed at an angle to the tank to cast a distinct shadow of each fish. The position of the shadow was used to calculate the depth of each fish in the school. In this way, records were obtained of the position in three dimensions of each fish.

Partridge et al. (1980) compared the schooling behavior of cod (*Gadus morhua*), pollack (also called saithe) (*Pollachius virens*), and herring (*Clupea harengus*). Each species tended to maintain a school of a particular three-dimensional shape. Viewed from above, herring schools were roughly circular and about three times as long (or wide) as they were deep. Their characteristic shape (the ratio of length: width: depth) was 3:3:1. In contrast, pollack schools tended to be twice as long as wide and about three times as wide as deep (shape ratio of 6:3:1). Cod schools were the most variable in shape, ranging from 10:4:1 to 2:4:1.

In addition to the different shapes of schools of different species, there are also characteristic arrangements of fish within the school. Fish avoid approaching closer than a certain minimum distance to one another and also have a preferred distance, elevation, and bearing relative to their nearest neighbors (Partridge 1982; Partridge et al. 1980). For example, herring and pollack tend to have a small proportion of their neighbors swimming at the same elevation (Figure 11.5a). As seen from above, herring most frequently have nearest neighbors at a bearing of 45° or of 135°, positions one would expect if herring schools were organized as cubic lattices. Cod have their neighbors most frequently at 90° while pollack are most likely to have nearest neighbors more uniformly distributed at bearings in the range of 60° and 140° (Figure 11.5b). Nearest neighbor distance depended not only on species but also on the size and speed of the school. Increasing the number of fish from five to twenty-five in schools of pollack or cod resulted in a decrease in nearest neighbor distance (Partridge et al. 1980). Nearest neighbor distances in pollack also decreased with increasing cruising speed (in the range 5.5 to 8.5 radians/min in the 10 m-diameter tank).

Adaptive Significance of Schooling Behavior

Many hypotheses have been offered to explain how the coordinated movements of fish in a school provide defense against predators and in some cases, increase foraging efficiency.

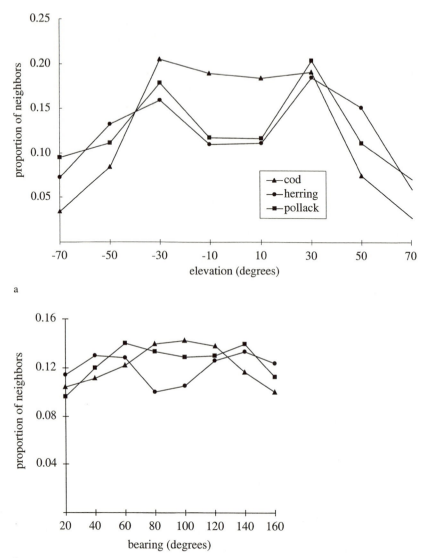

a

b

Figure 11.5 Proportion of nearest neighbors at different elevations (a) and bearings (b) for herring, pollack, and cod. In (a), the reluctance of fish to swim on the same level as near neighbors is especially clear for herring and pollack. (Data from Partridge et al. 1980)

These schooling behaviors may reduce predation in a number of ways. A large group of moving prey may confuse a predator trying to focus its attention on capturing a single individual. For example, squid (*Loligo vulgaris*), cuttlefish (*Sepia officinalis*), pike (*Esox lucius*), and perch (*Perca fluviatilis*) are more confused and have less success when hunting fish in large schools than in small schools (Neill and Cullen 1974). This may be due, largely to the coordinated movements of the school, such as unpredictable zigzagging maneuvers. Driver and Humphries (1988) call such behavior "united erratic display" and remark that such protean behaviors involve "special mechanisms for the close coordination of individuals in united movements. These mechanisms are clearly innate and must have been developed by natural selection." The confusing movements may be further enhanced by the pattern of body coloration, such as the flashing of silvery or iridescent patches, or shifting patterns of body stripes. Schools of fish (as well as flocks of birds) may further confuse a predator by starting out in a united erratic display and then breaking up into smaller groups that may cross over each other, recombine, and further separate. Examples among fish schools are flash expansion and the fountain effect discussed previously (Figures 11.3 and 11.4). Driver and Humphries (1988) also cite examples involving mullet, starlings, ducks, and wildebeest.

Another collective defensive maneuver involves individuals drawing together in a tight aggregation. Sometimes birds in a loose flock will snap into a tight group at the approach of a falcon and zigzag in close coordination. Driver and Humphries (1988) believe that such behavior prevents the predator from attacking, since the speed of a peregrine falcon may be more than 240 km/hr and at such speeds the falcon risks injury from a collision unless it strikes with its talons first. In such situations, the only recourse for the predator is to attack a bird that becomes separated from the flock. This may provide strong selection pressure on individuals to stay with the group and maintain close coordination with nearest neighbors. Flocks of ducks and waders (such as dunlin, turnstones, and oystercatchers) generally fly even faster than starlings, making the coordination of the individuals a more difficult task. Driver and Humphries (1988) suggest "that the special wing and tail markings of these two groups—particularly the wing bars and specula in waders and ducks respectively—have been developed at least in part to assist in the co-ordination of flock movements."

The coordinated movements of predatory animals in schools may also increase the harvesting efficiency of its members. A school of fifteen to twenty killer whales (*Orcinus orca*) was observed cooperatively hunting dolphins (*Delphinus bairdi*) by encircling their prey and gradually constricting the circle to crowd the dolphins (Martinez and Klinghammer 1970). Then one killer whale rushed into the middle of the school while the others continued to circle. After feeding on the dolphins, this individual returned to the circle and another whale entered to feed. This cooperative hunting technique, like that of the bluefin tuna described earlier, evidently confers higher per capita success than without such coordination.

In addition, some work suggests that the regular pattern of fish in a school functions to increase the hydrodynamic swimming efficiency of the group. Wiehs (1973) reported that the endurance of fish can be increased as much as six times when they travel in a school. However, Partridge and Pitcher (1979) present data suggesting that—for at least the three species of fish they studied—fish do not swim in the appropriate positions to gain a hydrodynamic advantage. In contrast, the evidence strongly supports an increased aerodynamic efficiency for flocks of Canada geese (*Branta canadensis*) flying in their typical V-shaped flight formation (Lissaman and Shollenberger 1970, May 1979). Theoretically, the energy savings allow a group of twenty-five birds to increase their flight range by about 70 percent over that of a solitary bird. The migrations of spiny lobster (*Panulirus argus*) provide another example of energy savings through group locomotion. In the autumn, mass migrations of thousands of lobsters have been observed near Bimini, Bahamas. As many as sixty-five lobsters line up in single-file queues maintained by tactile contact. Studies have shown that a queue of nineteen can maintain a pace of 35 cm/s with the same hydrodynamic drag as individual lobsters traveling only 25 cm/s (Bill and Herrnkind 1976).

To maintain its position in a school each fish must constantly gauge and monitor the positions and velocities of its neighbors. It seems likely therefore that the structural differences of schools reflect differences in the positions at which individuals are best able to sense and react to their neighbors' movements. These positional differences in turn may reflect underlying differences in the sensory capacities of individuals of a particular species. All these factors, both at the school and individual levels, may facilitate species-specific behavioral strategies for locomotion, defense, and predation. The fundamental problem facing a fish in a school is how to optimize its ability to respond rapidly to its neighbor's actions, to reduce its risk of colliding with a neighbor or being captured by a predator. These risks apparently exert a strong selection pressure for the evolution of coordinated group movements.

Thus a school acts as a group-level vehicle with traits that confer upon the members significant survival advantages. A school of fish is not merely a disorganized crowd, a "selfish herd" (Hamilton 1971), in which an individual is seeking cover within the crowd and minimizing its likelihood of predation by placing itself so that a few neighbors are between itself and a predator's jaws. A school is "a social organization to which the fish are bound by rigorously stereotyped behavior and even by anatomical specialization" (Shaw 1962). This organization has been carefully molded by natural selection acting upon individuals within the school.

Behavior of Individuals within the School

To understand schooling behavior we must understand how an individual fish coordinates its behavior with that of other members of its school. This requires examination of sensory inputs each individual acquires as it moves in the school and determination of how individuals use this information to adjust their behavior.

Most schooling fish rely on vision and the lateral line system. The need for visual input for schooling is clearly demonstrated by the inability of a group of experimentally blinded fish, to school, and by cessation of schooling at low light intensities (Pitcher et al. 1976; Shaw 1970). The sufficiency of vision has been demonstrated by placing individual fish on one side of a transparent glass barrier and observing their schooling behavior. Experiments such as these (reviewed in Shaw 1970) show that fish can maintain a parallel orientation to fish on the other side of a barrier. Moreover, a fish blinded in one eye approaches and lines up with its neighbor from its unblinded side (Shaw 1962). In these experiments, a fish can see its neighbors movements but obviously cannot sense other cues such as water displacement, sound, or chemical cues, thus strongly suggesting that a fish *visually* aligns itself with respect to its neighbor.

Although vision appears to be sufficient for schooling it cannot be the only sense involved. Vision certainly plays a role in the attraction and approach of a fish to others, but eventually a fish isolated from others by a glass barrier appears to lose interest and move away (Shaw 1970). Two other observations provide even stronger evidence for the role of other sensory input. Shaw (1970) described an experiment with carangids (*Caranx hippos*) in which individuals maintained a parallel orientation to neighbors separated by a one-quarter inch thick, distortion-free transparent or translucent partition. The fish responded regularly to each other's changes in positions and swam the length of the barrier almost in contact with it. Shaw noted that this was unusual behavior because under normal schooling conditions the fish would not swim so close to one another. In these experiments, the normal repulsion between fish that are too close together appeared to be blocked by the transparent partition. Parr (1927) hypothesized that schooling resulted from a balance between attraction and repulsion, both guided by vision alone. However, Shaw's experiments suggest that vision functions in the approach phase whereas another sense, blocked by the partition, is responsible for the repulsion of a fish when it gets too close to its neighbor.

That other sense involves the lateral line, gelatinous canals located internally along the side of the fish and connected to the external environment by a series of pores. Inside the canal are thousands of hair cells that function like sound receptors in the human ear. They respond to water displacements and provide the fish with information about the speed and direction of neighboring fish (Partridge 1982). The role of the lateral line was demonstrated conclusively in experiments in which individual, blinded pollack were introduced into a school of twenty-five normal fish (Pitcher et al. 1976; Partridge and Pitcher 1980). The blind fish were able to maintain their position in the school and respond to short-term movements of other fish in the school. When its lateral line was also cut it was unable to school.

Additional quantitative studies (Partridge and Pitcher 1980) using pollack with opaque blinders or with sectioned lateral lines have confirmed these results. Fish with severed lateral lines swim closer to their neighbors than usual, whereas blinding increased the nearest neighbor distance. Vision is most important for maintaining the attraction—when a fish strays too far from its neighbors it moves closer to nearby fish—whereas the lateral line is most responsible for repulsion. By means of the lateral line, a fish can sense whether it is too close to a neighbor and must move away to avoid collision. Vision appears to be most important for maintaining the position and angle between fish, while the lateral line is most important for monitoring the swimming speed of neighbors (Partridge and Pitcher 1980).

Even with an understanding of how visual and lateral line information allow a fish to maintain the proper spacing with respect to its neighbors, there is the further problem of how individuals, surrounded by many fish integrate information from many neighbors and decide on a plan of action. Each individual must use some scheme for weighting information coming simultaneously from several fish. That scheme may be understood by examining the correlation of the speed and direction of a particular fish with the speed and direction of other fish in the school a short time before. As one might expect, a fish generally is most strongly influenced by its nearest neighbors.

Alternative Hypotheses for the Mechanism of Schooling

Based on what we know about schooling and what we know about the behaviors of the individual fish involved, what mechanism can explain how these two levels of behavior—individual and the collective—are linked. Specifically, we wish to determine whether individuals possess behavioral instructions that directly code for collective schooling, or whether schooling is a self-organized activity, where coordinated motions arise in a less obvious, more indirect, way. Mechanisms involving templates, sophisticated leaders, plans, or recipes are all based on explicit instructions that directly specify the collective pattern of movement. For example, all the migrating birds in a flock "know" to fly south by reference to an external template provided by physical cues (which may include wind direction and the earth's magnetic field).

Hypotheses Based on Explicit Instructions

It is unlikely that an external imposed stimulus such as water currents can account for the coordinated motion of a school in the vast, apparently featureless expanses of the ocean. Nor is there any evidence that the myriad fish in a school coordinate their movements by responding to a common guiding stimulus, such as light or chemical cues.

Descriptive studies have also failed to show that schools of fish have leaders (Partridge 1982; Partridge et al. 1980). Within the school, fish continually change positions. Those at the leading edge often fall back and trade places with those behind. When the school changes direction, a fish that was following on the flank suddenly finds itself in front (Shaw 1962). In schools of fish each individual appears to adjust its speed and heading according to that of several of its nearest neighbors (Partridge 1982; Partridge and Pitcher 1980). Thus, there appears to be no leader and each fish bases its behavior on nearby fish.

It is even more difficult to account for schooling with other hypotheses based on direct coding for the pattern of movements of individuals. It seems exceedingly unlikely that thousands of fish in a large school could coordinate their movements by means of a common plan of action or step-by-step recipe specifying where and when to move. Schools of herring may contain a million members and migratory schools of Atlantic cod may number hundreds of millions of fish (Rose 1993). Even if these aggregations were to have some innate knowledge of where to go, there remains the mystery of how such an enormous number of fish precisely coordinate their movements relative to one another.

A Self-Organization Hypothesis of Schooling

The most likely explanation for schooling is a self-organization mechanism where each fish applies a few behavioral rules in response to local information from neighboring fish. Using this approach, several investigators have developed convincing mathematical models describing the movement of fish schools in two dimensions (Aoki 1982; Huth and Wissel 1992; Warburton and Lazarus 1991).

Huth and Wissel's (1992) model shares many features of Aoki's (1982) model and shows how simple attraction and repulsion rules used by individuals readily generate schooling patterns, without direct behavioral coding for the patterns. Instead, behavioral interactions among individuals are a balance between positive feedback—which brings individuals together as a group—and negative feedback that triggers repulsion when one fish comes too close to another. Neither feedback behavior explicitly codes for the positions or movements of the fish. Another feature of the model is that the rules rely on local information—the positions and headings of a few nearby fish—rather than a global plan or centralized leader. Since each fish must rely on limited, local information it cannot determine exactly where it should move. A determination would require instantaneous knowledge of all the fish in the school—an overview—and the ability to instantaneously calculate the velocity of the whole school. Rather than solve this complicated problem, a fish makes a series of approximate, moment-to-moment responses based on the limited knowledge it can gather over the short time interval in which it must act. Its solution to the problem of information processing is to rely on simple rules of thumb based on limited local cues gathered from near neighbors or the environment. The organism then acts without a complete overview of the situation. Its resulting behavior is adequate but not exact. Though imperfect, such self-organization mechanisms apparently have evolved because they work well.

A Model of Schooling Based on Self-Organization

The model incorporates the known behaviors and sensory capabilities of individual fish as they move in a school. Its main features and assumptions include:

1. Each fish in the simulated school follows the same behavioral rules so the school has no leaders and each fish uses the same behavioral rules.
2. To decide where to move, each fish uses some form of weighted average of the position and orientation of its nearest neighbors. Precisely what form this weighted average takes is poorly known, in part because experimental evidence is hard to obtain.
3. Each fish responds to its neighbors in a probabilistic manner. In other words, there is a degree of uncertainty in the individual's behaviors. This is incorporated into the model through an element of randomness or chance, that reflects both the imperfect information-gathering ability of a fish and the imperfect execution of the fish's actions.
4. There are no external influences, such as water currents or obstacles, affecting the fish.

We now examine the behavioral components of the individual fish that are incorporated into the model. In response to a neighboring fish with a particular position and orientation, a real fish can show one of four behaviors: repulsion, to avoid collision with other fish; attraction, or swimming toward the school; parallel orientation, where the fish orients itself by turning in the same direction as its neighbor; and searching behavior, where a fish that becomes isolated from the school (out of visual or other sensory range), or finds that its neighbors are within the blind spot behind it, turns randomly to restore contact with its neighbors.

The precise regions in which a neighboring fish elicits these four behaviors are species-specific and included as parameters of the model (Figure 11.6). Of course, a fish is influenced by several neighbors simultaneously. Therefore, the particular response of an individual fish at any moment must involve some means of prioritizing or averaging the mixture of influences (assumption 2 above). Since there is little empirical evidence to indicate how fish integrate and evaluate the mixture of influences from different neighbors, Huth

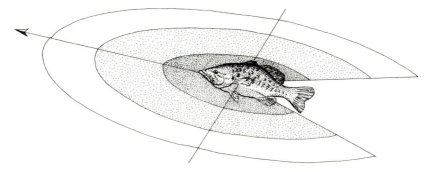

Figure 11.6 Ranges of the basic behavior patterns of an individual fish in the model of Huth and Wissel (1992). If another fish is in the central area (dark stipples), closest to the fish, the fish reacts with repulsion to avoid collision with its neighbor. If another fish is in the intermediate area (light stipples) the fish reacts by orienting parallel to its neighbor. In the next zone (white area), the fish reacts by being attracted to its neighbor. Beyond these areas and in the blind spot behind the fish, the fish cannot see any neighbors. Such an isolated fish will turn randomly to find another fish. (Illustration by Mark A. Klingler)

and Wissel (1992) developed alternative hypotheses that they incorporated in their model. One method, which they called decision models, involved the assignment of priority to one neighbor over others. Another method they applied (averaging models) used various algorithms to average the influence of different numbers of neighbors.

Results of the Model

Figure 11.7 shows the path of individual fish in an actual school (Partridge 1981) compared to the school that Huth and Wissel simulated. The simulation tracked the movement of eight fish, where each fish was initially placed in a randomly selected position and randomly selected orientation, and each fish traveled at an average velocity of 1.2 body lengths/s. Other parameter values used in the simulation can be found in Huth and Wissel (1992).

The mixture of influences of neighboring fish was calculated as the arithmetic mean of the responses to the four nearest neighbors. Within a broad range of parameter values, averaging models that combined the influences of several nearest neighbors resulted in a cohesive group with fish moving in a parallel orientation. In contrast, the decision models, which prioritized the influence of one neighbor over another, showed little resemblance to real schools. The group as a whole did not advance very well and all the fish were oriented in different directions.

Figure 11.8 compares the results of the decision and averaging models and shows that the decision model results in a nonpolarized school that hardly advances. Polarization is defined as the arithmetic average of the angular devi-

a

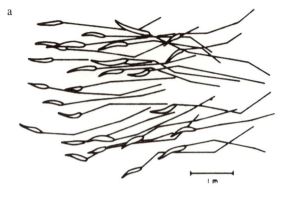

b

Figure 11.7 Paths of individual members of a fish school: (a) real fish (from Partridge 1981), where fish position was tracked from a videotape and plotted every 1.4 s; (b) simulated fish (from Huth and Wissel, 1992), where fish position was plotted every 0.5 s. Parameter values: time step = 0.5 s, number of fish in school = 8, radius of repulsion area = 0.5 body length (BL), radius of parallel orientation area = 2 BL, radius of attraction area = 5 BL, dead (blind zone) angle = 30°, maximum number of neighbors affecting a fish = 4. (From Huth and Wissel, 1992, figure 6, used with permission)

ation of each fish from the average swimming direction of the fish group; it ranges between 0 degrees and 90 degrees, where 0 degrees represents a group with optimal parallel orientation and 90 degrees a group that is maximally disoriented. The mean polarization for the six time steps shown was 73 degrees for the decision model and 12 degrees for the averaging model.

Lessons of the Model

The model based on averaging the influences of nearby fish captures the essential feature of schooling, namely the synchronous movement of a group of aligned fish. What can we learn from such a model?

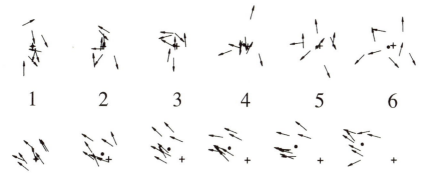

Figure 11.8 Comparison of decision and averaging-models of schooling. The top row (time steps 25–30) shows six consecutive time steps of a simulation based on a decision model where each fish adjusts its velocity based on the movements of one of four neighbors. The bottom row shows six consecutive time steps (16–21) of the same simulation except that each fish adjusts its velocity based on the average influence of its four nearest neighbors. Compared to the decision model, the averaging model results in much greater polarization and advancement of the school. For each time step, arrows represent the orientation of each of the eight fish in the simulation. The black dot marks the current center of mass of the school and the cross marks the initial center of mass at the first illustrated time step. (From Huth and Wissel, 1992, figure 5, used with permission)

Using realistic parameter values, the model generates a convincing simulation of schooling without the need for directly coding the collective behavior of the school. This offers some assurance that mechanisms based on leaders, plans, or templates are unnecessary for the observed patterns of movement. The proposed behavioral rules also are sufficient to generate the main features of schooling. This suggests that the mechanism represented in the model may actually be the one used by schooling fish. Furthermore, by testing various permutations of the behavioral rules and a range of relevant parameter values, the model can help determine whether each of the individual behaviors is necessary for schooling, and whether certain behaviors are more important than others in generating the collective pattern.

There is one caveat. One can always argue that many other models may be proposed that would generate convincing schooling behavior, yet not all of them are mechanisms that schooling fish actually use. We must stress that insofar as is possible, the self-organization models presented in this book are built from the bottom up. In other words, the models are not based on a set of arbitrarily chosen behavioral rules; they are built on a foundation of behaviors derived from careful experiments. It is true that the match between the predictions of a particular model and reality cannot *prove* that a particular mechanism is the one used by schooling fish, but the choice of a particular model—as it was in this case—should be based on experimental observations.

At the same time, models are a simple means of doing hypothetical experiments that are not easily done with real fish. The choice of a particular scheme for combining the influences of neighboring fish has a large effect on the behavior of the model. Mathematically, it is very easy to implement a variety of different schemes and to evaluate which scheme gave the most realistic schooling behavior. Since so little is known about how a fish actually integrates potentially conflicting sensory inputs, the model provides a simple means of doing hypothetical experiments which cannot easily be conducted with real fish schools. The results of such mathematical experiments can point to productive areas of experimental work.

The model is also useful in suggesting the relative importance of different features of the model and the effect of varying the magnitude of different parameters. Huth and Wissel present these data in the form of graphical sensitivity analyses showing how two measures of schooling behavior—polarization and expanse—vary under different conditions. Expanse is defined as the root mean square distance of each fish from the center of mass of the group and measures the compactness or cohesion of the group. The results suggest, for example, that polarization and expanse are not improved when each fish is influenced by more than its three nearest neighbors; however, a marked improvement occurs in these measures when two neighbors influence each fish's behavior (Figure 11.9). This confirms our observations of the sensory capabilities of fish. In terms of the spacing among individuals and the speed with which a fish in a school has to make its behavioral decisions, it is unlikely that an individual could simultaneously process the visual information derived from many fish. The model suggests that each fish must take into account more than a single neighbor, but only a small number of fish are required for coordinating each individual's movements with the movements of others.

Evaluating the Model

We should first ask ourselves whether the model's assumptions are realistic. Of the main features of the model, two details do not reflect actual fish behaviors. The first detail involves the mathematical simplification of analyzing fish movement in only two dimensions. However, there is no reason to believe that the results of the two-dimensional simulation would be qualitatively different from a three-dimensional simulation. For computational convenience, many biological models are presented, at least initially, in just two dimensions.

The second detail concerns how a fish simultaneously processes the influences of several neighboring fish. By presenting alternative mathematical hypotheses, the model allows an initial test of different approaches.

In evaluating the model, we might also determine that it fails to produce important features of schooling. Fish schools certainly exhibit greater complexity than cohesive parallel movement. Real fish schools respond to obstacles

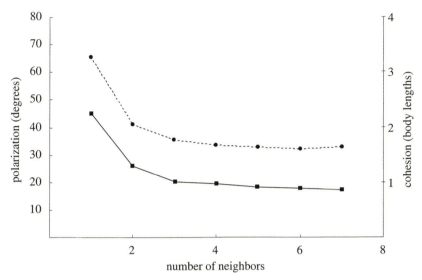

Figure 11.9 Graph showing the degree of parallel orientation (solid line) and the cohesion (dotted line) of simulated fish versus the number of neighboring fish influencing an individual. Note that up to about six neighbors, the more neighbors a fish takes into account the better the parallel orientation and cohesion of the school. (Data from Huth and Wissel 1992, figure 12)

in their path and sudden disturbances such as the approach of a predator. Of course the model presented here generates only the subset of schooling behavior characterized by motion in the absence of external influences as commonly found in a broad expanse of featureless ocean. This is to be expected of course, since the model was constructed with the assumption of no external influences. Refinements of the model could, no doubt, generate obstacle avoidance if that were considered a useful goal. In practice, goal avoidance would be simple to incorporate into the model, since the model fish already have repulsion behavior.

Similarly, the model does not produce sophisticated avoidance behaviors such as the fountain effect or flash expansion. One area of fruitful study would be an analysis, conducted both empirically and through modeling, of the behavioral changes required to go from the typical undisturbed schooling produced in this model to a behavior pattern such as flash expansion. Especially interesting would be a determination of whether the more sophisticated collective behaviors could be generated using similar behavioral rules based on self-organization, or whether these behaviors would require additional nonself-organization mechanisms. If it could be shown that only a few changes in individual behavioral rules are required to produce more complex schooling behaviors, it would demonstrate the versatility and power of mechanisms based on self-organization.

The main goal of this chapter has been to introduce the reader to a view of fish schooling based upon self-organization. Our review of the literature suggests that fish guide their movements within a school through the use of a limited number of behavioral rules based on local information gleaned from moment to moment as they swim. Within the past few years, additional studies have taken this work even further, addressing issues such as predator and obstacle avoidance, the fine structure of fish schools, and a number of different modeling approaches (Dagorn and Freon 1999; Dagorn et al. 2000; Doustari and Sannomiya 1995; Josse et al. 1999; Krause et al. 1996, 1998; Reuter and Breckling 1994; Stocker 1999; Vabo and Nottestad 1997).

Box 11.1 Simulating a School of Fish

Based on the model of Huth and Wissel (1992), we can outline the general form of a computer algorithm that simulates a school of fish or flock of birds. The program Flocking was written in Pascal for the Macintosh and is available at our website http://beelab.cas.psu.edu).

```
initializations:
specify number of fish in school
specify initial orientation angle, position (x, y coordinate) and
   speed of each fish
specify an array for the angle, x-coordinate, y-coordinate, and
   speed of each fish
specify number of neighbors which influence an individual fish
specify time step
specify radial distances and angles defining the areas for each
   type of behavior
  Repeat
   For each individual
      determine which neighbors influence the individual
      For each neighbor having an influence
        Case neighbor in
          repulsion area:
           generate new orientation based upon repulsion rule
          parallel orientation area:
           generate new orientation based upon alignment rule
          attraction area:
           generate new orientation based upon attraction
          searching area:
           generate new orientation randomly
          average the orientations for all the influencing
           neighbors
          update array with new orientation of the individual
          calculate new position based on new orientation,
           velocity & time step
          update array with new x, y coordinates of the individual
          erase old position and heading of the individual
          draw new position and heading of individual
          check if done = true
   Until (done = true)
```

Box 11.1 continued

A brief description of this algorithm may be helpful. The parameter values for the simulation are entered in initializations section. The main portion of the program is contained in the **Repeat—Until** loop. For every cycle of the loop, each fish is examined in turn. All neighboring fish are examined and those that influence the fish in question are identified. If, for example, the number of influencing fish was initialized to four and a nearest neighbor influence rule is used, the fish's four nearest neighbors are selected. For each of these influencing neighbors the program determines the area in which the neighbor lies, based on the parameters of distance and angle. Depending on whether the neighbor lies in the repulsion, parallel orientation, attraction, or searching area, a new orientation angle for the fish is generated. After a new orientation angle is generated based upon the location of each influencing fish, the program computes the average new orientation. The new orientation for the fish is entered in the array. New *x* and *y* coordinates for the fish are calculated based on the new orientation, the velocity of the fish, and the duration of the time step. The new *x* and *y* coordinates of the fish are entered in the array. The old position and orientation of the fish are erased from the screen and the new position and orientation are plotted on the screen. A check is performed to see whether the program is done, perhaps triggered after a specified number of passages through the main loop or a command by the user, such as clicking the mouse button. If done = **true**, the program ends, otherwise the program executes another cycle of the **Repeat—Until** loop.

This is only the basic outline of the main loop of the program. Separate subroutines (Pascal procedures) need to be created to determine which of the neighboring fish influence each individual; calculate the new orientation angle for each subcase; calculate the average of the orientation angles generated by each neighbor; update the array; calculate the new position; and provide graphic display of the calculations by means of a plotting procedure that erases the previous position of the fish and draws the new position.

A simulation of collective movement written in StarLogo is available at our website, http://beelab.cas.psu.edu. As discussed in Box 8.1, the StarLogo application and documentation can be downloaded from the MIT Media Laboratory website at http://www.media.mit.edu/starlogo. The StarLogo website contains a number of simulations demonstrating collective motion and self-organized behaviors. Also see the excellent Boids website at http://www.red3d.com/cwr/boids.

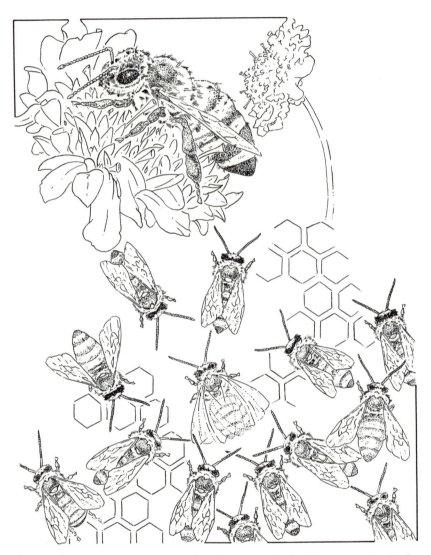

Figure 12.1 Behavioral castes of worker honey bees involved in nectar collection include foragers that collect nectar from flowers in the field, bring it back to the hive, and transfer it to food-storer bees that place it in the honey-storage combs. (Illustration © Bill Ristine 1998)

12

Nectar Source Selection by Honey Bees

How doth the little busy bee
Improve each shining hour,
And gather honey all the day
From every opening flower!
> —Isaac Watts, Divine Songs

On a typical summer's day a colony of honey bees dispatches some ten thousand bees—roughly a third of its members—into the surrounding countryside to gather the colony's food. Each of these forager bees travels up to 10 km from the hive, alights at a patch of flowers, collects a load of nectar or pollen, and then wings her way home. Back inside the hive, each bee promptly unloads her food and a short time later departs on her next collecting trip (Figure 12.1). The net effect of these behaviors, summed over the thousands of foragers in a colony, is a steady flow of food from the far-flung flowers back into the central hive. One can view a colony of honey bees therefore as a large and diffuse creature that can extend itself over great distances and in multiple directions simultaneously to tap a vast array of food sources. Over the course of a summer, this amoeboid entity will extract from its environment some 120 kg of nectar and 20 kg of pollen.

A colony that is to succeed in gathering such large quantities of nectar and pollen must closely monitor the flower patches within its foraging range and wisely deploy its foragers among patches so that food is gathered quickly and efficiently. In general, flower patches that are large and offer easy collection of nectar or pollen should receive many bees, whereas smaller patches with sparse forage should receive few foragers, perhaps none at all. In this chapter we will skirt the complexities in this labor allocation problem that arise when considering the joint collection of nectar and pollen (this is discussed in Seeley 1995), and will instead focus on the smaller but still sizable puzzle of how a colony concentrates its foragers among flower patches yielding only nectar.

The Colony-Level Pattern

A clear picture of a colony's ability to concentrate its nectar collection efforts on the more rewarding nectar sources emerged from an experiment in which a small colony was presented with two artificial food sources and the number of foragers allocated to the two sources was measured (Seeley et al. 1991). To accomplish this, the investigators worked with a colony with all 4,000 workers painstakingly labeled for individual identification. After labeling the bees over a two-day period, in 1989, the colony was moved to a special study site, the Cranberry Lake Biological Station. This field station is located deep in a heavily forested region of the Adirondack mountains in northern New York state. It is especially attractive for bee research because there are no feral colonies of honey bees to disrupt experiments and, owing to the dense forest cover, there would be little natural forage to entice bees away from artificial food sources or feeders. Each feeder consisted of a closed container of sucrose solution (artificial nectar) with grooves at the base where bees could insert their tongues to suck up the sweet liquid (Figure 12.2). Such feeders allow one to precisely control the quantity and quality of the food available at a food source.

To start the experiment, the investigators trained two groups of approximately ten bees each to two identical feeders positioned north and south of the hive, each one 400 m from the hive. During the training period, both feeders contained a rather dilute (1.0 M) sucrose solution that motivated the bees visiting each feeder to continue their foraging but did not stimulate them to recruit any additional hivemates to the feeders. A worker bee can inform her coworkers of the location of a desirable food source by performing a waggle dance (von Frisch 1967) that indicates the direction and distance of the source relative to the hive (see below).

The critical observations began at 7:30 on the morning of 19 June, following a ten-day period of cold, rainy weather. At this time the north and south feeders were loaded with 1.0 and 2.5 M sucrose solutions, respectively, and recordings were made of the number of different individuals visiting each feeder. By noon the colony had generated a striking pattern of differential exploitation of the two feeders, with ninety-one bees engaged at the richer feeder and only twelve bees working at the poorer one (Figure 12.3). The positions of the richer and poorer feeders were then switched for the afternoon, and by 4:00 P.M. the colony had fully reversed the primary focus of its foraging, from the south to the north. This ability to choose among forage sites was again demonstrated the following day during a second trial of the experiment. Thus when given a series of choices between two artificial nectar sources with different profitabilities, the colony consistently concentrated its collection efforts on the more profitable source.

The Individual-Level Processes

What are the mechanisms whereby a honey bee colony is able to organize adaptive patterns of forager allocation like those in Figure 12.3? Seeley and his colleagues have been exploring this question since 1980, and the answer was recently reviewed by Seeley (1995). Here we will examine these mech-

a

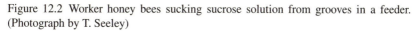

b

Figure 12.2 Worker honey bees sucking sucrose solution from grooves in a feeder. (Photograph by T. Seeley)

anisms with the view that they include a process of self-organization. That is, the mechanisms appear to arise entirely from the actions and interactions between forager bees rather than on guidance from an external, directing influence. Studies have failed to find any supervisory leader of the foragers.

To function as a leader a bee within a colony would have to acquire synoptic knowledge of the foraging opportunities and the foragers' deployment among the food sources, calculate the proper allocation of foragers among the food sources, and issue instructions to the foragers whenever their distribution

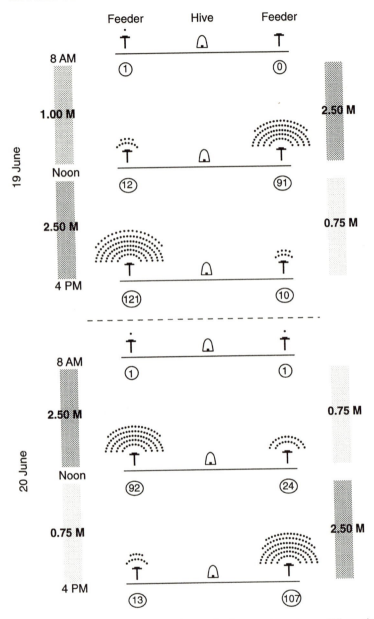

Figure 12.3 A honey bee colony can choose the best nectar sources. When given a choice between two feeders with different profitabilities, the colony consistently directed most of its foraging effort to the richer one. The number of dots above each feeder denotes the number of different bees that visited the feeder in the half hour preceding the time shown on the left. (From Seeley 1995, figure 3.8)

among the food sources needed adjustment. No bee, including the queen bee, appears to possess such abilities of information acquisition, processing, and communication. Also, there is no blueprint to which foragers could refer to know what spatial distribution they should create. A blueprint is a static plan, whereas the allocation pattern of a colony's foragers is highly dynamic as the ever-changing array of flower patches in the surrounding countryside presents the bees with a kaleidoscopic array of food sources. These dynamics also preclude the bees from following a stereotyped sequence of actions, or recipe, to create the proper allocation pattern. As we shall see below, a forager bee must be able to respond flexibly to her food source, sometimes advertising it to nestmates, other times exploiting it quietly, and still other times abandoning it altogether. Such flexibility in behavior is the antithesis of executing a rigid behavioral recipe.

Finally, it seems clear that the fourth alternative to self-organization, namely a template mechanism of pattern formation, cannot explain how foragers manage to distribute themselves properly in the environment. There is no template, no preexisting guide, that indicates how many foragers should be working in each subregion around the hive. In the experiment depicted in Figure 12.3, for example, nothing about the physical structure of the two feeders determines how many bees should exploit each feeder. This is clearly demonstrated by the fact that the two feeders remained identical (except in their contents) throughout the two-day experiment, yet on both days the number of bees at each feeder changed dramatically between morning and afternoon.

In summary, we find no evidence of the involvement of any leader, blueprint, recipe, or template controlling the spatial distribution of a colony's foragers. Thus we turn to an examination of decentralized, self-organized mechanisms by which a colony's foragers distribute themselves among food sources. In part, we base this belief on arguments of plausibility, and thus may be criticized for not providing stronger evidence against alternative mechanisms of nectar-source selection. Nonetheless, we believe that both empirical evidence, and the modeling approach overwhelmingly favor the conclusion that self-organization underlies nectar-source selection by honey bees.

Employed and Unemployed Foragers

An important first step toward understanding how the allocation process works is to distinguish between *employed* and *unemployed* foragers; between foragers currently engaged in exploiting a patch of flowers and those that are not. The members of these two groups have markedly different behaviors. Employed foragers generally bring home food and may provide information about the location of a food source, while unemployed foragers receive information about food sources and then search for one of the advertised work sites. We will first consider the behavior of employed foragers, placing special emphasis on their ability to express their knowledge of desirable food sources.

Every time an employed nectar forager flies home from a patch of flowers, she brings home not only nectar stored in her honey stomach but also information stored in her brain about the food source. She can share this knowledge with the unemployed foragers by means of the waggle-dance, a remarkable behavior in which a bee, deep inside her colony's hive, performs a miniaturized reenactment of her recent journey to some flowers. Bees observing the dance learn the distance, direction, and odor of these flowers and can convert the information into a flight to the specified flowers (von Frisch 1967).

To understand how bees communicate using waggle dances, let us follow the behavior of a bee upon her return from a rich nectar source. The bee is returning from a large cluster of flowers located a moderate distance from the hive, say 1500 m, and along a line 40 degrees to the right of the line running between the sun and its hive (Figure 12.4). Excited by its experience at the flowers, the bee scrambles into the hive and immediately crawls onto one of the vertical combs. Here, among a massed throng of unemployed foragers, she performs the recruitment dance. This involves running through a small figure-eight pattern: a waggle run followed by a turn to the right to circle back to the starting point, another waggle run, followed by a turn and circle to the left, and so on in a regular alternation betweeen a right or left turn after each waggle run. The waggle run portion of the dance is the most striking and informative part of the bee's performance and is given special emphasis both by the vigorous waggling—the lateral vibrating of the body, with sideways deflections greatest at the tip of the abdomen and least at the head—and by the dorso–ventral vibrating of the wings at approximately 260 Hz. Usually several unemployed foragers will trip along behind a dancer, their antennae always extended toward her to detect the dance sounds (Dreller and Kirchner 1995).

The direction and duration of each waggle run is closely correlated with the direction and distance of the flower patch the dancing bee is advertising. Flowers located directly in line with the sun are represented by waggle runs in an upward direction on the vertical comb, and any angle to the right or left of the sun is coded by a corresponding angle to the right or left of the upward direction. In the example illustrated in Figure 12.4, the flowers lie 40 degrees to the right of the sun and the waggle run is correspondingly directed at an angle of 40 degrees to the right of vertical. The distance between nest and flowers is encoded in the duration of the waggle run: the farther the flowers, the longer the waggle run. Bees observing the dance can detect the buzzing sound producing during a waggle run, so it seems likely that the observers measure the duration of a waggle run by sensing the duration of the sound associated with each waggle run.

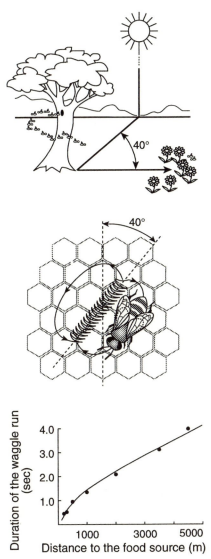

Figure 12.4 In this example of the honey bee waggle dance, the patch of flowers (top picture) lies 1500 m away at an azimuth of 40 ° to the right of the sun as a bee leaves the colony's nest inside a hollow tree. To report this rich food source when inside the nest (middle), the bee runs through a figure-eight pattern, vibrating its body laterally as it passes through the central portion of the dance, called the waggle run. Each waggle run is oriented on the vertical comb by transposing the azimuth angle (between the food and the sun) to the angle between the waggle run and vertical. Distance to the flowers is coded (bottom) by the duration of the waggle run (based on data in Table 13, von Frisch 1967). (From Seeley 1995, figure 2.9)

Besides information about the direction and distance to her patch of flowers, a dancing bee also expresses information about the desirability of the flowers. This is done by adjusting the strength of the dance. In principle, the adjustments could involve modulation of the dance duration (the number of waggle runs per dance) or the dance intensity (the vigor of each waggle run), or both. The bees, however, rely primarily on modulation of dance duration (Seeley 1995). The richer the food source, the greater the number of waggle runs performed, as shown in Figure 12.5. If one monitors the dances of bees exploiting natural flower patches, one sees that dance duration varies by about two orders of magnitude, from 1 to about 100 waggle runs, which implies that natural flower patches vary greatly in their rewards to the bees. It is important to note too, however, that the duration of dances varies greatly for a single food source (Figure 12.5) and that considerable overlap occurs in the distribution of dance duration for different food sources, even ones that differ markedly in quality. This implies that the duration of any one dance cannot provide precise information about the quality of the food source it represents. The significance of these facts will become clear when we consider how the unemployed foragers follow dances.

Whereas the waggle dance actively displays the direction, distance, and desirability of a food source, a dancing bee passively conveys information about the odor of the flowers at her forage site. This occurs as the dancing bee incidentally exudes the floral scent that was absorbed in her waxy cuticle while she worked in the flowers. The audience of unemployed foragers evidently draws upon its knowledge of a food source's odor to help pinpoint the location of the food after using the dance's distance and direction information to arrive in the correct general vicinity.

Although every employed forager brings home information about her food source's location and quality, only bees returning from high-quality sources perform dances and share their information with their unemployed nest mates. This is shown in Figure 12.5, where we see that few bees danced when the feeder yielded a 1.0 M sucrose solution, and no bees danced when it yielded only a 0.5 M solution. Bees returning from a mediocre flower patch will simply unload and head back out to the flowers. If the flowers offer only very meager rewards, it is likely that the bee will not even leave the hive and, instead, joins the ranks of the unemployed foragers standing around inside the hive. In nature, the fraction of returning foragers that perform a dance is low, generally less than 10 percent (Seeley and Visscher 1988). Clearly, an important feature of the waggle-dance communication process is a strong filtering out of information about low-yield food sources. This implies that the pool of shared information within a hive consists mainly of information about rather high quality food sources.

Employed foragers do not report on their food sources throughout the hive but concentrate their announcements in the area just inside the hive entrance,

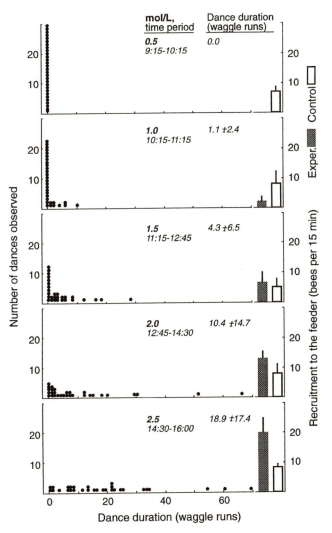

Figure 12.5 Dance duration corresponds to nectar-source profitability. Thirty bees were trained to a feeder from each of two colonies—one experimental and one control—and over the course of one day the sugar solution in the experimental colony's feeder was raised in stages from 0.5 to 2.5 M. At each concentration, the duration of thirty dances and the rate of recruitment (filled bars) to the feeder were measured. Bees that did not dance were assigned a zero dance-duration. The sugar solution in the control colony's feeder was held constant at 1.5 M and the rate of recruitment (open bars) to it was also measured. The absence of significant changes over the day in recruitment rate for the control feeder implies that ambient conditions were stable throughout, hence the dramatic changes in dance duration and recruitment rate for the experimental feeder were caused solely by changes in feeder profitability. (From Seeley 1995, figure 5.7)

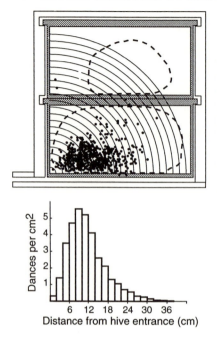

Figure 12.6 The locations of 437 dances on the dance floor in an observation hive are shown in the top figure. The locations were observed with scan sampling at 2-min intervals between 9:00 and 10:30, on July 12, 1989. Regions of beeswax comb inside the hive are white while wooden surfaces (comb frames) are shaded. The dashed lines mark the boundaries of brood-filled cells on each comb. Quarter-circle lines centered on the entrance mark the 2-cm-wide bands used to measure dance density. The density as a function of distance from the hive entrance is shown in the lower figure. Dance density was calculated based on the data shown at the top plus comparable data gathered on two other days; $n = 1224$ dance locations. (From Seeley 1995, figure 5.4)

the so-called dance floor (von Frisch 1993), as shown in Figure 12.6. Such conspicuous clustering may be merely a by-product of foragers minimizing their time and travel inside the hive between foraging trips, but it may also benefit the overall process of a colony's food collection by facilitating the flow of information. In particular, an unemployed forager's task of finding a dancer inside the dark hive is surely simplified by having a special region of the hive where the density of dancing bees is high.

Thus we see that every time an employed forager returns to the hive and unloads her food, she must decide whether or not to perform a waggle dance. Moreover, we see that if she decides to dance, she must decide how many waggle runs to produce based on an assessment of the desirability of the food source. The collective effect of this decision-making by the employed foragers is that the dance floor functions as a labor clearinghouse where employed foragers advertise high-quality work sites and unemployed foragers listen to their nestmates announcements to learn about these work sites. Employed foragers visiting highly desirable forage sites persist at these sites and produce vigorous dances which arouse unemployed foragers to join them, whereas employed foragers visiting less desirable sites will tend to refrain from announcing them to the unemployed foragers, and may even cease working at these sites.

Finally, it should be noted that the foraging decisions that each employed forager makes are based only on very limited information, namely a knowledge of her own particular patch of flowers. For example, an employed forager's assessment of the quality of her food source (whether low, high, or some point in between) does not involve acquiring knowledge of other food sources and then comparing her food source to these others. A foraging bee knows about only her particular flower patch and evidently is able to assess its quality through reference to an internal scale of food-source quality that is built into the bee's nervous system during development (discussed in detail in Seeley 1995). In short, each employed forager functions with extremely local, not at all global, knowledge of her colony's foraging opportunities.

The Behavior of Unemployed Foragers

Unemployed foragers have an opportunity to become well informed about the various food sources exploited by their colony whenever they attend the dance floor to observe dances. Certainly the dance information is arrayed in the hive in a manner that should make it possible for a bee to acquire an overview of the foraging opportunities. Throughout the day, dances representing multiple flower patches are performed close together in the hive. Also, food-source quality is coded in the dances—in dance duration—so there is the possibility that dances for richer and poorer nectar sources are distinguished by the bees. Furthermore, the strongest dances for natural food sources are more than a hundred times longer than the weakest ones, which implies that the profitability of natural food sources varies greatly. Conceivably, unemployed foragers could follow numerous dances on the dance floor and acquire broad knowledge of the available food sources, including their relative qualities. Having acquired this information, they could compare the various food sources and choose the best one as their next work site. Do unemployed foragers do so, or do they actually adopt a new forage site based on much more limited information?

There is now strong evidence that an unemployed forager does not conduct a thorough survey of the information available on the dance floor before leaving the hive to search for a new food source (Seeley 1995). First, direct observations of bees watching waggle dances have revealed that each bee follows just one dancer closely for several waggle runs before leaving the hive to search for a food source. These observations also reveal that it is exceedingly rare for a dance-following bee to watch a dance from start to finish. This indicates that dance-followers do not acquire information about food-source quality from dances since, as was previously mentioned, information about quality is strongly expressed only in dance duration.

The question of how unemployed foragers sample the information on the dance floor has also been addressed experimentally. A critical experiment involved presenting a colony with two sucrose-solution feeders equidistant from the hive but different in profitability. If each unemployed forager follows multiple dances, compares them, and selectively responds to the strongest dance,

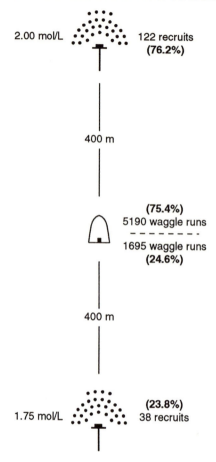

2.00 mol/L   122 recruits
(76.2%)

400 m

(75.4%)
5190 waggle runs
—————————
1695 waggle runs
(24.6%)

400 m

(23.8%)
38 recruits

1.75 mol/L

Figure 12.7 Experimental results show that waggle runs are equally effective for richer or poor sources. Thirty bees were trained to each of two feeders 400 m from the hive and in opposite directions. On July 11, 1990, the feeders were loaded with 2.00 and 1.75 M sucrose solutions, and both the number of waggle runs performed for each feeder and the number of recruits arriving at each feeder were recorded between 10:00 A.M. and 3:00 P.M. Recruitment was found to be proportional to the number of waggle runs for each feeder. The estimated number of recruits at each feeder, determined from the proportion of waggle runs for each feeder, was not significantly different from the observed number of recruits ($P > 0.78$). The result implies that the waggle runs for the two feeders were, on average, equally effective in recruitment; however the richer food source provokes a greater number of waggle runs. (From Seeley 1995, figure 5.9)

then dances for the more profitable feeder will be disproportionately effective per waggle run and the proportion of recruits to the richer feeder will exceed the proportion of waggle runs for that feeder. If, however, each unemployed forager follows just one dance chosen more or less at random, then the proportion of recruits to each feeder will match the proportion of waggle runs for that feeder. Eleven trials of this experiment were performed and in each trial it was found that the proportion of waggle runs for each feeder accurately predicted the proportion of recruits arriving at the feeder (Figure 12.7). Thus it is clear that unemployed foragers do not follow multiple dances and do not selectively respond to those advertising the best food source.

    These findings indicate that each unemployed forager samples just one dance, chosen essentially at random, before exiting the hive to search for a new food source. This in turn implies that many bees will fail to learn about the best food source. Why do unemployed foragers ignore all the dancers but one? We can suggest several possible answers. First, recall that dance-followers generally do not attend a dance from beginning to end and, as far as is known, information about quality is strongly expressed only in dance duration. Thus the information for comparison is not readily available to a dance follower. Second, it may be better for a colony if its foragers are not broadly informed and selective, for this could lead to the colony concentrating its entire foraging efforts on one or a few food sources. Such a response pattern is inappropriate for a honey bee colony that distributes thousands of foragers among multiple food sources rather than crowding them on one best source. A third explanation is that sampling multiple dances takes time. In order for such behavior to evolve, the time spent comparing information from multiple dancers would have to yield sufficient benefit in foraging efficiency to offset the time lost foraging. As we show in the model of forager allocation in the next section, there appears to be little to be gained by such comparison. Finally, for a system of comparison to be workable requires that foragers have the cognitive ability to compare and simultaneously keep track of information from multiple foraging sites. Even if comparison were advantageous, it may be beyond the cognitive capacities of an insect.

Model of Collective Wisdom in Forager Allocation

We have seen that an employed forager's knowledge of the array of food sources exploited by her colony is limited to her own particular source, that she makes her own independent assessment of this source, and that based on this assessment she chooses an appropriate foraging response. This involves deciding whether to continue foraging at the source and, if so, deciding how strongly to advertise the source by dancing. We have also seen that an unemployed forager follows just one dance, chosen more or less at random, each time she attempts to locate a new food source. Thus we now know that each forager possesses only extremely limited knowledge of her colony's foraging opportunities. But at the same time we know that the colony as a whole responds in a way that takes account of the full array of foraging opportunities outside the hive. How are these two seemingly contradictory facts reconciled?

An answer is suggested by the information in Figure 12.8, which depicts some of the experimental results shown in Figure 12.3 and also includes information about the per capita rates of recruitment and abandonment for both feeders. This figure shows that the process of allocating foragers among nectar sources may be viewed as akin to natural selection; that is, a decision-making process in which the distribution of individuals among alternative states (food sources, in this case) is determined not by a high-level, well-informed supervisor but simply by the differential survival (persistence at a food source) and reproduction (recruitment to a food source) of the individuals, each responding

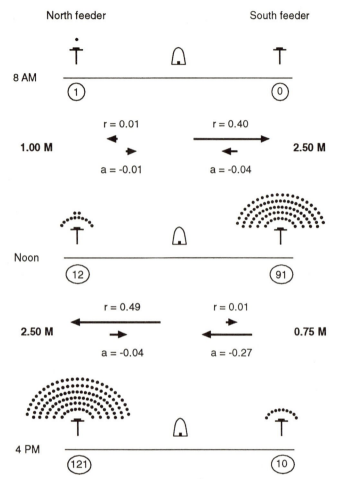

Figure 12.8 The distribution of bees exhibits the preferential exploitation of the richer of two food sources (as observed on 19 June 1989) as well as the pattern of recruitment (r) and abandonment (a) for each feeder. The number of dots above each feeder denotes the number of different bees that visited the feeder in the half hour preceding the time shown on the left. The variables, r and a, denote the average per capita rates of recruitment or abandonment for a feeder, measured in recruits (or deserters) per 30 min per bee visiting the feeder. (From Seeley 1995, figure 5.32)

to its own, immediate set of circumstances. For example, on the afternoon of 19 June, after the relative qualities of the north and south feeders had been reversed, the colony produced an appropriate redistribution of its foragers simply by means of each employed forager responding to the changed conditions at her own feeder. The south-feeder bees responded to the deterioration of their feeder by lowering their recruitment rate and increasing their abandonment rate, while the north-feeder bees responded to the improvement of their feeder by increasing their recruitment rate (their abandonment rate remained low). The net result was that the percentage of the colony's forager population associated with the richer food source automatically increased while that for the poorer food source inevitably decreased. In short, through a process analogous to natural selection, the colony built a globally correct response out of the locally controlled actions of its members.

A Mathematical Model

The idea that the foraging behavior of honey bees can be viewed as the process of natural selection among poorly informed foragers can be tested by means of a mathematical model that endows each forager with strictly limited information and expresses a detailed hypothesis of how the allocation process works (Camazine and Sneyd 1991). We will describe such a model and assess the model's relevance to what actually happens in a beehive by comparing the exploitation pattern of two unequal nectar sources predicted by this model with the exploitation pattern observed during the experiment illustrated in Figure 12.8.

The model deals with the situation of a colony choosing between two nectar sources, A and B, under a fixed set of foraging conditions (no changes in weather, etc.). To explain how the model operates, let us first consider a flow diagram which shows the behavior of one unemployed forager bee as she begins her day of foraging (Figure 12.9). The bee enters the dance floor area of the hive where employed foragers are dancing for different nectar sources. The forager selects a dancer advertising nectar source A. After following the bee's dance, she flies to that nectar source. Upon arrival, the forager gathers a load of nectar and returns to the hive. After relinquishing her nectar to a food-storer bee, the forager may do one of two things, as indicated by the branch points (diamonds) in Figure 12.9. First, she may abandon the food source and return to the pool of unemployed dance followers. If she decides to continue to forage at the nectar source, she may either perform recruitment dances before returning to her patch of flowers or continue to forage at the food source without recruiting nestmates.

We can model the process of foragers selecting one or another of the two nectar sources by combining two flow diagrams, one for each food source, as shown in Figure 12.10. In this model each forager is in one of seven distinct compartments, each of which is characterized by an activity:

A: foraging at nectar source A
B: foraging at nectar source B
D_A: dancing for nectar source A

D_B: dancing for nectar source B
F: unemployed foragers observing a dancer
H_A: unloading nectar from source A
H_B: unloading nectar from source B

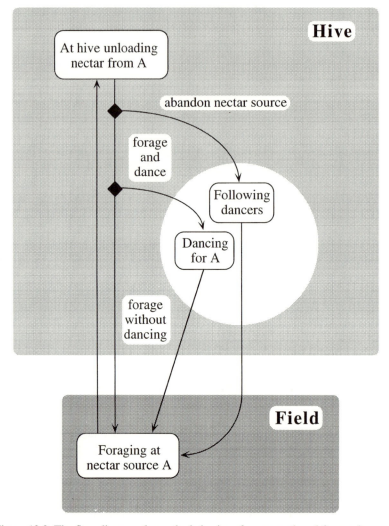

Figure 12.9 The flow diagram shows the behavior of an unemployed forager bee as it begins its work day. The forager follows a dancer for nectar source A, flies to the food source, and then returns to the hive. The black diamonds represent branch points for the decision whether or not to abandon the food source and whether or not to dance for the food source. (After Camazine and Sneyd 1991)

Note that the variables A, B, D_A, D_B, F, H_A, and H_B refer to both the name of the compartment and the number of bees within each compartment. A forager bee is in one of these seven compartments until the moment she enters her next compartment. Thus, for example, the time spent in D_A includes both the time spent dancing for and returning to nectar source A. Note that the dance floor (shaded area in Figure 12.10) contains three separate compartments: bees dancing for source A, bees dancing for source B, and bees following a dancer. Note too that Figure 12.10 consists of two separate cycles, one for each food source, with the follower compartment (F) the only intersection point for the two cycles. Thus bees from one nectar source can switch over to the other source only by passing through the dance floor and following a dancer for the other nectar source.

Two factors affect the proportion of the total forager force in each of the seven compartments: the rate at which a bee moves from one compartment to another, and the possibility that a bee takes one or the other fork at the five branch points (black diamonds) in Figure 12.10. The fraction of bees leaving a compartment in a given time interval is denoted by the appropriate rate constant r_i, with each r_i, equal to $1/T_i$, where each T_i is the time to get from the relevant compartment to the next. Thus each r_i, has the units min^{-1}. For example, the rate constant for bees leaving compartment A is r_3 and the fraction of the bees at nectar source A that leave in time interval Δt is equal to $r_3 \Delta t$. The values of the rate constants r_i for a particular experimental situation will be presented in a later section.

Let us consider what determines the probabilities of the different behaviors at each of the five branch points in Figure 12.10. The first branch point occurs after a bee has unloaded her nectar in the hive. Here, she may abandon the nectar source and return to the dance floor to observe another dancer. The probability that a bee does so is denoted by the abandoning function, P_x. Its value will depend on the profitability of the nectar source, thus P_x^A denotes the probability that a bee leaving compartment H_A will abandon nectar source A and become a follower bee (F). Abandonment, of course, diminishes the number of bees committed to a nectar source and provides a pool of unemployed foragers that will observe a dance for one source or another.

The second branch point applies to bees that did not abandon their nectar source. It determines what proportion of these bees will dance for their nectar source. Although at this branch point there is no filtering of bees away from the nectar source to which they are committed, this branch point affects the probability with which a follower bee follows dances for one or the other nectar source. The probability that a bee performs a dance for her nectar source is denoted by the dancing function, P_d. As with the abandoning function, its value depends on the profitability of the nectar source, with P_d^A denoting the probability that a bee foraging at nectar source A performs a dance.

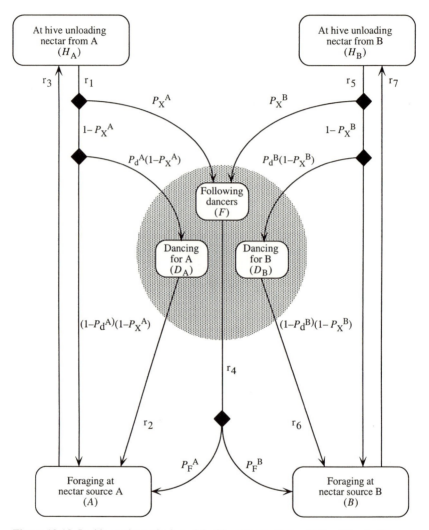

Figure 12.10 In this mathematical model of how honey bee colonies allocate foragers between two nectar sources (A and B), at any given moment each forager can be in one of the seven compartments shown (H_A, H_B, D_A, D_B, A, B, F are the compartments as well as the number of foragers in the compartments). The rate at which bees leave each compartment is indicated by r_1–r_7. The functions P_X^A, P_X^B, P_d^A, P_d^B, etc. indicate the probability of taking one or the other fork at each of the five branch points (black diamonds). (After Camazine and Sneyd 1991)

The third branch point occurs on the dance floor when bees follow dancers for one or another nectar source. The probability of a follower bee following dances for nectar source A and then leaving the dance floor to go to this nectar source, is denoted by the following function P_F^A. As we have seen, each follower bee follows just one dancer, chosen essentially at random, before leaving the hive to search for a new food source. Hence in the situation of just two nectar sources, A and B, the probability of following a dancer for nectar source A (P_F^A) is roughly estimated by $D_A/(D_A + D_B)$. However, since only a portion of a bee's time in the dance area is actually spent dancing, it is necessary to weight D_A and D_B in the above expression by the proportion of time that the foragers actually dance. These fractions are denoted by d_A and d_B. Thus

$$P_F^A = \frac{D_A d_A}{D_A d_A + D_B d_B}, \tag{12.1}$$

$$P_F^B = \frac{D_B d_B}{D_A d_A + D_B d_B}. \tag{12.2}$$

Each function takes into account the number of dancers for each food source as well as the time spent dancing, and so indicates the proportion of the total dancing for each nectar source.

For simplicity in making calculations with the model, we make two further assumptions: All foragers go to either one of the two nectar sources, and the total number of foragers (employed and unemployed) is constant.

The Model's Equations

From Figure 12.10 we can write the following set of differential equations:

$$\frac{dA}{dt} = (1 - P_d^A)(1 - P_x^A)r_1 H_A + r_2 D_A + P_F^A r_4 F - r_3 A, \tag{12.3}$$

$$\frac{dD_A}{dt} = P_d^A(1 - P_x^A)r_1 H_A - r_2 D_A, \tag{12.4}$$

$$\frac{dH_A}{dt} = r_3 A - r_1 H_A, \tag{12.5}$$

$$\frac{dB}{dt} = (1 - P_d^B)(1 - P_x^B)r_5 H_B + r_6 D_B + P_F^B r_4 F - r_7 B, \tag{12.6}$$

$$\frac{dD_B}{dt} = P_d^B(1 - P_x^B)r_5 H_B - r_6 D_B, \tag{12.7}$$

$$\frac{dH_B}{dt} = r_7 B - r_5 H_B, \tag{12.8}$$

$$\frac{dF}{dt} = P_x^A r_1 H_A + P_x^B r_5 H_B - r_4 F. \tag{12.9}$$

A detailed derivation and discussion of these equations is given in Camazine and Sneyd (1991).

Testing and Using the Model

We can test the model and the hypothesis it embodies by comparing its predictions with the distribution of foragers to two food sources that was observed in a particular experimental situation. It is important to note that correct predictions by the model are not guaranteed either by the assumptions underlying it or by the correct determination of the parameters (P_x^A, r_2, . . .). Only if both the model's parameters are accurately measured *and* it's structure accurately represents the mechanisms underlying a colony's forager allocation will the model's predictions resemble what is actually observed in a real colony of bees.

We will examine the model's ability to predict the allocation dynamics for the situation shown in Figure 12.8, namely a colony given a choice between two sucrose-solution feeders equidistant from the hive and identical except that one contained a 0.75 M solution and the other a 2.50 M solution. This requires an estimate for each of the rate constants, r_i, $i = 1$ to 7. Each is equal to $1/T_i$, where each T_i is the time required to get from the relevant activity to the next. Values of T_i appropriate to the foraging situation shown in Figure 12.8 are

Table 12.1 Parameter values for the model of a colony allocating foragers between two nectar sources, as shown in Figure 12.1.

Parameter: definition	Value
T_1: time from start of unloading to start of following, dancing, or foraging, A foragers	1.0 min
T_2: time from start of dancing to start of foraging, A foragers	1.5 min
T_3: time from start of foraging to start of unloading, A foragers	2.5 min
T_4: time from start of following dancers to start of foraging, A and B foragers	60 min
T_5: time from start of unloading to start of following, dancing, or foraging, B foragers	3.0 min
T_6: time from start of dancing to start of foraging, B foragers	2.0 min
T_7: time from start of foraging to start of unloading, B foragers	3.5 min
P_x^A: probability of abandoning A, per foraging trip	0.00
P_x^B: probability of abandoning B, per foraging trip	0.04
P_d^A: probability of dancing for A	1.00
P_d^B: probability of dancing for B	0.15

Note: A and B correspond to the 2.50-mol/L and the 0.75-mol/L feeders, respectively.
Source: Based on Table 2 of Seeley et al. 1991.

given in Table 12.1. Values of the abandoning function (P_X), and the dancing function (P_d), have also been measured for the situation shown in Figure 12.8, and are likewise shown in Table 12.1. Values of the following function (P_F) are calculated for the Figure 12.8 foraging situation with the formulae shown in equations 12.1 and 12.2, using $d_A = 0.38$ and $d_B = 0.02$ as estimates for the proportion of time that bees actually dance to advertize food A and B, respectively (see Seeley et al. 1991 for a discussion of how the model's parameters were estimated based on empirical studies of bee foraging behavior).

Using the parameter values shown in Table 12.1, we can assess how well the model's predictions correspond with actual field observations. The top section of Figure 12.11 shows in detail the colony's response in the two-feeder test depicted in Figure 12.8, while the bottom section shows the computed solutions of the model.

The starting conditions for the computer simulation were chosen to match the real-world example where approximately twelve bees were committed to each feeder, and during the course of the day a total of approximately 125 different bees visited the two feeders. A comparison of the top and bottom sections of Figure 12.11 indicates that in the computed solutions of the model, as in reality, the colony exploits the most profitable nectar source and rapidly responds to changes in location of the richer food source. In the experiment the south feeder showed a rapid buildup of bees between 8:00 A.M. and 12:00 noon, when it was loaded with a 2.5 M sugar solution, followed by a declining number of bees visiting the feeder over the next four hours, when the feeder was switched to 0.75 M. The computed solutions of the model show a similar pattern of rapid rise in bees at the south feeder (loaded with a 2.5 M solution) and a marked decline four hours later when the feeder was switched to 0.75 M.

The north feeder initially contained a 1.0 M sugar solution, rather than 0.75 M, to prevent total abandonment of this feeder. Therefore a slight buildup of bees was observed at the north feeder during the morning. In the simulation, however, the north feeder initially contained a 0.75 M solution (to keep the simulation simple) and subsequently showed a slight decline in the number of bees. But during the afternoon, when for both the experiment and the simulation the north feeder was loaded with a 2.5 M solution, in both settings the number of bees visiting the north feeder rose rapidly with virtually identical trajectories.

The results show that the model has the correct qualitative behavior, and even provides a remarkably good quantitative match between the model and observation. This demonstrates that the proper allocation of honey bee foragers among nectar sources can arise by a process of self-organization in which each bee has only limited knowledge of the array of available nectar sources in the field. Indeed, the results of this model show unequivocally that each employed forager needs only knowledge of the nectar source at which she is currently

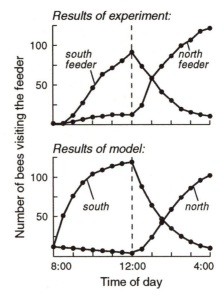

Figure 12.11 The results of an experiment (top) and a mathematical model (bottom) show how a colony selectively exploits the richer of two food sources (see text). For the first 4 h of the experiment on June 19, 1989, the south feeder contained a 2.5 M sucrose solution while the north feeder contained a 1.0 M solution, so the number of bees visiting the feeder increased in the south and stayed constant in the north. (The number of bees visiting the feeder denotes the number of different individuals that visited a feeder during the previous half hour.) During the next 4 h, the sugar solution at the south feeder was switched to 0.75 M and that at the north feeder to 2.5 M. As a result, the number of bees in the south declined and the number in the north correspondingly increased. The computed solutions of the model (bottom) are qualitatively similar to the experimental results. The number of simulated bees visiting the feeder is defined as the sum of the number of bees at the feeder, the number of bees at the hive unloading from that feeder, and the number of bees dancing for that feeder. The strong similarity between the actual observations and the computer simulation suggest that the mathematical model (Figure 12.10 and equations 12.3–12.9) accurately describes the essence of the forager-allocation process. (After Camazine and Sneyd 1991)

foraging, and that each unemployed forager needs to follow only one randomly encountered dancer.

With the confidence that we have developed a valid model of forager allocation, the model can now be used as a tool to explore more fully the mechanism of forager allocation. First, we can help answer the question asked earlier in this chapter: "Why don't unemployed foragers obtain more complete information about potential nectar sources in the field by attending multiple dances?" Figure 12.12 shows the results of the model in which we compared the dynamics of recruitment to a nectar source under two hypothetical conditions:

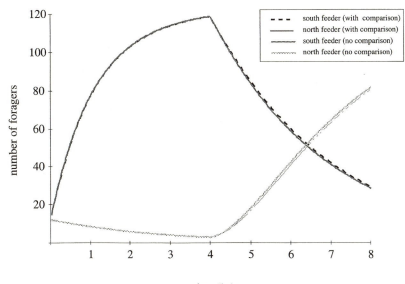

Legend:
- - - south feeder (with comparison)
——— north feeder (with comparison)
~~~~ south feeder (no comparison)
········ north feeder (no comparison)

number of foragers (y-axis)

time (hr)

Figure 12.12 Results of the model comparing the dynamics of recruitment to a nectar source under two hypothetical conditions: Each unemployed forager follows a randomly selected dancer on the dance floor or compares four different dances and selects the best. The strong similarity between the two curves suggests little is gained by a forager comparing dances and selecting the one advertising the best food source. (After Camazine and Sneyd 1991)

in one, each unemployed forager followed a randomly selected dancer on the dance floor, and in the other each unemployed forager compared four different dances and selected the best. If dancers were able to communicate not only location, but also the quality of the food source, then dance-followers might benefit by comparing dances and heading for the most profitable nectar source. Despite the intuitive expectation that the most efficient exploitation of a food source would occur when dance-followers make comparisons, this is not the case. Even with as many as four comparisons, the maximum recruitment rate to the feeder is hardly increased. This suggests that one reason why dance-followers do not compare dances is that there is little to be gained in terms of increased foraging efficiency.

The process of forager allocation also illustrates the utility of interactions between modeling and experimentation. A model such as this can both guide research questions and provide additional confirmation of experimental results. In this particular case, prior to the experimental demonstration that dance followers do not compare dances, it was suggested that comparison among dances would be unlikely due to the predicted lack of increased foraging efficiency (Camazine and Sneyd 1991).

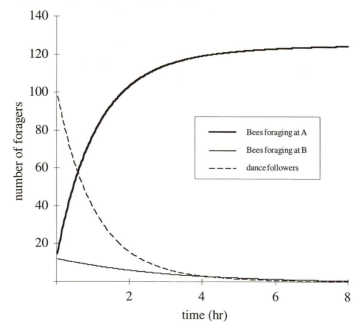

Figure 12.13 Results of a model identical to that in Figure 12.11 and which also shows the dynamics of the dance followers. (After Camazine and Sneyd 1991)

Another use of the model is to predict behaviors. Figure 12.11 (top section) shows the number of foragers at the two feeders for the experiment shown in Figure 12.8. These data were obtained experimentally by marking foragers as they visited the feeder. Figure 12.13 illustrates the portion of the model run depicted in Figure 12.10 in which there is a buildup of foragers at the south feeder (with 2.5 M sugar solution), and a decline in the number of foragers at the north feeder (with 1.0 M sugar solution). If we were interested in the dynamics of the unemployed forager population, it would be difficult to obtain that information experimentally, as there is no way to identify these bees a priori. However, our model easily provides such information and shows that the number of unemployed foragers declines exponentially as these foragers become committed primarily to the higher quality south feeder.

Finally, we can use this model to ask theoretical questions that would be difficult or impossible to answer experimentally. For example, what is the relative importance of abandonment, dancing, and dance-following (shown in the model of Figure 12.10) on the dynamics of forager allocation? Each branch point for abandonment, dancing, and dance-following affects the distribution of foragers exploiting the two nectar sources. In the field, a change in nectar-source quality simultaneously affects the partitioning of bees at *all*

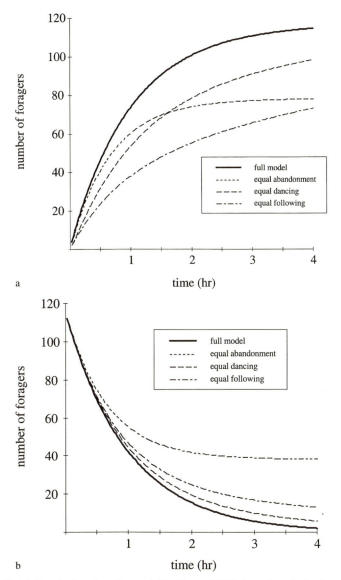

Figure 12.14 Results based on the model in Figure 12.10 show the relative importance of the functions of abandoning, dancing, and dance-following on the allocation of foragers to the two nectar sources. Dynamics of the foragers at A and the dance followers are shown in (a) and (b), respectively. The solid line shows the results of the original model (this is the same curve as shown in Figure 12.11). (After Camazine and Sneyd 1991)

these branch points, making it impossible to assess the relative importance of each branch point independently. In the model, however, we can change each parameter independently and observe the resulting behavior of the system. To compare the effect of abandonment and dancing on the rate of recruitment to a feeder, for instance, we can set the probability of abandonment at the better feeder (A) equal to that of the poorer quality feeder (B), and keep all the other parameters fixed. Intuitively one might expect that increasing the probability of abandonment of the better feeder would have a significant effect on the recruitment rate to that feeder. However, the initial recruitment rate to feeder A over the first half hour (Figure 12.14a) was nearly the same as in the complete model. The most significant effect of increasing the abandonment of feeder A was a decrease in the long-term (steady-state) buildup of bees using feeder A, and a corresponding increase in the number of bees in the follower compartment (Figure 12.14b). Increasing the abandonment rate at both feeders does not result in increased utilization of the poorer food source (B), but only leads to a decreased steady-state number of bees at the better food source (A). This is largely due to the length of time spent as a follower (60 min, Table 12.1). The time required to find a food source after following a dancer is the bottleneck in the foraging process. An individual bee often follows a dancer yet fails to find the indicated nectar source, repeatedly returning to the dance floor to follow another dancer before discovering a source (Seeley and Visscher 1988).

Next we examined the effect of eliminating different rates of dancing for the two nectar sources by having all bees dance for their nectar source. This resulted in a decrease in the maximum recruitment rate to feeder A and a temporary increase in the number of bees using the poorer nectar source, B. However, the long-term (steady-state) distribution of bees at nectar source B is still zero.

Finally, we modeled the situation in which bees had an equal probability of choosing either nectar source after visiting the dance area by setting $P_F^A = 0.5$ for both A and B. This change removes the nonlinearity from the model. The result is a more gradual increase in the initial recruitment rate of new bees to the better nectar source, A (Figure 12.14a).

The results of these models can be understood in a qualitative sense by considering abandonment as a process that creates opportunities for bees to switch from one food source to another, while differential following selectively filters bees to the best food source. The differential dancing for one food source or another determines the degree of selectivity that occurs in this filtering process. In other words, abandonment provides a pool of uncommitted bees on which the rules of dancing and following rules act to selectively filter foragers to the best food source.

This example of forager allocation among nectar sources demonstrates how detailed field experiments, coupled with a simple differential equation model, can be used to elucidate the self-organized processes among social insects.

Careful observations and detailed experimental field work are indispensible prerequisites to model building, and the recommended starting point for understanding how such a system works (Figure 6.2). Models follow later and provide the means to dissect and tease apart the various components of the system. The end result is an understanding of how the self-organized process of food gathering by honey bees can generate a satisfactory solution to a problem without any of the participants having broad knowledge of the problem.

Figure 13.1 In collective foraging by fire ants, *Solenopsis*, workers recruit their nest-mates with pheromone odor trails by dragging the extruded sting along the ground. This is one of the species for which trail recruitment has been extensively studied (see, for example, Wilson 1962). (Illustration © Bill Ristine 1998)

13

Trail Formation in Ants

Go to the ant, thou sluggard; consider her ways and be wise:
Which having no guide, overseer or ruler,
Provideth her meat in the summer, and gathereth her food in
the harvest.

—Book of Proverbs

Trail Formation in the Field

A common ant in northern Europe is *Lasius niger*, also known as the black garden ant. It is a highly opportunistic ant that exploits diverse food sources and will be the main species discussed in this chapter. One can easily observe foraging in *L. niger* by setting out a dish of sugar solution in the vicinity of a nest. After some time, a forager will discover the sugar and shortly thereafter through a recruitment process, numerous additional foragers will appear at the food source. If one plots the number of ants at the sugar as a function of time, one usually obtains a logistic curve that eventually reaches a plateau (Figure 13.2). Observation reveals ants trafficking between the nest and the food source as if the creatures were following an invisible highway on the ground (Figures 13.1 and 13.3). We shall see later that the ants indeed are moving along a chemical trail deposited by the ants and termed pheromone. The ants reinforce this trail with additional pheromone both after they have ingested food and are returning to the nest, and when they are following the pheromone trail to return to the food source (Hölldobler and Wilson 1990; Traniello and Robson 1995).

The logistic growth curve, introduced by Pierre-François Verhulst in 1845, is typical of populations in which an initial exponential growth phase gives way to a phase in which further growth is self-inhibited. The initial exponential growth phase suggests a snowballing positive-feedback process. Each ant returning from the food source can stimulate many other nestmates to forage, these ants in turn stimulate still others, and so on. Of course, this phase of rapid population growth cannot continue indefinitely for the simple reason that when all the available foragers have been recruited and are participating in active foraging no new ones are left to join the foraging system. The population size reaches a plateau (stable population size) when the "birth rate" (rate of

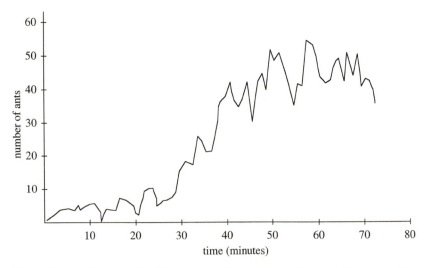

Figure 13.2 Logistic growth of the number of *Lasius niger* ants at a 1 M sucrose solution food source. The food is discovered at time $t = 0$.

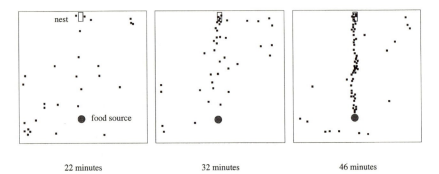

Figure 13.3 Trail between nest and 1 M food source (*Lasius niger*): each dot represents one ant. At the beginning of the experiment the ants are homogeneously distributed. As the recruitment proceeds a well-defined trail emerges.

new arrivals at the food source) equals the "death rate" (rate of abandonment of the food source) so that net change occurs in the forager population size.

Recruitment is often regarded merely as a mechanism that enables an ant colony to assemble rapidly a large number of foragers at a desirable food source. However, in this chapter we will explain how a trail's recruitment system also enables efficient decision making, as it collectively selects the most profitable among an array of food sources. We will also explain how a trail's recruitment system enables it to select the shortest route to a food source.

Considering the tiny size of an individual ant in comparison to the obstacles—such as stones, branches, and crevices—in its pathway, the task of selecting an efficient route to a distant food source may seem too complicated to be solved by the limited intellect of an individual ant. Nonetheless a colony of ants can complete the task and may do so several times in a day. For example, colonies of garden ants often live under flat paving stones and dispatch conspicuous columns of foragers, that follow pheromone-scent trails from the entrance of their nest more or less directly to a persistent food source, sometimes several meters away. A typical example of a natural food source is a cluster of aphids. Foragers tend clusters of aphids from which they obtain a secretion, honeydew, rich in sugar. When a food source is initially discovered by more than one ant, each one will try to find its own way back to the nest, negotiating various obstacles in its path. The result is a number of different routes home. Remarkably, even when all the routes initially have similar amounts of forager traffic, within an hour or two the longer paths will be abandoned in favor of the shortest one.

Trail Formation in the Laboratory

The majority of recent experimental work on the ability of ants to lay trails to implement informed foraging decisions has been undertaken in the laboratory, where the ant colony's environment can be more carefully controlled than in the field. This is essential because the foraging movements of ants have a significant random component that requires repeated identical experiments to yield statistically significant results. Furthermore, in the field many different food sources may be present simultaneously, key orientation cues may change, competitors and predators may interfere with the activity of the colony, and temperature and other climatic changes may influence the outcome of the experiments. Although laboratory experiments may appear somewhat artificial and abstract, they provide the uniform conditions needed to elucidate the mechanisms underlying the phenomena under investigation.

The following experiments are laboratory equivalents of the discovery of different food sources by an ant colony in the field.

Selecting the Richest of Two Foods in an Arena

This experiment examined the foraging behavior of ants in a flat 80×80 cm arena containing just two food sources. The ants can form foraging trails anywhere within the arena. In a typical experiment of this design, the food source provides an unlimited supply of sugar solution. In experiments with *L. niger*, these food sources are placed 60 cm apart with each one positioned 60 cm from the nest (Beckers et al. 1990). In different experiments, two different pairings of food sources are used. In one setup both food sources have equal quality

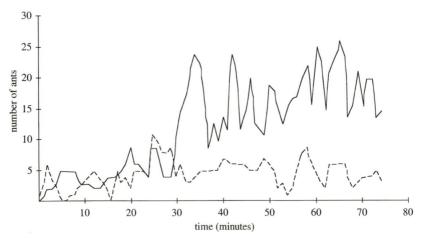

Figure 13.4 Temporal dynamics of the number of *Lasius niger* workers around two 1 M food sources offered simultaneously.

(sucrose concentration). Since 1 molar sucrose solution was used, we refer to these as the 1 M vs. 1 M experiments. In the second setup, one food source has 1 M sucrose solution and the other a 0.1 M sucrose (1 M vs. 0.1 M experiments).

When a colony of *L. niger* was offered two identical 1 M sources *at the same time*, after a short period of equal exploitation a bifurcation was observed so that one of the sources was exploited more strongly than the other (Figure 13.4). When a second source was presented after recruitment to the first source was well under way, the second source tended to remain grossly underexploited. Pasteels et al. (1987) obtained similar results with *Tetramorium* species as did Traniello and Robson (1995) with *Monomorium viridum*, *Pheidole hyatti* and *Liometopum apiculatum*.

When a *L. niger* colony was offered two sources with *different* sugar concentrations (1 M vs. 0.1 M) and the richer source was discovered before or at the same time as the poorer one, the richer source was always most heavily exploited (Figure 13.5a). When the richer source was discovered *after* the poorer one, however, the richer source was only weakly exploited (Figure 13.5b). As before, the colony exploited the *first* source it discovered, even when this meant neglecting a richer source for a poorer one.

Selecting the Richest of Two Foods in a Y-Maze

The second experiment employed a more restricted foraging space. Here, the ants were confined to a Y-shaped cardboard bridge (Figure 13.6a). The branches of the Y are separated by an angle of 60 degrees and each leads to a

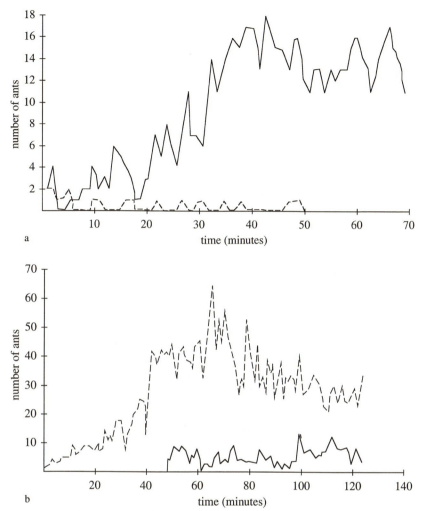

a

b

Figure 13.5 Temporal dynamics of the number of *Lasius niger* workers around two food sources 1 M (solid line) and 0.1 M (dotted line). In (a) the two food sources were offered simultaneusly, and in (b) the 1 M food source was introduced after the 0.1 M food source.

platform. The distance between the nest and each food source was 20 cm. The ants were allowed to explore the bridge for two hours before a small dish of sucrose solution (2 cm diameter) was placed on each platform. Both platforms were loaded with food simultaneously. In this experiment, colonies of *L. niger* with between 2000 and 3000 workers were starved for four days before the

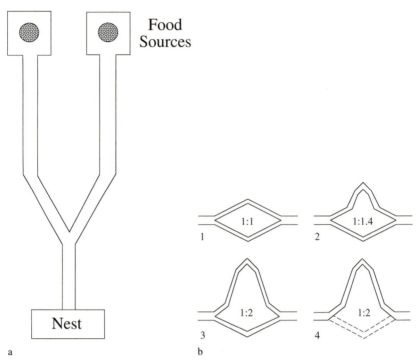

Figure 13.6 A Y-shaped cardboard bridge is used in experiments with two food sources
(a). Various types of bridges may be used in the studies of the selection of the shortest
path (b). The ratios 1:1, 1:1.4, and 1:2 refer to the length of the short path divided by the
length of the long path. The fourth figure refers to the experiment with a bridge-length
ratio of 1:2 where the short path is introduced after a delay.

food was set out (Beckers et al. 1993). The following pairs of food sources
were tested: (1) 1 M vs. 1 M; (2) 0.5 M vs. 1 M; (3) 0.1 M vs. 1 M; (4) 0.05 M
vs. 1 M.

In tests (1) and (2) the ants chose one or the other food source with equal
probability. With a greater difference in concentration between the two simultaneously introduced sources, the majority of foragers exploited the 1M source
in an increased proportion of the trials. In (3) and (4), respectively, 86 percent
and 100 percent of the experiments led to the selection of the 1 M source.

Selecting the Shortest Path

The following laboratory experiments can be thought of as representing
the discovery of a single food source by ant foragers that find two different
paths to the food and eventually select the shortest one (Beckers 1992; Beckers

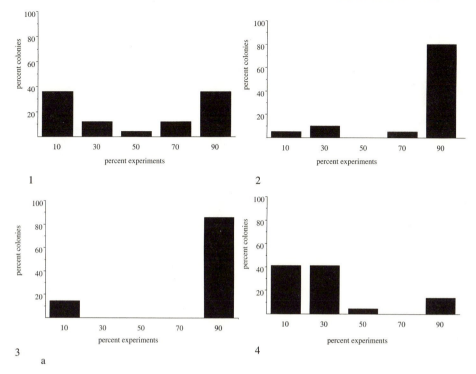

et al. 1992b). In these experiments starved colonies of *L. niger*, also containing 2000–3000 workers, were given access to a plastic box containing sugar water that was connected to the nest by a flat cardboard bridge that provided two alternative routes to the food. Four different bridges were used:

1. The 1:1 series, a bridge with two equal 14 cm-long paths (Figure 13.6b,1).
2. The 1:1.4 series, a bridge with two unequal routes, one a short path (14 cm long), the other 1.4 times longer (20 cm) (Figure 13.6b,2).
3. The 1:2 series, a bridge with two unequal routes, one a short path (14 cm long), the other 2 times longer (28 cm) (Figure 13.6b,3).
4. The same experiment as in (3), except that the long path was the only route available during the first part of the experiment. When the ants were heavily engaged in exploiting the food source, the short path was made available (Figure 13.6b,4).

After several hours during which foragers had time to explore the bridge and the plastic box, 2 ml of 1 M sucrose solution was placed on a microscope slide in the box. Between 30 and 40 minutes after the first ant had discovered the food, the number of ants on the two paths had stabilized. Figure 13.7 shows

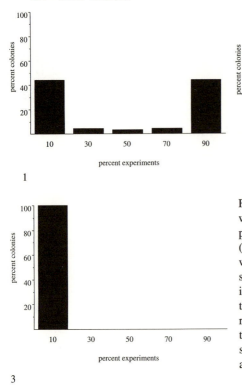

Figure 13.7 The percentage of experiments with up to 100 percent traffic on the shorter path for each experimental series is shown in (a). Parts 1, 2, and 3 refer to the experiments with a bridge ratio of 1:1, 1:1.4, and 1:2, respectively. The fourth graph refers to the experiment with a bridge-length ratio of 1:2, where the shorter path is introduced after a delay. Corresponding percentages of computer simulations with 0–100 percentage of traffic on the shorter path are shown in (b) for the ratio 1:1 and 1:2 (with and without a delay).

the overall outcomes of these experiments and reveals the ants' ability to select the shortest route to the food. In the case of paths of unequal length, clearly a colony is most likely to adopt the shortest path; in most of the experiments more than 80 percent of the foragers used the short path. In the case of a bridge with two equal paths, the choice was random; in approximately 50 percent of the experiments the ants selected the left path and 50 percent the right. When we introduced the two paths with a time delay (the shorter path being introduced later), the colony failed to select the shortest path in most of the experiments. Similar experiments with similar results have been performed with the Argentine ant, *Linepithema humile* (Goss et al. 1989). Later in this chapter we will discuss why the ants fail to utilize the shortest path when it is introduced after a delay.

Adaptive Significance of Ant-Trail Patterns

The selection of the richest food source or of the shortest path is likely to be adaptive for the obvious reason that such a choice, all else being equal, is probably the most profitable option. The shortest path enables ants to minimize the time spent traveling between nest and food source, takes less time to complete, and therefore allows ants to consume their food more quickly, minimizing the risk that a good source of food will be discovered and monopolized by a larger or more aggressive neighboring colony. Shorter paths also mean lower transportation costs.

The use of one path rather than two is also likely to be adaptive. A strong, heavily marked pheromone trail is probably easier to follow and fewer ants will lose their way. There is likely to be safety in numbers as well. Competition from other ants may influence the effective profitability of a food source. Thus defense against potentially interloping foragers from other colonies may be important at a food source. A single dense traffic column is probably better able to defend itself against predators than a sparse column, since isolated ants can easily fall prey to predators (Hunt 1983). The self-defense of ants on a trail or the defense of a food source are cooperative phenomena that probably depend for their effectiveness on concentrated numbers (Franks and Partridge 1993). Roughly speaking, the probability that a colony will monopolize a food source grows nonlinearly with the numbers of its foragers at the food source, on the trail, or readily available to be recruited. An asymmetrical exploitation of two equal food sources therefore may facilitate cooperative defense.

The Behavior of Individual Ants during Trail Formation

In this section we present individual behaviors that contribute to the collective process of trail formation. As in the previous chapter, no evidence is available to suggest that leaders, templates, recipes, or blueprints play any role in the development of foraging patterns in the species of ants described in this chapter. Instead, the process appears to be based on local competition among information (in the form of varying concentrations of trail pheromone on the substrate), and is used by individual ants to generate the observed collective foraging decisions. Note, however, that in certain other ant species (Verhaeghe 1982; Hölldobler and Wilson 1990) foragers that have found a food source do play an active role in leading other ants to that source.

The goal of this section is to explain the bridge experiments with *Lasius niger* that were reported above, and to formulate a model of trail formation compatible with both the results of the bridge experiments and the known ant behaviors. The process of trail formation can be divided into two steps: trail-laying and trail following. The behavior of individuals during trail recruitment begins when a forager ant, on discovering a food source, returns to the nest and lays a chemical trail all the way home. Other foragers, waiting in the nest, are then stimulated to leave the nest under the influence of either the pheromone trail alone or the trail plus other stimuli created by the recruiting ant. In some cases the recruiter runs around in the nest contacting her nestmates, apparently exciting the foragers (Hölldobler 1971; Traniello 1977). The recruits follow the trail to the food source and load up with food. On their return journey to the nest, they add their own pheromonal secretions to the trail and may even provide further stimulation to other foragers in the nest. Although the main function of the recruitment pheromone is to define the trail, the pheromone actually serves both as an orientation signal for ants traveling outside the nest and as a general arousal stimulus inside the nest.

Our purpose now is to review the quantitative aspects of ant behavior pertinent to the modeling of trail systems. For a more detailed treatment of recruitment behavior in ants the reader is referred to the classic works of Sudd (1957) and Wilson (1962), and for reviews, to Wilson (1971), Passera (1984), Hölldobler and Wilson (1990), and Traniello and Robson (1995). Unless stated otherwise, trail laying behavior discussed in this chapter is that of *L. niger* observed during the previously described bridge experiments. A detailed description of these behaviors can be found in Beckers et al. (1992a,b) and Beckers et al. (1993).

Trail-Laying

When a scout ant discovers a desirable food source it feeds and promptly returns to the nest. During the return trip and during subsequent trips to the food source, it usually deposits small amounts of chemical markers, or trail pheromones. These chemicals are low molecular weight substances produced in special glands or in the gut. In the case of *L. fuliginosus*, for example, the chemical trail is produced in the hindgut and two components have been identified (Kern et al. 1997). Although the trail pheromones of *L. niger* have not been identified, the mean lifetime of this chemical signal is estimated from the behavior of the ants to be between thirty and sixty minutes. The biological activity of the trail depends on the amount of pheromone laid down, the rate of pheromone loss due to evaporation and other factors, and the sensitivity of the ants. A trail remains detectable over a period much greater than the mean lifetime of the pheromone since even after many half-lives the ants are able to detect the phenomone remaining on the trail.

During trail-laying, a *L. niger* worker periodically curls its gaster towards the ground (the normal gaster position being horizontal) and interrupts its forward walk for a fraction of a second by "backing-up," thereby making a clear contact between the tip of its gaster and the ground and creating a so-called pheromone mark. This characteristic marking behavior is used as the criterion for trail-laying; it is never performed by individuals that have not found food.

Of course, observations of marking behavior record only the number of trail-marking events, not the quantity of pheromone deposited, which is the value of interest in the formation of a trail. Workers modulate the frequency of marking behavior, which suggests that the frequency of marking rather than the amount of pheromone laid down per mark is the primary means by which an ant varies the strength of the pheromone trail.

When exactly do ants deposit their trail pheromones? Most individuals start trail-laying either as soon as they begin returning to the nest after loading up at a food source, or at the start of their next trip back to the food source. Specifically, when foraging at a 1 M food source, more than 80 percent of individuals begin trail-laying during their first trip back to the nest or during their first return trip back to the food source. Approximately half of these ants begin trail-laying on the return to nest, and the other half begin on the return to the food. The remaining 20 percent of ants show variable trail-laying behavior. Some start during later trips back and forth from the food source, and a few never lay trails. Nevertheless, most ants that begin trail-laying continue to do so for several trips. The probability of trail-laying diminishes rapidly with the number of trips made by an individual. Once an ant has stopped trail-laying, it rarely starts again.

There are also differences between recruiters and recruits. In this chapter, a recruiter is defined as an ant that was in the foraging area at the introduction of the food source, found the source, fed at it, and then returned to the nest. A recruit is defined as an ant that left the nest after the return of the first recruiter and exhibits behavior easily distinguishable from that of ants leaving the nest spontaneously. A recruiter is more persistent in its foraging, making more trips back and forth to the food source. Furthermore, a greater percentage of the trips conducted by recruiters, relative to those by recruits, is accompanied by trail-laying.

The intensity of trail-laying—marks per trail-laying passage—is essentially equivalent for recruits and recruiters, as well as for trips to and from the nest and for each sequential trip. For example, in one set of observations foragers made an average five marks per 20 cm of trailing passage to or from a 1 M food source. Thus *L. niger* ants lay trail at a fairly constant intensity.

Some individuals mark much more than others, however. Indeed, the first and second most active trail-layers in a group of five to seven individually marked ants contributed respectively 60 percent and 20 percent of trail-laying. This was true for both recruiters or recruits, even when they made no more passages than the others. This same variability in recruitment behavior was also seen among dancers for a single food source, as discussed in Chapter 12.

L. niger workers also modulate their trail-laying intensity as a function of the quality of the food source. The ratio between the trail-laying intensity for a 1 M food source vs. that for a 0.05 M source is 1.5. However, the intensity difference between a 0.1 M and a 0.05 M food source is negligible. A counterintuitive finding was that the frequency of trail-laying in *L. niger* (i.e. the proportion of all passages in which trail-laying occurs) is not modulated in relation to the food-source quality.

A final feature of trail-laying behavior in *Lasius*—and probably in other ants as well, such as *Tetramorium* (Verhaeghe 1982)—is the modulation of trail-laying in relation to the orientation of movement towards the nest. It appears that ants keep track of the general direction to the nest, and when forced to return home along a circuitous trail, a portion of which is oriented at a large angle from the homeward direction, the ants show a lower intensity of trail marking. This was seen in certain of the bridge experiments (B, in Figure 13.6), where a long flat branch was used. The angle between its first section and the direct axis between nest and food was 30 degrees, whereas its second section was at 75 degrees from the direct axis between nest and food. The number of marks per passage by each ant was approximately twice as great on the section where the angle with the nest-food axis was only 30 degrees, compared to the bridge section where the angle was 75 degrees. Thus there appears to be a strong orientation-based modulation of trail laying. Similarly, data suggests that once ants have learned the direction to the food source, they show a decreased intensity of trail marking if they are moving at a large angle away from the food source (Beckers et al. 1992b).

Trail-Following

In this section we review how an ant moves over a substrate in response to pheromone trails laid down by itself or other ants. The basic layout of the experiments described previously, a bridge with a Y-shaped branch, allows us to simplify the problem of how an ant decides to go right or left at a bifurcation in the trail. An important assumption of the experiments previously described, well supported by experimental data, is that ants preferentially follow the stronger of two pheromone trails. In the bridge experiments described previously, the foragers showed no tendency to follow the path they had used previously. Therefore, in the experiments that we will now examine, we can disregard the potential complication that an ant will learn and remember a particular path to the food. The only additional detail about trail-following that must be considered is the tendency for ants sometimes to retrace their steps and make U-turns along the trail. This will be discussed in a later section.

Binary Choice

We begin our analysis of trail-following by developing a mathematical relation describing how an ant chooses the right or left branch of a trail in relation to the concentration of pheromone on each branch (Deneubourg et al. 1990b). The following equations are simple, general choice functions consistent with experimental data (Beckers et al. 1993). They quantify the probability of taking the left branch, P_L, or the right branch, P_R, in relation to the absolute and relative pheromone concentrations on the left and right branch, C_L and C_R, of a bifurcation in the trail:

$$P_L = \frac{(k + C_L)^n}{(k + C_L)^n + (k + C_R)^n},$$ (13.1a)

and

$$P_R = 1 - P_L.$$ (13.1b)

The parameter, n, determines the degree of nonlinearity of the choice; a high value of n means that if one branch has only slightly more pheromone than the other, the next ant that passes will have a high probability of choosing it. The parameter k corresponds to the attractivity of an unmarked branch. In other words, the greater k, the greater the pheromone concentration necessary for the choice to reflect a tendency rather than pure chance. Figure 13.8a shows the probability of an ant choosing the left-hand branch ($n = 2$) as a function of the concentration of trail pheromone on the left-hand branch. Note that when the trail pheromone concentration on the left branch is less than that on the right ($C_R = k$ units), the chance of moving to the left is less than pure chance (0.5), but as the trail pheromone concentration on the left branch exceeds that on the right branch, the chance of moving left exceeds 0.5.

Data obtained from *L. niger* are best matched when the general choice function is fitted most accurately with $n = 2$ and $k = 6$. Note that this function is analogous to the dance-following function (equations 12.1 and 12.2) in the nectar-foraging model for honey bees (Chapter 12). The rate of evaporation of the trail pheromone was also measured and the mean life time of the trail pheromone is around 2000 s for *L. niger*.

U-Turns

In our observations, we noticed that ants following trails often make U-turns and double back on a trail. For example, an ant leaving the food source may take one branch, then make a U-turn and return to the common branch point, and then follow the second branch. It may also make a U-turn and return to the food source. When an ant backtracks after a U-turn it does not lay down trail pheromone.

Two hypotheses may explain why an ant makes a U-turn: trail-based U-turns and orientation-based U-turns. The first hypothesis assumes that an ant makes a U-turn on sensing a decrease in the pheromone on the trail. Such trail-based U-turns are expected to occur just beyond a branch point because here would be the greatest change in the concentration of the trail pheromone. If a forager accidentally leaves the well-marked trail used by most of the ants, it will find itself on a weakly marked branch and is likely to make a U-turn.

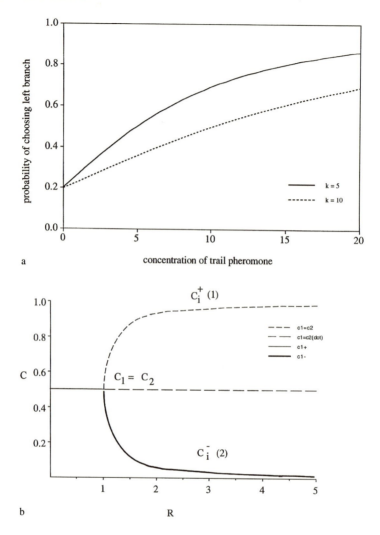

a   concentration of trail pheromone

b   R

Figure 13.8 (a) The probability of an ant choosing the left-hand branch as a function of the concentration of trail pheromone on the left-hand branch for $k = 5$ (solid line) and $k = 10$ (dotted line). $C_R = k$ and $n = 2$. (b) A bifurcation diagram shows the stationary states of the pheromone concentration $C_i/(C_1+C_2)$ (equation 13.3) for two equal food sources versus R, where $R = q\Phi/2fk$. A switch from symmetrical to asymmetrical food source exploitation occurs when $R > 1$. In small colonies ($R < 1$) the workers remain dispersed equally over both sources ($C_1 = C_2$, branch 1), but when $R > 1$, the colony concentrates its activities on one food source (branch 2 or 3). The greater the colony size, the greater the asymmetry of exploitation. Branch 1 corresponds to equation 13.4a and branches 2 and 3 correspond to equation 13.4b.

The second hypothesis assumes that an ant makes a U-turn on sensing an unattractive orientation of a branch of the bridge. It is known that ants are aware of the general direction to the nest and of the general direction of the food source, once they have learned it. Thus when ants are given a bridge such as the one shown in b Figure 13.6, where the long branch makes an angle of 75 degrees with the base of the Y, an ant finding itself in the middle of the long branch may realize that it is heading nearly perpendicularly to the direction required to reach the food source. This may induce it to make a U-turn.

These two hypotheses are not mutually exclusive. Indeed, ants may be stimulated to make U-turns for both these reasons. If the second hypothesis is correct, then one would expect the proportion of orientation-based U-turns to be higher on the long branch. This prediction is fulfilled; early in an experiment, before a significant amount of pheromone has accumulated on either branch, 50 percent of ants make U-turns on the long branch compared to approximately 15 percent on the short branch. But if the first hypothesis is correct, then one would expect trail-based U-turns to increase in frequency over time as the short branch becomes more strongly marked with pheromone and captures more of the traffic. Experiments show that the proportion of trail-based U-turns on the long branch increases on average by 150 percent during the first 15 min of an experiment (Beckers et al. 1992b). On the short branch this proportion is always small and decreases over time. Hence, the occurrence of both orientation-based U-turns and trail-based U-turns favor the shorter branch.

These dynamics can be described mathematically. Assume that the probability, P_0, specifies the initial likelihood that an ant turns back in response to a particular orientation of a branch on the bridge. This baseline probability of U-turns (P) will decrease if there is a significant concentration of pheromone on the branch, since the presence of pheromone indicates that the branch has been well used. This can be expressed by the equation,

$$P = \frac{P_0}{1 + \alpha C},$$
(13.2)

where C is the pheromone concentration on the branch and the parameter α specifies the extent to which trail pheromone diminishes the importance of orientation-based U-turns. Larger values of α give a greater weighting to pheromone-based trail following. Over time the proportion of U-turns falls almost to zero on the shorter branch as the pheromone trail on this branch gets stronger.

Models of Collective Decision Making through Recruitment

This section is devoted to the discussion of mathematical models that link the behavior of individual ants to trail-formation behavior of the colony, drawing on the biology presented earlier in this chapter. We will discuss a model for selection of food sources and the selection of shorter path—both expressed in the foraging trails of a colony—based on studies of the foraging behavior of *L. niger* (Beckers et al. 1992a,b; 1993).

Because of the extensive experimental data available, this modeling provides an unrivaled opportunity to determine the relative contribution of these different behavioral components to the pattern of trail formation by colonies. The effects of different trail-following strategies and of variations in their parameter values are examined through Monte Carlo simulations and differential equation models.

Selecting the Richest of Two Food Sources

The initial goal is to describe a simple model that is based on empirical findings about the behavior of the individual ants and includes the following assumptions:

1. There are two food sources, designated equivalently as R and L, or 1 and 2, equidistant from the nest at the two ends of a Y-shaped bridge.
2. Ants leave the nest at a constant rate (Φ, the flux), ingest food, and promptly return to the nest.
3. C_L and C_R are the concentrations of trail pheromone on the left and right branches of the bridge, respectively.
4. Φ_L and Φ_R are the flux of ants on each branch of the bridge. The traffic flow over each branch is affected by the choice function (equation 13.1).
5. q_L and q_R are the amounts of pheromone deposited by a returning ant.
6. On its return journey from a food source, each ant lays a quantity of pheromone, q_L or q_R, depending on which food source it visited. There is no trail-laying during travel from the nest to the food source. Although ants are known to deposit pheromone in both directions, we make this assumption for simplicity and to demonstrate that bidirectional trail-laying is not an essential ingredient in certain types of collective decision making. In a later section we present a more complete model that does include bidirectional trail-laying.
7. The evaporation rate of the trail pheromone is proportional to the pheromone concentration and to a constant, f, that is the inverse of the mean lifetime. When the mean lifetime is relatively long compared to the duration of the experiment, evaporation can be neglected ($f \cong 0$).
8. Since the time needed to visit the food sources and to return is equal for both sources, we neglect the time delay between the choice (left or right source) and the ant's return.

These relationships can be expressed in the following equation describing the rate of change in concentration of pheromone on trail i:

$$\frac{dC_i}{dt} = q_i \Phi P_i - f C_i \quad i = 1, \ldots, m \tag{13.3}$$

where

$$P_i = \frac{(k + C_i)^n}{\sum\limits_{j=1}^{m} (k + C_i)^n}.$$

In the case of a simple Y-bridge with 2 branches, i refers to either the left or right branch of the bridge and the expression for P_i is identical to equation 13.1.

Modeling Trail-Formation to Two Identical Food Sources

When food sources are identical each ant will deposit the same amount of pheromone on returning from either food. Thus $q_1 = q_2$. Figure 13.9 shows the change over time of the population on each branch produced by a Monte Carlo

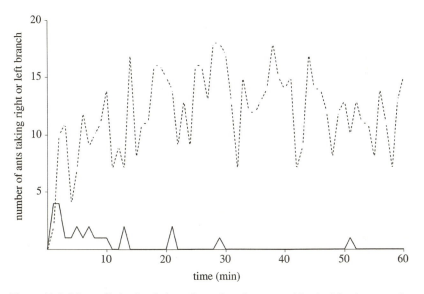

Figure 13.9 Monte Carlo simulation of ants foraging at two identical food sources indicates that, for the first few minutes, both sources are exploited equally but a break suddenly occurs after which one source is exploited more than the other. Parameter values: $\Phi = 0.2$ ants s^{-1}; $q_1 = q_2 = 1$; $k = 5$; $n = 2$; $f = 0.0005$ s^{-1}.

simulation corresponding to equation 13.3 using parameter values corresponding to *L. niger* (see legend for Figure 13.9). Notice that, initially, both sources are exploited equally but a break suddenly occurs, after which one source is exploited more than the other. This temporal pattern is similar to the pattern seen with real ants and discussed at the beginning of the chapter. The asymmetrical utilization of identical food sources arises as one trail becomes, by chance, slightly stronger than the other, with the result that it is followed more accurately and thus becomes further reinforced in a positive feedback cycle. Note, however, that this asymmetry results, not only from the positive feedback in pheromone deposition but also from the inherently stochastic nature of the ants' trail-following behavior. In the simulation, as in actual experiments, the amplification of random behavioral events leads, at a critical moment, to one trail becoming slightly stronger than the other (see Figure 13.4). Since this happens by chance, it cannot be predicted which trail will dominate.

In lieu of using a Monte Carlo simulation to examine the behavior of this system, one can also take an analytical approach to determine the general solution for the predicted stationary-state number of ants on each branch of the bridge. The stationary states of the system are found by setting the derivatives, dC_i/dt, of equation 13.3 equal to zero and solving algebraically. After simplifying, we arrive at three stationary states. The first is the symmetrical solution in which both food sources continue to be utilized equally:

$$C_1 = C_2 = \frac{q\Phi}{2f}. \tag{13.4a}$$

However, when $q\Phi/(2f) > k$, two new stationary states appear:

$$C_1 = \frac{q\Phi}{2f} \pm \sqrt{\left(\frac{q\Phi}{2f}\right)^2 - k^2} \quad \text{and} \quad C_2 = \frac{k^2}{C_1}. \tag{13.4b}$$

These results are shown in Figure 13.8b and point up a classical bifurcation phenomenon. The figure shows the distribution of ants on the two branches of the bridge as a function of a parameter, R, equal to $q\Phi/2fk$. A switch from a symmetrical to an asymmetrical food source exploitation occurs as R is increased (for example, by increasing q or Φ, or decreasing f). When R (or Φ) is small (in a small colony, for example), the workers remain dispersed equally over both sources. This is the stationary state characterized by equation 13.4a.

When R is above the critical value (R = 1) the colony suddenly switches from equal exploitation to asymmetrical exploitation. The greater the colony size, the greater the asymmetry of the exploitation and the more quickly the exploitation becomes asymmetrical. In the vicinity of R = 1 equal exploitation of both food sources still exists as a solution but is unstable; any fluctuation is quickly amplified and the system moves toward an unequal exploitation as illustrated by the two branches labeled, 2 and 3, in Figure 13.8b.

If one wants to observe these phenomena experimentally, it is easiest to tune the parameter Φ. This can be done by varying the number of foraging ants, or using colonies of different sizes. One can also experimentally vary q by changing the quality of the food source and changing the intensity of trail-laying by the ants. Of course in the simulation f and k can also vary. It is difficult to vary these parameters experimentally, however.

To the extent that the model accurately expresses the behavior of ants, it allows one to perform experiments that would be difficult or even impossible to do in the laboratory. For example, we can use the model to assess the role of each of the parameters in the system, by varying each parameter independently. Doing this experimentally may be impossible, since a change in one parameter often affects the value of another.

Consider, for example, the situation in which two food sources of the same sugar concentration are at different distances from the nest. In this case, the closer food source is of higher value because it is closer to the nest. Ants will deposit a greater quantity of pheromone when returning from this food source. In addition, ants will be able to reach the closer food source in a shorter time and less pheromone will evaporate from the trail between trips back and forth from the nest. In this situation, it would be difficult to separate the relative contributions of greater pheromone deposition and less pheromone evaporation. It would be interesting to know whether the shorter trip time alone is sufficient to establish preferential utilization of the closer food source; that is, even if the amount of pheromone deposited by the ants were the same for both food sources. Although, it is difficult experimentally to uncouple these two contributions to trail-formation, since one cannot control the amount of pheromone an ant releases, it would be a simple exercise to simulate such a situation.

Modeling Trail-Formation to Nonidentical Sources

Here we assume that the ants simultaneously find two nonidentical food sources and that the only difference in behavior between ants is that those ants visiting the better food source deposit a greater amount of pheromone, q. The results obtained with the corresponding Monte Carlo simulation are shown in Figure 13.10. The percentage of colonies preferring the richer source is graphed as a function of the ratio of the quantity of the pheromone deposited by an ant returning from the poorer source to that deposited by one returning from the richer source (q_1/q_2). The selection that ants make during each trial is generally unequivocal, meaning that for most of the colonies the vast majority of forager traffic is to one of the two food sources. The previously described

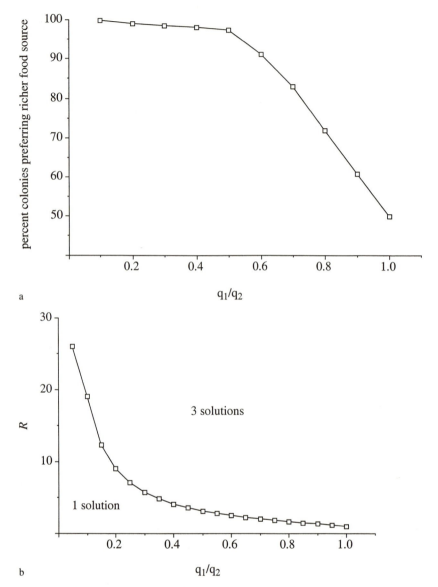

a

b

Figure 13.10 The percentage of colonies preferring the rich source is plotted in (a) as a function of the ratio, (q_1/q_2), of the quantity of pheromone laid by an ant return-ing from the poor source to that laid by an ant returning from the rich source. Each point represents 10,000 trials. (b) The plot in (b) shows the number of solutions of the equation 13.3 with respect to R and the ratio. $q_1/q_2 = 1$; $k = 5$; $n = 2$; $f = 0.0005$ s^{-1}.

experiments showed a weak modulation of trail-pheromone release as a function of food-source quality. The ratio between the intensity of trail-laying for a 1 M vs. a 0.05 M food source was found to be 1.5. Our simple simulation results show that even with such weak modulation the selection of the better food source occurs in at least 80% of the cases when q_1/q_2 at least 0.7.

The selection of a source is a collective process. No individual ant visits both food sources, compares them, and decides which is best. The elegance of this foraging system is that a collective decision to exploit the better food source arises automatically from the trail-following behavior, without the need for leader ants that would need a synoptic view of the situation.

What happens in the model when the food sources are not discovered simultaneously? Simulations with the same model, but with an initial delay so that the better food source is discovered later, show that the food source that is found first generally captures the greatest amount of traffic. Only when R is large, for example, when the amount of trail deposition (q) is much greater for the better food source, are the ants able to select a better food source that is introduced later. This is summarized in Figure 13.10b where we see that the model (equation 13.3) can exhibit one or three stationary solutions. When R is small and or the ratio q_1/q_2 is an intermediate value, the model exhibits only one stationary state. When R is large and or the ratio q_1/q_2 is intermediate in value, the model exhibits three stationary states. Only two such states are stable and observed: one with the rich source largely exploited, the other with the poor source largely exploited. Which stable state is selected depends on random events and the time-delay between the discovery of the first and second food source (see Figure 13.5). For example, if the poor source is discovered first it is the one exploited. The third stationary state is unstable and not observed. In the case of two identical food sources, this unstable state shows an equal exploitation of both.

The Significance of the Number of Food Sources

We can use this model to examine a slightly different situation in which we consider m identical sources (where $q_1 = q_2 = \ldots = q_m$) (Nicolis and Deneubourg 1999). Our purpose is not to discuss all the properties of this model. However, using equation 13.3 it can be shown that the behavior of this system is more complex than the one in which there are only two food sources. Figure 13.11 summarizes the behavior of the model. Keeping R constant (and equal to $q\Phi/2fk$), for $m < m_{c1}$, there exists stable stationary states in which one food source is preferentially exploited and the others are less, but equally, exploited. For $m_{c1} < m < m_{c2}$, there exists two stable stationary states: the previous one and the homogeneous state in which all food sources are equally exploited. When $m > m_{c2}$, the only stable stationary state is the homogenous state. Similarly, keeping m constant, when R grows the foraging

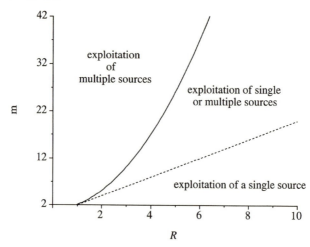

Figure 13.11 Pattern of exploitation of multiple sources as a function of R and the number of sources (m) (see text).

pattern switches from an equal exploitation of food sources to the state where one source is preferentially exploited. The critical values m_{c1} and m_{c2} are:

$$m_{c1} = 2R, \qquad (13.5a)$$
$$m_{c2} = 1 + R^2, \qquad (13.5b)$$

with $R = q\Phi/2fk$.

This plasticity (change of exploitation pattern) is significant because a change in an environmental parameter (the number of food sources) led to the adoption of different foraging patterns with no change in the workers' behavioral rules of thumb. Hahn and Maschwitz (1985) conducted experiments with the granivorous ant, *Messor rufitarsis*, and found that this species modified its exploitation pattern in relation to the number and the distribution of food sources. A well-marked trail can be observed for one source and a diffuse exploitation when numerous sources are present. It is difficult to imagine how an individual ant could perceive or count the number of sources and modify its behavior accordingly. Equations 13.5a and b suggest, however, how such plasticity can occur at the colony level. Similarly, the change of foraging patterns of army ants, discussed in Chapter 14, may be another example of such self-organized regulation of foraging patterns.

Selecting the Shortest Paths

In the previous model, ants deposited trail pheromone only when returning to the nest after visiting an attractive food source. For all species studied so far, however, it is believed that ants deposit pheromone when moving both to and from a food source (see pages 226–228). In this model we introduce bidirectional trail laying and modify the foraging arena to simulate the bridge arrangements in Figure 13.6b. The aim is to determine what features of trail-laying and trail-following are required for the colony to select the shortest path between nest and food. The basic assumptions of the model include:

1. Ants lay trails when traveling in both directions; when walking from the food source to nest and from nest to food source.
2. Foragers exhibit exactly the same behavior on both the long and short branches of the bridge, depositing pheromone at equal rates when traveling on either branch. Thus we assume that individual ants have no knowledge of the relative lengths of branches.
3. The selection of one branch or the other is determined solely by the choice function described previously in equation 13.1. Individual ants have no memory of the routes they have taken, but choose one branch or the other based on the relative concentrations of trail pheromones on the two branches.
4. Once an ant selects a branch, it always traverses it completely.
5. The evaporation of the pheromone is proportional to the trail pheromone concentration and to a constant, f, equal to the inverse of the mean lifetime.

Figure 13.12a summarizes the scenario underlying the model. Ants going back and forth from the nest to the food source arrive at forks 1 and 2 (where 1 is the fork nearest the nest, 2 is the fork nearest the food source, and Φ_1 and Φ_2 are the corresponding fluxes of ants). Each ant chooses a branch, S or L (where S = short and L = long branch) and lays a trail along the branch by depositing pheromone at a constant rate.

P_{S1} and P_{L1}, respectively, are the probabilities that an ant at fork 1 moving towards the food source takes the short or the long branch. Similarly P_{S2} and P_{L2} are the corresponding probabilities at fork 2 for an ant returning to the nest. These probabilities (equation 13.6a) have the same nonlinear form as equation 13.1:

$$P_{Sj} = \frac{(k + C_{Sj})^n}{(k + C_{Sj})^n + (k + C_{Lj})^n} \quad j = 1, 2 \tag{13.6a}$$

and

$$P_{Lj} = 1 - P_{Sj} \quad j = 1, 2$$

Four variables are needed to describe the concentrations of pheromone at the two forks: They are, respectively, the concentrations, C_{S1}, C_{L1}, C_{S2} and C_{L2}.

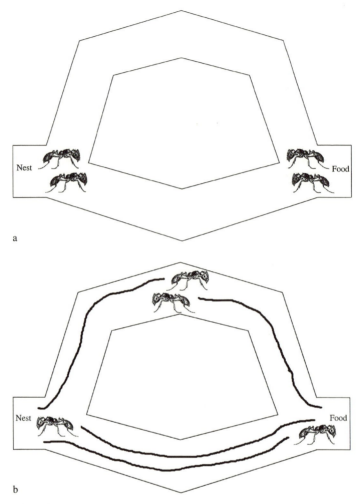

Figure 13.12 The model of bidirectional trail-laying is shown in (a) and (b). Ants arrive at forks 1 and 2, where fork 1 is nearest the nest and fork 2 is nearest the food source. Each ant chooses a branch, S or L (short or long), and lays a trail along the branch by depositing pheromone at a constant rate.

The equations describing the dynamics of trail of pheromone concentration are:

$$\frac{dC_{si}}{dt} = \Phi_j P_{sj}(t - T_s) + \Phi_i P_{si}(t) - fC_{si} \qquad (13.6b)$$

$$\frac{dC_{Li}}{dt} = \Phi_j P_{Lj}(t - T_L) + \Phi_i P_{Li}(t) - fC_{Li} \qquad (13.6c)$$

with either $i = 1$, $j = 2$, $i = 2$, $j = 1$, and T_S and T_L equal to the time to cross over the short or long branch, respectively.

The delays, T_S or T_L, in equations 13.6 are key to how the colony selects the short branch. Figure 13.12b illustrates how this model works. The two delays are small compared to the mean life time of the pheromone ($1/f$). These delays lead to an initial period in which the short branch at both forks has ants moving in both directions while the long branch at both forks has ants moving in only one direction. Consider the fork nearest the nest. As ants first arrive and there is little or no pheromone at the fork, ants travel along their randomly chosen branch, laying down pheromone as they go. After T_S, however, the short branch at the second fork will be marked by these ants. This phenomenon is symmetrical for ants moving from the food source to the nest; that is, the same events occur later, at T_L seconds, for the long branch. Between these two moments, the short branch accumulates a slight advantage over the long branch at both ends. This is amplified by the autocatalytic nature of the choose-and-mark process, and the short branch becomes preferred.

The necessary condition for this selection procedure is that the ants *must* lay trails in both directions. With trail-laying in only one direction, no unequivocal selection is possible. In between the two extremes of one-way and two-way trail-marking, we find that the greater the difference in trail-laying between ants leaving and returning to the nest, the more closely the choice between the two branches approaches pure chance.

To make comparisons between the model's predictions and the results of experiments, we must specify a flux of ants that develops over time. As a result of recruitment, this flux will not be constant. From experiments, we know that the flux shows an initial increasing phase due to recruitment, followed by a plateau when all the foragers willing to forage have been recruited. From the data of Beckers et al. (1992b), we have the following general expression for the flux, where Φ_1 is the flux arriving at the choice point 1, and Φ_2 is the flux arriving at point 2. We assume that Φ_2 is equal to the flux Φ_1 after a time delay corresponding to the time needed to cross the bridge (T_S or T_L) and the time spent at the food source, ingesting food (T_1):

$$\Phi_1 = \frac{k_0(k_1 + C_{L1} + C_{S1})^2}{k_2 + (k_1 + C_{L1} + C_{S1})^2}, \tag{13.7a}$$

$$\Phi_2 = \Phi_1 P_{S1}(t - T_S - T_1) + \Phi_1 P_{L1}(t - T_L - T_1). \tag{13.7b}$$

Equation (13.7a) gives the time evolution of the flux Φ_1 of ants leaving the nest. The parameters k_0, k_1 and k_2, are fitted from experimental data. In this equation we assume that the total quantity of pheromone at the choice point 1 ($C_{S1} + C_{L1}$) governs recruitment of ants from the nest.

In the case of *L. niger*, there was poor agreement initially between the model's predictions and our experimental observations. The model predicted a weak preference for the shortest path, whereas the ants expressed a strong preference for the shortest path (Figure 13.7). Apparently, simple bidirectional trail laying is not a complete explanation for how *L. niger* selects the shortcut. When using the Argentine ant, *Linepithema humile*, however, agreement between the model and the experiments is much better. The selection of the shortest path predicted by the theoretical models almost matches the level of selection shown in the experimental results (Goss et al. 1989).

How can we explain this difference? In the model, the greater the difference in the length of the two branches, and thus in the difference in the time it takes to cross the two branches (T_S versus T_L), the more likely it is that the short branch is chosen. This highlights the importance of running speed in the selection of the shortest path. Since the running speed is higher for *L. niger* than for *L. humile*, for the same bridge design the time difference between T_S and T_L would be smaller for *Lasius* than for *Linepithema*. It is therefore more difficult for *Lasius* to select the shortest path under experimental conditions in which the branches of the bridge differ in length by only 14 cm. It so happens that for the bridges used in the experiments the model predicts that *Lasius* will choose the shortest path only 55 percent of the time, but in actual experiments *Lasius* does much better, selecting the shortest path 80 percent of the time.

These discrepancies suggested that we had failed to incorporate in the model an important aspect of the ant's behavior. Indeed, more detailed, observations of *Lasius niger* revealed that U-turns, could have an important affect on the ability to select the shortest path. In the next section, we present a more detailed model that includes U-turns. With this refinement of the model, the results of the model agree more closely with those of the experiments.

A Model Incorporating U-turns

Based on the discussion of equation 13.1, we define the probabilities, U_{S1}, U_{S2}, U_{L1}, and U_{L2}, of an individual making a U-turn on the short and long branches at forks 1 and 2, respectively. The probability decreases from an initial baseline value U_{S0} for the short branch and U_{L0} for the long branch as a function of the amount of pheromone on that branch. Pheromone amount is represented by a parameter α. Recall that these baseline values refer to the probability of making a U-turn based solely on the orientation of the branch, as there is little or no pheromone on each branch at baseline. As the amount of pheromone on a trail increases, the likelihood that an ant on that trail will make a U-turn decreases. Thus when C_{Sj} and C_{Lj} are the trail pheromone concentrations for each branch at the choice point $j = 1, 2$, we have:

$$U_{Sj} = U_{S0}/(1 + \alpha C_{Sj}), \qquad U_{Lj} = U_{L0}/(1 + \alpha C_{Lj}). \tag{13.8}$$

Recall that roughly half of ants making a U-turn go back to the nest without further marking and the other half take the alternate branch and mark it as they proceed. Taking account of these U-turns, we obtain a new, more complicated set of equations describing the change over time of the trail pheromone concentration at the four sites just beyond the two forks:

$$\frac{dC_{Si}}{dt} = \Phi_i\left(P_{Si} + 0.5P_{Li}U_{Li}\right)(t) + \Phi_j\left[P_{Sj}(1 - U_{Sj}) + 0.5P_{Lj}U_{Lj}\right]$$

$$\underset{\text{I}}{} \qquad\qquad\qquad \underset{\text{II}}{}$$

$$\underset{\text{III}}{(t - T_S) - fC_{Si}} \qquad\qquad\qquad\qquad (13.9a)$$

$$\frac{dC_{Li}}{dt} = \Phi_i\left(P_{Li} + 0.5P_{Si}U_{Si}\right)(t) + \Phi_j\left[P_{Lj}(1 - U_{Lj}) + 0.5P_{Sj}U_{Sj}\right]$$

$$\underset{\text{I}}{} \qquad\qquad\qquad \underset{\text{II}}{}$$

$$\underset{\text{III}}{(t - T_L) - fC_{Li}} \qquad\qquad\qquad\qquad (13.9b)$$

with either $i = 1$, $j = 2$, or $i = 2$, $j = 1$.

The first term (I) of equation 13.9a refers to the pheromone added by ants entering the short branch at fork 1 (or 2) at time t. This is given by the flux of ants past the site multiplied by their probability of entering the short branch (whether or not they turn back, since in either case they lay pheromone), plus the flux multiplied by the probability of choosing the long branch and turning back to take the short branch. Recall that only half of the ants that make a U-turn will take the short branch; the others will return to the nest. The second term (II) refers to pheromone added by ants that pass in the opposite direction on the short branch at fork 1 (or 2) at time t, having entered the short branch at the opposite fork 2 (or 1), T_S seconds earlier at time $t - T_S$. This is given by the flux multiplied by the probability of selecting the short branch and not turning back $(1 - U_{Si})$, plus the flux multiplied by the probability of selecting the long branch and turning back to take the short branch. The final term (III) refers to the loss of pheromone by evaporation as in equation 13.3. Equation 13.9b is an analogous equation to equation 13.9a and yields the change in pheromone concentration on the long branch.

This model illustrates the synergistic (cooperative) effect of orientation-based U-turns on pheromone-based trail-following. In other words, differences in the rate of U-turns (due to the orientation of the branch) are amplified by pheromone-based trail-following.

Consider, for example, the situation with no trail pheromone. If there were no trail-laying, the majority of ants would still take the shorter branch since the probability of making a U-turn on the shorter branch is less than that on the longer branch $(U_{S0} < U_{L0})$. With trail laying, however, the difference between U_{Si} and U_{Li} becomes even greater. This positive feedback amplifies the preference for the shortest path.

Table 13.1 Comparison between experimental and theoretical results.

Experiments	86
Simulation with bidirectionality	56
Simulation with U-turns	89
Simulation with bidirectionality + U-turns	91

Percentage of the experiments or simulations correspond-ing to the selection of the shortest path (from Beckers et al. 1992b). Parameter values: $T_S = 7$; $T_L = 21$.

The magnitude of this effect can be calculated algebraically by assuming that in the absence of any trail pheromone the ants choose both branches with the same probability, that the rate of U-turns remains constant, and that half of those that make a U-turn take the other branch and are therefore counted twice (the other half returning to the common branch). Hence:

$$\text{proportion taking short branch} = (2 + U_L)/(4 + U_S + U_L). \qquad (13.10)$$

For the bridge design with a 1:2 ratio of branch lengths, for example, the pro-portion of ants taking the short branch was 85 percent of the total traffic in the experiments. Based on equation 13.10 (and recalling that the initial probability of U-turns is 0.50 versus 0.15 on the long and short branches, respectively), we would expect a value of only 55 percent in the absence of trail-laying.

We are now ready to use this more sophisticated model to compare the relative contributions of bidirectional trail-laying and orientation-based U-turns. With the appropriate selection of parameters, we can independently es-timate the expected contribution of each mechanism of path selection, even though one cannot do this experimentally. To evaluate the contribution of U-turns alone, for instance we can equalize the time taken to cross each branch ($T_S = T_L$). Thus bidirectional pheromone deposition occurs, but we eliminate the difference in pheromone concentration on the two branches that arises because ants cross the shorter branch more quickly and thus increase the pheromone concentration on that branch at a greater rate.

Similarly, to evaluate only the contribution of bidirectional trail laying and its effect on pheromone concentration on the two branches, we can let $U_{S0} = U_{L0}$, and thus eliminate the effect of U-turns on the selection of the shortest path. In Table 13.1, the results of these simulations is compared with experimental results (Beckers et al. 1992b) and demonstrates the relative con-tributions of the two mechanisms. As in the experiments, most of the simu-lations resulted in more than 80 percent of the forager traffic focusing on the shorter of the two branches. For parameter values corresponding to *L. niger*,

U-turns make a larger contribution (89 percent) to path selection than bidirectional trail-laying (only 56 percent). With both mechanisms operating, the selection is slightly more efficient (91 percent) (see Figure 13.7). We should not expect a large gain in efficiency by combining the two mechanisms, since including U-turns alone achieves the high level of shortest-path selection seen in the experiments.

A final test of the model is shown in Figure 13.7a and b, in which a delay occurs before the introduction of the shortest path branch. The model predicts that ants will always remain faithful to the long branch and not select the shortest path. However, experimental results are not clear cut. This may be the result of disturbances to the system when the short branch is added to the bridge.

On a more general level, it now seems probable that other trail-laying ant species, using only one of these mechanisms, will be capable of selecting the shortest path. Species that lay bidirectional trails, however, will have a strong ability to select the shortest path only if a large time difference exists between alternative paths. This mechanism will also work better with a greater flux of ants.

Species with small colony sizes (low ant flux during foraging) are therefore not expected to utilize a shortest path mechanism based on bidirectional trail-laying but depend on the mechanism of U-turns. Species with larger colony sizes may tend to rely more on collective chemical orientation. In these species, colonies may nevertheless have the capacity to exploit orientation-based U-turns. This may be especially true in species with colonies in environments with many obstacles.

Conclusions: Recruitment in Ants and Bees

The examples discussed in this chapter and Chapter 12 demonstrate how rather simple rules, with no individual possessing global knowledge, can lead to diverse collective responses that efficiently solve particular foraging problems.

This analysis has revealed that accurate collective decisions can arise through the selective amplification of slight preferences among various alternatives. Such collective decision making can be as accurate and effective as decision making by solitary individuals, including certain vertebrates (Krebs and Davies 1984), even though vertebrates have a greater capacity for information processing.

The models described in this chapter and the previous one are similar in that both are based on mathematical expressions describing the relative attractiveness of one source or one trail compared to others. In the case of the ants, for example, the expression takes the form of a general choice function $A_i/(\Sigma A_i)$, where A_i is the attractiveness of trail i leading to source i. This attractiveness is proportional to C_i^n where C_i is the concentration of trail pheromone and n (n > 1) expresses the nonlinearity of the response to a trail pheromone concentration.

For honey bees, we showed that the attractiveness of a food source A_i can be represented by the similar form $(D_i d_i)^n$, where D_i is the number of bees dancing and d_i is the duration of their dancing for the source i. In the case of honey-bee foraging, $n = 1$, indicating that dancers do not cooperate and a bee is influenced only by a single dancing bee at any one moment. In the two systems, the trail concentration is analogous to the number of dancing bees and both represent the attractiveness of one foraging site over another. The primary difference between the two systems is at the level of the power, n. If, for example, the bees were successively able to sample dancers to compare the relative profitability of food sources, this would introduce an equivalent nonlinearity in the bee-foraging system. But as Camazine and Sneyd (1991) predicted and Seeley et al. (1991) demonstrated experimentally, bees do not compare dancers.

This seemingly small difference is responsible for the rather different properties of the honey-bee recruitment system versus the recruitment system of many ants. As has been shown, bees are always able to select the most rewarding nectar source even if many weaker food sources have been discovered previously. Furthermore, it can be predicted that a honey-bee colony presented with an array of identical sources will not focus on a single source. This is not the case for the ants, which are not always able to select the most rewarding source but often focus their efforts on one among a group of identical sources. Moreover, among bees the solution is independent both of colony size and of the number of food sources. As was seen, such is not the case for the ants.

The different results for bees and ants are not primarily due to differences in the sensory or physiological capacities of the individuals involved; they are a by-product of the properties of the decision-making system operating at the collective level. Most of these collective decisions arise as a result of competition between different sources of information flow that can be amplified in different ways. In each case, however, the underlying logic is that the food source attracting the most recruits is the one selected.

Both chapters consider competition between food sources which have either a similar or dissimilar concentration of sucrose. However, several factors affect the amplification of recruitment to such sources and the collective responses take into account particularities of the environment, such as, food quality, the number of food sources, and the distance from the nest to the food source, as well as characteristics of the colony, such as size and nutritional state.

These parameters can be divided into two categories depending on how they affect the rate of recruitment amplification (Detrain et al. 1999). One set of factors relates to the parameters involving various external or environmental constraints such as the distance between the nest and a food source and the time of discovery of a food source. In these cases any constraint that reduces the rate of recruitment or diminishes the trail concentration to a food source can cause that food source to be less attractive than other food sources. Thus an efficient decision can be made without particular modulation of individual behavior. The second group of parameters, in contrast, are intrinsic to the behavior of the individual. For example, as discussed above, an individual scout bee or ant modulates its dancing or trail-laying activity depending on its perception of the profitability of a particular sugar source.

We have also examined situations in which many sugar sources are present and it is clear that plasticity in the pattern or recruitment trails may be driven simply by an increase in the number of food sources, with no requirement for changes in the communication system or the individual behavior of the ants. A colony may select only one source if the number of sources is small and/or its forager population is large. Alternatively, a colony may exploit all sources equally if they are numerous or the colony itself is small. Such switching behavior, which is not present in the model for honey bees, may facilitate a response that is a good adaptation for defense in terms of the distribution of foragers among food sources (Franks and Partridge 1993).

The simplicity of the rules involved in production of a response appears even more clearly if the recruitment dynamics are taken into account along with other colony dynamics, such as changes in food consumption and demand. Such dynamics can play important roles in the recruitment system, but for pedagogical reasons were ignored in our presentations of recruitment mechanisms. We expect that the integration of these other dynamics processes will lead to response patterns of varying levels of complexity. For example, when there is a choice between sources of different quality (e.g., proteins vs. sugars), workers will selectively forage to meet the requirements of the colony. In these cases, we suspect that nonforaging workers within the colony can provide important information to foragers by means of trophallactic and behavioral interactions, and that this information plays an essential role in the emerging choice between foods (Howard and Tschinkel 1980; Sorenson et al. 1985; Cassill and Tschinkel 1999; Camazine et al. 1998).

Influences such as changing patterns of crowding at a food source or variations in the characteristics of a food source due to changes in exploitation or temporal variation can lead to far more complicated foraging dynamics than the simple choice of one food source over another. Wilson (1962), for example, showed that ants arriving at a highly profitable but crowded food source go back to the nest or explore the foraging ground without trail-laying. Such effects can lead to oscillations in exploitation (Verhaeghe and Deneubourg 1983) or equal exploitation of food sources of quite different quality.

Finally, a set of spatial factors relates the global configuration of the system (for example, the food distribution, or the trail network) and the particularities of its history to the exploitation of a new source. If, for instance, a novel food source is present near a source that already possesses recruits, this new food source may be exploited in preference to other sources that have higher nutritional potential or are closer to the nest simply because of its proximity to other sources and the history of its discovery. It has been seen, for example, that the sequence of discovery of different sources influences the colony's choices. Recruits that get lost during recruitment are not locked into a particular food source but wander around and may discover new food sources ripe for exploitation. In essence, the colony may have an automatic mechanism for modulating its level of exploration. If the food source is not high quality, the lack of a strong trail (many ant species vary the quantity of trail pheromone deposited according to food quality) may promote trail-following that is more diffuse, and may promote a more extensive search of the foraging area. This may increase the probability that new sources of higher value will be discovered (Deneubourg et al. 1983).

Associated with such issues is the pattern shown in Box 13.1, where the exhaustion of food as a result of foraging, together with trail-laying behavior, can lead to foraging trails that rotate systematically around a nest.

A number of other models (differential equations, simulations) have been developed to study the growth of trail networks in two-dimensional systems. This subject has been explored by Edelstein-Keshet and her coworkers (Edelstein-Keshet 1994; Watmough and Edelstein-Keshet 1995a, b; Edelstein-Keshet et al. 1995). The next chapter discusses how, in army ants, such patterns characteristic to one species emerge from the interplay between food-distribution and trail-recruitment.

Recruitment and collective decision-making systems (including those based on trail-laying) are not restricted to the foraging of honey bees, ants, and termites. Numerous other examples are described in the literature. Some involve chemical and or mechanical trails, such as those in stingless bees (Michener 1974). Other recruitment systems involve chemicals that diffuse into the air, or mechanical vibration or visual communication (Hölldobler and Wilson 1990; Wilson, 1971). Recruitment systems are also used for other purposes, such as nest defense (Hölldobler and Wilson 1990; Wilson 1971), swarming (Jeanne 1991; Lindauer 1955), exploration (Deneubourg et al. 1990b; Detrain et al. 1991; Fourcassié and Deneubourg 1992) or inter-nest communication (Aron et al. 1990). Recruitment has also been shown in other gregarious species such as social caterpillars (Fitzgerald 1995), cockroaches (Rivault and Cloarec 1998), mollusks (Focardi et al. 1985), and vertebrates such as Norway rats (Galef and Buckley 1996) and naked mole-rats (Judd and Sherman 1996).

These examples of recruitment address the fundamental question of how the most rewarding site can be selected by a group of individuals. The answer depends in part on whether a selective advantage accrues to aggregation and concentration of the colony's efforts on a single site, or whether it is better to distribute one's work force more widely. At the most fundamental level, however, essential similarities are found in the mechanisms used to arrive at such collective decision making. In these chapters on honey bees and ants we have explored how individuals make advantageous colony-level decisions by actively exploiting competition between semi-isolated positive feedback loops in which different pools of recruits attempt to recruit more and more of their comrades.

Box 13.1 Self-Organized Trail Rotation in Harvester Ants

The examples and models presented in this chapter were relatively simple. In most cases ants traveled on trails that presented binary choices and the environmental substrate was uniform. This is obviously unrealistic. In this section, we discuss an example of a more elaborate foraging pattern that can be understood as an extension of the problems previously discussed. The main difference between this example and the previous ones is that here we consider variations in food sources over time that result from the ants' foraging activity. The aim is to suggest that a richer variety of ant-foraging patterns may arise as self-organized processes involving the same simple rules of thumb for trail-laying and trail-following.

The Pattern

Field observations by Bernstein (1975) and Rissing and Wheeler (1976) on the harvester ant, *Messor pergandei*, have revealed spectacular collective behavior. Figure 13.13 gives a typical map of the direction of a nest's main foraging trail over more than a dozen sequential foraging periods. This trail can be seen to make a more or less complete revolution about the nest with a period of several days to about three weeks. As the ants forage for seeds, the pattern of trails sweeps around the nest like one of the hands of a clock. The pattern is not, however, clocklike in its regularity. Rissing and Wheeler (1976) reported that occasionally bouts of random foraging occur, mostly when food is scarce, and rarely in years of high seed density. In addition, the foraging columns change direction more slowly in years or in regions when food is abundant. Rissing and Wheeler (1976) also observed that double columns may develop just before a major change of foraging direction. The newer of the two trails becomes dominant as the old one is progressively abandoned.

This unusual behavior can be modeled with a scenario based on just two factors. The first is trail-laying and trail-following behavior as described in the previous portion of this chapter. The second is the changing

Box 13.1 continued

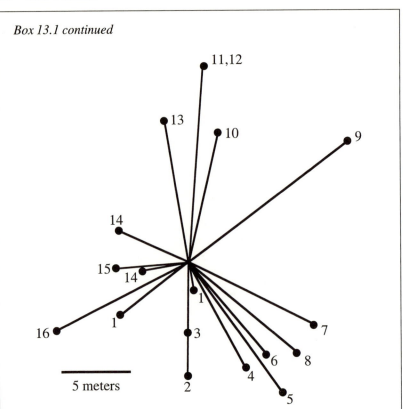

Figure 13.13 In consecutive foraging columns of the harvester ant, *Messor pergandei*, the numbers identifying each trail refer to different foraging periods (based on Rissing and Wheeler 1976).

distribution of food in the environment through the ants' own foraging efforts.

The Model

Our model of *Messor* foraging behavior is based on the assumption of a circular foraging area divided into m sectors, with the nest at the center (Figure 13.14). Each sector is characterized by a seed number, S_i, a trail concentration, C_i, and the number of foragers, F_i, where $i = 1, \ldots, m$.

The seed number, S_i, in each of sector is influenced by three factors: the arrival of new seeds to the sector, the loss of seeds by foraging ants, and the loss of seeds by external factors, such as competitors or decay.

Box 13.1 continued

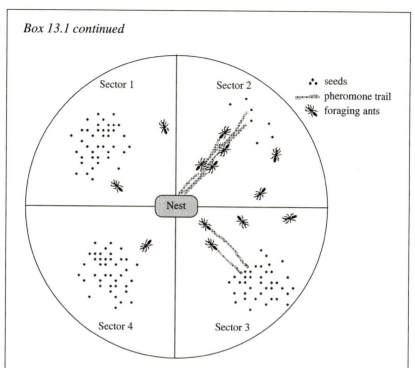

Figure 13.14 The layout of the simulation of harvester-ant foraging includes a circular foraging area. The nest at the center divided into m = 4 sectors. Each sector is characterized by a seed number, S_i, a trail concentration, C_i, and the number of foragers, F_i, in the sector. Sector 2 is exhausted and foragers explore mainly sectors 1 and 3. In sector 3, after discovering seeds two foragers go back to the nest laying a trail.

Let Θ_i be the rate of seed arrival per unit time in each sector. The number of seeds found by ants in a sector can be expressed as the product of the number of foragers, F_i, in the sector, a constant, g, and a function of the number of seeds in that sector: $S_i/(\alpha + S_i)$. The constant, g, describes the foraging behavior of the ants. The higher the value of g, the easier it is for an ant to collect seeds, once the ant has found seeds. A higher g suggests, for example, that the seed is easier to handle.

The function $S_i/(\alpha + S_i)$ increases monotonically from 0 to 1 as S_i increases; α is a constant that specifies the seed-finding capacity of the ants and increases as ants find it harder to locate seeds. This expression describes the ease with which seeds are discovered as a function of the number (or density) of seeds in the sector. If the density of seeds

Box 13.1 continued

is very low, then $S_i/(\alpha + S_i)$ is close to 0, and seed removal is low. The disappearance rate, rS_i, is assumed to vary with external factors, such as competitive removal of seeds by other animals or the decay of the food. The rate constant, r, specifies the magnitude of this effect.

The trail leading to each sector is characterized by C_i units of pheromone. This quantity is under the control of trail reinforcement and trail evaporation, as described earlier. Each successful forager returns to the nest laying a unit of pheromone. Because such return journeys are initiated by finding a seed, this term is equal to the rate of finding seeds, $gF_i S_i/(\alpha + S_i)$. The evaporation rate of the pheromone is given by fC_i.

Each sector contains a population of F_i foragers. Four factors govern the number of foragers. At each time step, N foragers leave the nest. A fraction, P_i, choose the sector i according to a nonlinear function of the trail pheromone associated with each sector (equation 13.11d). Note that this is the equivalent to equation 13.1. Of this fraction a small quantity η diffuses into each of the two adjacent sectors (lost ants). At the end of each step, all the foragers that found food return to the nest adding one pheromone unit to the trail leading from that sector. Those foragers that did not find seeds return to the nest without laying trail pheromones. The complete system of equations for the process are thus:

$$\frac{dS_i}{dt} = \theta_i - gF_i \frac{S_i}{(\alpha + S_i)} - rS_i, \quad i = 1, \ldots, m \qquad (13.11a)$$

$$\frac{dC_i}{dt} = \frac{gF_i S_i}{(\alpha + S_i)} - fC_i \qquad (13.11b)$$

$$F_i = N\big((1 - 2\eta)P_i + \eta P_{i+1} + \eta P_{i-1}\big), \quad i = 1, \ldots, m \qquad (13.11c)$$

$$P_i = \frac{(k + C_i)^2}{\sum\limits_{j=1}^{m} (k + C_j)^n} \qquad (13.11d)$$

where $j = 1, \ldots, m$, and P_i has the periodic conditions, $P_{n+1} = P_1$ and $P_0 = P_1$.

Figure 13.15 shows the results of a simulation exhibiting periodic behavior. As in the bridge experiments with *L. niger*, ants choose one sector, i, after an initial fluctuation during the formation of a strong recruitment

Box 13.1 continued

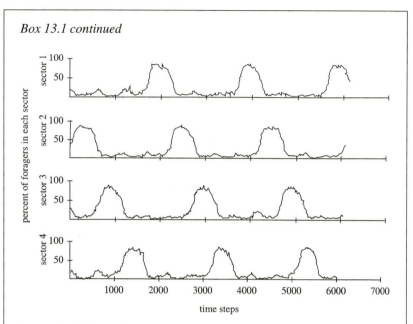

Figure 13.15 The results of a Monte Carlo simulation with 4 sectors show the percentage of foragers in each sector as a function of time. The foragers form a trail that starts in sector 2 and rotates clockwise to sectors 3, 4, 1, etc., with a regular period. $N = 100$; $f = 0.03$; $g = 0.1$; $r = 0.001$; $\eta = 0.05$; $\Theta = 2$; $\alpha = 1000$; $k = 20$.

trail. Thereafter, the number of seeds in sector i diminishes through the foragers' activity, while they accumulate in the other less frequented sectors. Eventually, the foragers have difficulty finding seeds in sector i, and so the trail to that sector is no longer reinforced but diminishes by evaporation of pheromone. Ants then begin randomly to explore neighboring regions on either side of the trail, eventually find seeds, and return to the nest depositing trail pheromone. Now, the trails to its two neighboring sectors $i - 1$ and $i + 1$ have a higher concentration of pheromone than the other sectors. Thus, rather than collectively selecting at random the $(m - 1)$ remaining sectors, the foragers collectively choose between sectors $i - 1$ and $i + 1$.

Suppose that the foragers choose sector $i - 1$. When sector $i - 1$ is nearly empty, again weak trails lead to the two adjacent sectors, i and $i - 2$. However, sector i is still more or less depleted of seeds, not having had enough time to become restocked from the environment. Therefore, the foragers collectively choose sector $i - 2$, since they find seeds there

Box 13.1 continued

more easily than in sector i. Thereafter, they continue to choose the next sector in the *same* direction as before, and thus the trail rotates around the nest. By the time the trails have done a full circle, the sectors originally exploited have been restocked and the rotation continues indefinitely.

The elegance of this pattern of trail rotations is that there is nothing in an individual ant's behavior that specifies that it should forage in one sector one day and in the neighboring sector on the next. The rotating pattern arises in a self-organized manner from interactions between the ants' simple trail-following behavior and their interaction with the environment (in this case the distribution of seeds).

This behavior is similar to that described in Chapters 8 and 10 (devoted to slime mold waves and synchronized flashing in fireflies), where we discussed the periodic activities emerging at the group level. An important hypothesis is that such periodicity is generated through the interplay between a period of amplification and a refractory period. We find the same idea here. Amplification occurs in trail-laying and trail-following behavior, while the refractory period corresponds to local exhaustion of seeds. The only difference here is that this refractory period is a by-product of the foragers' interaction with the environment and not part of the individual's intrinsic behavior or physiology, as in the case of fireflies.

Trail rotation is not the sole property of the model. As in previous models, a number of tunable parameters can switch the system into other regimes. Other patterns can arise, depending on the number of foragers and the amount of food in the environment. With increasing food abundance, the model predicts a switch from homogeneous foraging to the formation of a trail that rotates about the nest. The greater the abundance of the food the more slowly the trail rotates. With extremely high food abundance the trail can become fixed on one sector (a permanent trail). These theoretical results are in close agreement with experimental observations described earlier.

While *M. pergandei* exhibits rotating trails, other harvester ant species develop other foraging patterns. Do these different patterns necessarily require different behavioral rules on the part of the individuals? The model suggests that a diversity of individual behaviors is unnecessary. The results that we have presented show that different patterns can result simply from different seed densities, which is a function not of the ants but of ecological conditions.

The model also makes the weaker prediction that diverse patterns may result from slight differences in the values of parameters that specify

Box 13.1 continued

the communication behavior of these ants (e.g. trail-laying intensity and trail-following ability). By modulating these parameters, we are also able to generate different patterns without resorting to qualitative modification of individual behavior. Until controlled experiments unravel the various factors involved, however, we will not be able to assess the relative influences of the environment and behavior on the generation of these patterns. A more detailed discussion of the factors affecting these patterns is offered by Goss and Deneubourg (1989).

Finally, we suggest that the logic of the model is not only applicable to trail-following social insects. The potential to generate such collective patterns of foraging is inherent in all group-living centrally nested foragers that have mechanisms to coordinate their direction of foraging, or in solitary foragers able to self-reinforce their successive foraging trips.

Figure 14.1 *Dorylus (Anomma) nigricans* workers in Africa attack an earthworm. Colonies of this army ant species may contain 20 million workers, each utterly blind. Shown beside the workers is the raiding trail used by the colony. (Illustration © Bill Ristine)

14

The Swarm Raids of Army Ants

It is remarkable that notwithstanding the high pitch of
excited raiding and the variety of individual activities which
might suggest a lack of organization, the huge body of ants
behaves as a mobile unitary organism.
> —T. C. Schneirla, *Further studies on
> the army-ant behavior pattern:
> mass-organization in the swarm-raiders*

Introduction

The group activity considered in this chapter is the army ant swarm raid
and includes the spatial distribution of the ants, their pheromone trails, and
the dynamic pattern in space and time of ant-traffic flow within this foraging
system. Consider, for example, the huge raid of the African army ant, *Dorylus (Anomma) nigricans*, composed of millions of completely blind workers.
This chapter will explore how the army ant swarm raid is created literally by
the blind leading the blind. Imagine trying to organize vast numbers of blind
automata so that they can cross unknown and dangerous terrain, fight battles
against foes more powerful *per capita* than themselves, sustain few casualties
in their own ranks, retrieve the corpses of their prey, and find their way home
efficiently without getting lost. This is not fiction from H. G. Wells's *War of
the Worlds*. It is what army ants do almost every day of their lives.

The Collective Structure of Swarm Raids

The swarm raids of army ants are one of the wonders of the insect world.
(For recent reviews of army ant biology and behavior, see Franks [1989], Gotwald [1995] and Hölldobler and Wilson [1990]). The largest raids of neotropical army ants are those of *Eciton burchelli*, which may contain up to 200,000
voracious and virulent workers. Such a raid may be 15 m or more wide, proceed at 0.2 m/min, and sweep over an area of more than 1500 m^2 in a single
day (Willis 1967; Franks 1982; Franks and Bossert 1983). The raid system is
composed of a swarm front—a dense phalanx of foragers carpeting the ground
that extends for approximately 1 m behind the leading edge of the swarm—and
a large reticulate system of trails, including one principal trail (Figure 14.1).

The swarm front is served by a series of these looping trails along which new reinforcements join the battle lines while others among their forces retreat with captured prey, mostly other social insects and arthropods.

The trails in the wake of the swarm front form larger and larger but fewer and fewer loops progressively back from the front. This delta-shaped area of anastamosing columns, between the principal trail of the raid and the swarm front, has been termed the *fan* by Schneirla (1940). About 10 m or so back from the killing field of the raid a last single loop of trails leads into the one principal trail—the main highway for the ants—which permanently links the foraging ants to their bivouac, the location of the queen, brood, and brood-tending adult ants. The complex structure of the swarm raid is dynamic yet stable. Through 10 hours of raiding in a single day the raid system can grow to a length of 150 m, but its basic design remains essentially the same (Franks 1989). Also, each *E. burchelli* raid maintains the same general compass bearing throughout each day as it proceeds for 100 m or more through the depths of the rain forest (Franks and Fletcher 1983).

Each army ant swarm raid, with its principal trail, its fan of anastamosing columns, and its swarm front is created entirely anew each day. How this structure is built is the concern of this chapter.

Adaptive Significance of Army Ant Raids

As indicated by the quotation at the start of this chapter, Schneirla clearly saw the swarm raid as an adaptive phenomenon at the group level. Group raiding enables army ants to capture prey unavailable to solitary-hunting ants (Chadab and Rettenmeyer 1975). Army ants typically attack large arthropods (such as crickets, cockroaches, and scorpions) and social insect colonies (such as wasp nests and other ant colonies) that are either too large and powerful or too numerous to be overcome by small numbers of foragers.

Recently, Franks and Partridge (1993), invoking classical mathematical models from human warfare, suggested that very large numbers of army ant workers raiding in concert minimize their own casualties by concentrating their attack. In this way, their prey—which are almost always stronger *per capita*—can be quickly overcome before they kill or injure many of the army ant workers. This is essential because *E. burchelli* colonies, for example, would dwindle and vanish if more than 1 or 2 percent of the foragers in a swarm raid were lost during each day's raid (Franks 1985). Franks and Partridge (1993) suggest that to minimize the number of casualties it is likely that army ant colonies have evolved to possess immense numbers of workers even though many of them are rather small and relatively weak. This hypothesis, that concentrated numbers are of paramount importance, has deep implications for army ant evolution because it suggests that army ants rely on having very large numbers of workers instantly available wherever a prey item is discovered. This would explain the adaptive significance of a high density of army ant workers across the whole of the swarm front.

The army ant colony as a whole is the vehicle on which natural selection acts in all activities except sexual reproduction. *Eciton* workers are completely sterile, indeed they have no ovaries at all (Oster and Wilson 1978). Their inclusive fitness depends on their colony being able to reproduce and grow to a size at which it can split into two daughter colonies and produce males that inseminate queens in other colonies (Franks 1985; Franks and Hölldobler 1987).

The workers in an army ant raid appear completely selfless, with few limits to their apparent bravery in attacking well-defended prey. The extreme altruism of army ant workers is seen especially clearly in the majors (the largest caste of workers), who do not participate in killing prey but defend the colony against would-be predators. When an *E. burchelli* major bites into a vertebrate that gets too close to the colony, it sinks its ice-tonglike mandibles deep into the flesh of its enemy and cannot withdraw them after biting an opponent. Its mission is a suicidal one, but it sacrifices life for the good of the group, the colony. Similarly, raiding workers that die in a successful attack on a well-defended ant colony are also sacrificing themselves so that copies of their genes have a better chance of getting into the next generation, via their colony. The swarm raid is like a foraging pseudopod and is one of the key adaptive traits in the extended phenotype (see Dawkins 1982) that is the army ant colony.

Basis in Individual Behavior

In the context of this book it is particularly noteworthy that Schneirla (1940), while recognizing the complexity of the group-level phenomenon of the swarm raid, was adamant that it could only be understood by analyzing the behavior of individual workers. In this section we embrace this viewpoint and consider how individual behaviors contribute to the formation of the swarm raid. We will group these behaviors under two main headings: the behavior of ants leaving the bivouac, and the behavior of ants returning to the bivouac. Particularly important topics are the behaviors involved in trail-laying and trail-following, and the relationship between running speed and trail strength. We will also consider other topics, such as the influence of prey distribution and the dynamics of the trail pheromone.

Departure from the Bivouac

Most *E. burchelli* raids begin at dawn as increasing numbers of ants leave the bivouac and start to forage. Initially, the raid is small and amorphous, but within an hour or two it becomes the recognizable structure seen in Figure 14.2. The raid may contain between 100,000 and 200,000 ants (Willis 1967), most of which concentrate at the slow-moving swarm front. Throughout some ten

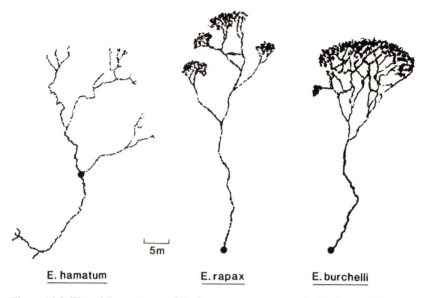

E. hamatum   **E. rapax**   **E. burchelli**

Figure 14.2 The raiding patterns of *E. hamatum* are compared with those of *E. rapax* and *E. burchelli*. (From Burton and Franks 1985)

to twelve hours of raiding a continual flux of ants leave the bivouac to join the raid while others return to deliver captured prey.

Figure 14.3 shows the development of a raid as recorded by Schneirla (1940). The movements of workers in a raid initially appear highly haphazard. But, after an hour or two, instead of apparently milling about at random, more and more ants run either from the bivouac fairly directly towards the swarm front or, when they have captured food, back down the raid system towards the bivouac. The outgoing ants have very different running speeds in different parts of the raid. They run at their highest speed in the emerging principal column that links the bivouac to the swarm front, but walk slowly and hesitantly at the swarm front.

The hundreds of thousands of army ants that eventually form an *E. burchelli* swarm raid are virtually blind; their eyes appear to possess only a single ommatidium on either side of the head. It is not surprising therefore that the army ants use other senses than sight to find their way through the raid system. Many experiments, which we will review briefly below, have shown that these army ants lay pheromone trails as they depart from the bivouac. The trails influence the movements of ants that enter the raid later and eventually enable all the surviving army ants to find their way back to the bivouac.

Schneirla (1940) clearly showed that chemical (pheromone) trail substances influence the flow of traffic through a raid system. It is possible to divert traf-

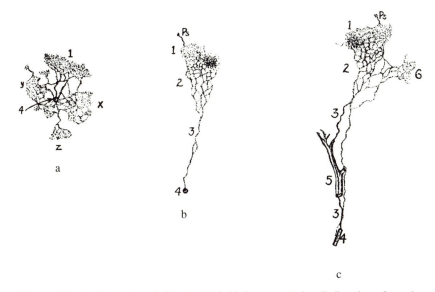

Figure 14.3 An *Eciton burchelli* raid (a) initially expands in all directions from the central bivouac site. The picture corresponds to 1.5 h after the start of raiding. Also shown are an *E. burchelli* raiding system (b) at an intermediate stage between (a) and (c) and an *E. burchelli* raiding system at about 11:00 A.M. after about 5 h of raiding (c). In (c) the swarm front (1) is approximately 55 m from the bivouac (4) and about 12 m wide. An auxiliary swarm (6) was produced by secondary division of the main body shortly after 10:00 A.M. The swarm raided through a fallen tree trunk (5). (All from Figure 1 in Schneirla 1940)

fic in an army ant raid by transferring part of the substrate, such as a leaf over which they have been running, to another part of the raid system. The ants precisely follow the path taken previously by their nestmates across such a leaf. An unmarked leaf, used as an experimental control, simply causes a disturbance when transferred into an army ant raid system and there is no systematic trail following. The entire area over which the swarm passes has greater or lesser amounts of trail pheromone depending on how many individuals have passed over it, how much pheromone they have laid, and how rapidly the pheromone evaporates. We will consider all of these factors in the following sections of this chapter. First, however, we will consider the behavior of individuals as they move into virgin terrain ahead of the swarm. Such ground initially bears no trail pheromones and the ants react characteristically to such virgin terrain.

Schneirla (1940) noted that individual workers make slow and hesitant progress into virgin terrain at the swarm front and after a few centimeters exhibit a characteristic "rebound pattern," in which they quickly return to their nestmates in the rest of the slowly progressing swarm. This rebound pattern probably occurs because the ants have a threshold distance for venturing into terrain that bears no trail pheromones, and at that distance they turn around and follow the trail they themselves have laid back towards the rest of their nest mates in the dense swarm front. Such rebound patterns and the slow movement of ants at the swarm front mean that the forward movement of the swarm front itself is slow compared to the running speeds of individual ants. Indeed, the swarm front proceeds an average of only 0.2 m/min (Willis 1967), whereas individual ants often run at an average rate of about 3 m/min in the principal trail behind the swarm front (Franks 1985, 1986). As we will see below, workers run at different speeds partly as a function of the strength of the pheromone trails they encounter. Before we explore the details of experiments showing relationships between running speed and trail pheromone concentrations, we will first consider the number of individuals involved in *E. burchelli* raids and their rate of departure from the bivouac.

The traffic flow along the principal trail of a swarm raid changes very slowly over the day (Franks et al. 1991). It seems that incoming foragers do not strongly recruit others at the bivouac to join the raid. In addition, there may be a form of negative feedback controlling traffic flow in the raid. Schneirla (1933, 1938, 1971) suggested that forager activity may be stimulated by a discrepancy between the amount of incoming prey and the colony's own energy requirements, especially that of the colony's voracious brood within the bivouac. Topoff and Mirenda (1980a,b) have verified this idea for the army ant, *Neivamyrmex*. The greater the number of prey-laden foragers returning to the nest, the greater the influx of prey and the *fewer* the number of foragers that leave the nest.

Trail-Laying

Little is known about the trail-laying behavior of army ants during their outbound phase of movement, before finding prey. Outbound ants probably continually deposit small amounts of pheromone and deposit larger amounts of pheromone as they return to the bivouac loaded with prey. During the course of a typical raid, large numbers of *E. burchelli* workers return to the bivouac carrying the prey they have killed and dismembered. The presence of prey and their capture has a major influence on the behavior of individual army ants and hence the structure of the swarm raid. Schneirla's (1940) suggestion that army ants that are returning with prey may lay particularly attractive recruitment trails was confirmed decades later by Chadab and Rettenmeyer (1975). They showed that *Eciton hamatum* workers discovering prey can lay highly influential recruitment trails between the place where prey are discovered and the massed numbers of their nest mates either in front of the raid or on established trails. One *E. hamatum* worker can recruit fifty to a hundred nestmates within the first minute of its trail crossing their path. There are no comparable data for *E. burchelli* workers.

Schneirla's (1940) observations suggest that *E. burchelli* foragers retrieving prey lay highly influential trails at the start of their return journey through the raid system. However, it appears that extremely strong trail-laying by returning ants only occurs between sites of prey discovery and the rest of the raiding swarm, or at most in the first few meters or few minutes of the return journey.

Running Speeds

Measurements have shown that the average running speed of *E. burchelli* workers on the principal trail is about 3 m/min (Franks 1985). The running speed of individual workers is not constant, however, but depends on the strength of the pheromone trail (see below), as well as the caste to which workers belong, and whether they are carrying prey.

E. burchelli have four distinct worker castes. In order of decreasing size, they are majors, submajors, medium workers, and minims. We have already touched on the specialized defensive role of the majors with their massive ice-tong like jaws. These ants (which comprise only 1 percent of the worker population) never carry prey and, loaded down by their huge heads and clumsy jaws, they move slowly. They are present in low density at the swarm front but sometimes appear numerous at trail junctions in the fan of the raid. The submajors have the longest legs in proportion to their body size of any *E. burchelli* workers and constitute a high-speed, road-haulage caste. Submajors are only 3 percent of the worker population but make up 26 percent of all of the workers observed carrying prey (Franks 1985, 1986). Submajors are also able to carry disproportionately large prey. Medium workers are the ubiquitous jacks-of-all-trades in the colony, able to function as both foragers and nurses. Minim workers are more common in the bivouac of an army ant colony than in its raid system and may be somewhat specialized as nurses. More than 97 percent of the ants in a raid system are members of the two smallest castes, the medium and minim workers. When unburdened with prey, the submajors run more quickly than the · medium workers that, in turn, run more quickly than the minims. When carrying prey, all three of these castes run at a slower and almost identical speed, the standard prey-retrieval speed (Franks 1985, 1986). This can be explained in terms of the larger ants carrying disproportionately heavier prey items (Franks 1986).

Evidence that an individual's running speed depends not only on which caste it belongs to (and whether it is burdened by prey), but also on the strength of the pheromone trail it is following, comes both from the experiments of Schneirla (1940) and recent experiments using circular mills (Figure 14.4a) (Franks et al. 1991). Schneirla (1940, p. 440) described the first controlled experiments

a

Figure 14.4 A circular mill of army ants is made to march endlessly in circles by placing a dish in their midst (a) (from Farb 1962). The average initial velocity (m/s) of
E. burchelli workers in an artificial circular mill is plotted as a function of the time (b)
after the mill was last saturated with trail pheromones. The average velocity falls exponentially as a function of the age of the trail: ln(velocity) $= -2.77 - 0.005/$age (min).
(From Franks et al. 1991)

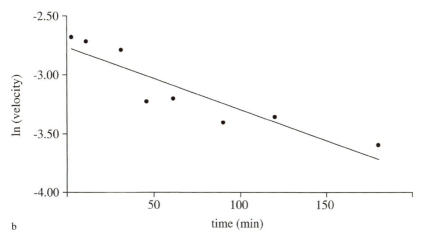

b

on *Eciton* foraging which strongly suggested the importance of quantitative variation in trail strength.

Circular mills have been used to establish quantitatively the relationship between trail strength and running velocity (Franks et al. 1991). An artificial mill-arena was used for this bioassay. A running track 2 cm wide at its base and 0.75 m in circumference was used to measure the running velocity of the ants as function of the amount of traffic flow. Two types of experiments were performed. In the first, groups of forty *E. burchelli* workers were introduced to the mill and their running behavior was recorded with a video camera. The running speeds of a particular physical caste of the workforce, the submajors (that are speedier than their nestmates [Franks 1985, 1986]), were noted 5 min after the start and then 20–30 min after in four separate mill experiments. In each of these time intervals, the time taken for ten submajors each to complete a full lap of the mill was noted. The running speeds of such submajors increased by about 20 percent from the first 5 min of the circular mill to when it had been in operation for 20–30 min. This suggests that as more pheromone is laid down by the ants, they respond by increasing their running speed.

In the second set of experiments, a group of forty army ants was given at least 30 min to form and saturate a circular-mill trail. These workers then were taken out and different groups of forty workers from the same colony were introduced to the mill after intervals ranging from 1 to 840 min, and their running speeds were noted. In this procedure, all ants were rapidly reintroduced to the trough at a point and the time recorded for the ants traveling clockwise to meet those running counterclockwise. This method was used so that the behavior of a group of ants could be assessed as they ran over an old trail that they themselves had not influenced. Their meeting point was almost always diametrically opposite to the reintroduction point. Hence, the average speed of the group of ants traveling over a trail for the first time since it was saturated could be calculated from the average distance traveled (half the circumference of the trough) divided by the time taken for the mill of ants to reform. For the studied intervals, the speed of the reintroduced ants declined exponentially as such trails decayed over time. The half-life of the ants' running speed as a function of trail age was between 2.0 and 2.5 h (Figure 14.4b). We term this the *trail-response half-life* since it denotes a change in the ants' behavior rather than direct information on the amount of pheromone present.

Choosing the Left or Right Trail

Not only does the strength of a trail influence the running speed of ants but the relative strength of two intersecting trails seems to influence the choice of which trail the ants take (as described in detail in Chapter 13). Although little experimental work has been done showing that army ants select the most heavily marked branches of a trail, Franks et al. (1991) examined the behavior of individual ants at natural Y-junctions in the fan of a swarm raid. In a first series of experiments, video recordings were made by filming branching points for extended periods of time. The Y-junctions filmed were those where the trail of ants moving outwards branched symmetrically; or, the two branches reunited symmetrically. In other words, naturally looping trails that were filmed formed both a Y-junction for outbound ants moving toward the raid front and a similar Y-junction for inbound ants moving toward the nest. Suitable diamond-shaped loops are common in *E. burchelli* raids and form a natural analogue to the double-bridge experiments that Deneubourg et al. (1990) used to examine the ability of other ants to choose shortcuts (see Chapter 13).

The video recordings of *E. burchelli* diamond loops were analyzed by counting the number of ants running outwards or inwards along both branches every minute. In this procedure, the left and right turns of tens of thousands of army ants were recorded. Thus, the number of ants choosing to take the left or right branch at a Y-junction could be determined as a function of the numbers that had earlier made such decisions on their way out to the swarm front, and had previously returned along these trails, presumably laying down pheromone in the process.

The results were that ants preferentially chose the branch of a Y-junction over which the most traffic had passed. As in the case of the choice experiments on branching trails in Chapter 13, the traffic along a bifurcating branch was generally quite asymmetrical (Figure 14.5).

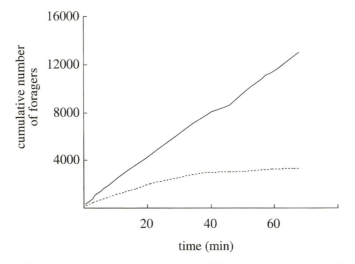

Figure 14.5 Analyses of traffic flow at a natural Y-junction in an *E. burchelli* swarm raid. The cumulative numbers of all foragers, includes all traffic on the winning (solid line) and losing (dotted line) trails over time. (From Franks et al. 1991)

Prey Recovery After a Raid

Franks (1982) estimated that an average of about 30 prey/m² (of all types) are present in the leaf litter through which *E. burchelli* colonies raid, and that a colony can retrieve 30,000 or more prey items in a single day's raid. Fifty percent of the total weight of prey retrieved are social insects, such as ants and wasps mainly captured in their nests. The other 50 percent consists of more homogeneously distributed larger arthropods, such as crickets, cockroaches, and spiders. Populations of large arthropods recover relatively rapidly, in about seven days, after a raid simply by walking back into a raided area from the surrounding areas. However, the social-insect prey take much longer to recover from a raid, about 200 days to regain their former abundance, since they largely grow *in situ* (see Franks and Bossert 1983). It can therefore be safely assumed that any recovery of prey in an area raided by an *E. burchelli* colony during the same day is trivial.

Self-Organization of Army Ant Raids

In the above account we have concentrated on the behavior of individuals within the *Eciton* swarm raid. Now, we will begin to consider how the swarm raid is organized by the individual behavior of the tens of thousands of participants. As discussed in Chapter 4, we have considered several alternatives to self-organized pattern formation, including leadership, blueprints, templates and recipes.

Schneirla (1940, p. 448) was convinced that *E. burchelli* raids do not arise through predetermined plans or any particular workers acting as leaders:

> We find the *Eciton* worker population a homogenous body so far as the social contributions of workers in the raiding situation are concerned: all workers are subject to the conditions of the rebound pattern and consequently all make similar contributions to group function under equivalent conditions.... The "leader-ship" of the sections and the factors "controlling" the entire system are at all stages functions of collective behavior.... The behavioral adaptiveness is... not [attributable] to any ontogenetic factors even remotely approaching "ingenuity" or "planning" in the foraging workers themselves.

We agree with Schneirla's conclusions that a special caste of leaders does not appear to exist in *Eciton*. It is important, at this point, to also consider the other possible alternatives to self-organization and understand why they are not reasonable. First, building the raid by means of repeated consultation of a predetermined blueprint or plan would require prior surveys, in this case of the terrain over which the raid would occur. Organization based on blueprints requires some global knowledge of the overall foraging activity even as it is emerging. It is extremely unlikely, that any one of the millions of totally blind workers in a *Dorylus* raid, for example, has global knowledge of the huge foraging pattern of which it is just a minuscule part.

Not only is there no evidence for any long distance scouting, or survey activity, independent of the direct progress of the raid itself, but such activity is unlikely to evolve. Scouting would require the movement of isolated army ants into hostile terrain where they are likely to be killed by their aggressive prey, which are almost always much stronger *per capita* than the army ants. The complex, capricious, and dangerous environment through which army ants blindly raid would seem to dismiss the possibility that they use any form of predetermined plan.

Templates for the organization of the raid as a whole can also be ruled out. It is clear from observations that existing structures in the environment through which army ants raid have some influence on the local geometry of the raid. For example, principal trails often run along fallen tree trunks (see Figure 14.3c). However, army ants use such paths of least resistance only when the front of the swarm raid has passed purely by coincidence along the length of such a fallen tree trunk. The ants may use this path simply because they are able to run most quickly along it. More traffic passes along easier paths per unit time and such paths get stronger pheromone trails and thus prevail over more difficult routes. The use of such natural pathways probably only indicates the army ants' ability to take advantage of easy paths that occur randomly along the direction already taken by the raid system. The stability of the geometry of the swarm pattern by itself suggests such physical templates in the environment can serve only a secondary, local role.

Clearly, army ant raids are not predetermined but are event-driven systems in which past foraging traffic directs future foraging traffic through the influence of trail pheromones. External conditions, including prey distribution and key unpredictable events such as prey capture seem to strongly influence the structure of a raid. This is because ants returning with prey from particularly rich foraging grounds lay trails that are initially strong and exert a major influence on the future course of the raid.

Schneirla (1940) appeared to have an intuitive understanding of emergent properties in complex, multicomponent systems: "Any consideration of the behavior of the isolated individual, however detailed, could not have prepared us for analysis of the general phenomenon."

A Self-Organization Model

We will now review a computer model that explains how the complex structure of an *E. burchelli* raid is generated by thousands of workers. The model shows how the characteristic patterns of army ant raids could be self-organizing, generated from large numbers of interactions among many identical foragers, each with simple trail-laying and trail-following behavior (Deneubourg et al. 1989). After reviewing the model and its predictions, we will examine the field experiments designed to test those predictions.

Basic Mechanisms

The space over which army ants move is represented in the model by a lattice converging toward the bivouac-nest (Figure 14.6). The lattice presents a sequence of binary choices to individual ants. In the simulation of the model we will consider the distance between two nodes in the lattice to be 5 cm. Time is treated as a series of discrete, four second steps, with each ant having a specified probability of moving from one node to the next at each time step. Ants are assumed to exhibit trail-following and trail-laying behavior; to change their speed in response to trail-pheromone concentration; and to return to the nest when they capture prey.

Other components of the model take into account physical constraints and initial conditions, such as the distribution of food in the environment. We present the components of the model by following, step by step, what an individual ant does, from leaving the bivouac to returning to it with a prey item.

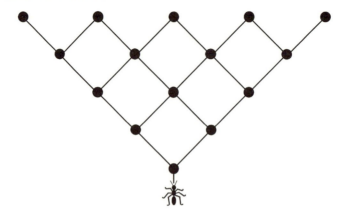

Figure 14.6 The representation of the space over which ants move in the simulation. The lattice represents a sequence of binary choices to each ant. The distance between two lattice nodes is 5 cm.

Departure from the bivouac

Franks et al. (1991) found that the number of ants leaving the bivouac per unit time was relatively constant. Here we assume that two to ten ants leave the bivouac every second. This is based on an estimate that twenty ants leave the bivouac every second during an emigration (Franks 1985). During foraging, the departure rate from the bivouac is certainly lower.

Trail-Laying

We assume for simplicity that each ant moving from the bivouac toward the swarm front lays one unit of pheromone in each time-step. The trail-laying of these outgoing ants is inhibited if the quantity of pheromone is greater than a saturation value (S_1). This response to saturation seems reasonable because ants moving quickly, on a strong trail may have less opportunity to lay trail than do ants moving slowly. Also, empirical observations have suggested an explicit inhibition of trail-laying behavior on a strong trail, in other ant species (Aron and Deneubourg, unpublished). A further reason to expect inhibition is that ants presumably have little need to continue laying pheromone on an already well established trail.

A returning ant lays q units of pheromone per step moved (rather than one unit for the outgoing ants). In the absence of any quantitative data, we have assumed that $q = 10$. We also assume that returning ants stop depositing pheromone when the quantity of pheromone is greater than a saturation value, S_2.

We suspect that *E. burchelli* workers have different thresholds for ceasing to lay trails, depending on whether they are departing towards the swarm front or returning to the bivouac. If returning ants stop laying trails at a lower level of trail pheromone, this might explain why the ants lay a strong trail between sites of prey discovery and their nest mates at the swarm front (Schneirla 1940; Chadab and Rettenmeyer 1975), whereas trail-laying in the fan of the raid seems to be no greater than that of outgoing ants in this part of the foraging system. An explanation for this difference may be that ants returning with prey lay strong trails only through virgin terrain or the first couple of meters or minutes of established trails during their incoming journey, and are inhibited by well-established trails further back in the raid system.

Running Speed as Modulated by the Strength of the Pheromone Trail

In the model, again for simplicity, we assume that all ants are identical. This is a reasonable approximation given that 97 percent of the ants in a raid are minim and medium workers and that the latter predominate (Franks 1985).

Based on the empirical observation that running speed is a function of trail pheromone concentration, we have used the following mathematical relation for the probability P_M that forager, M, will move at each step:

$$P_M = \frac{1}{1 + e^{-(C_L + C_R)/C^*}}, \tag{14.1}$$

C_L and C_R are the trail-pheromone concentrations on the left and the right branch, respectively, and C^* is the concentration of trail pheromone for which there is a 50 percent probability of moving at each step. This functional form has been chosen since it incorporates the known biology of army ants and corresponds to one of the classical relationships for a stimulus-response reaction. We know there is an upper limit to an ant's velocity, that its speed is a function of trail-pheromone concentration, and that this function is nonlinear. At each step, an ant decides whether or not to move forward by evaluating the total amount of pheromone ahead of it $(C_L + C_R)$.

In the model, C^* values of approximately 40 have been used, meaning that 40 passages by trail-laying ants are needed to reach 50 percent of the maximum speed. This number seems reasonable since 40 ants in circular-mill experiments can produce a strongly marked trail after only a few revolutions. The maximum speed of ants along a well-marked principal trail during an emigration is 300 cm/min (Franks 1985). In a typical foraging raid, however, the speed of the ants is much less, since the trail is not as well marked, and ants occasionally collide with one another moving in opposite directions. We assume therefore that the maximum speed during foraging is approximately 75 cm/min. If the distance between two nodes is 5 cm, one time-step of the simulation corresponds to 4 s.

Probability of Going Left or Right

Once an ant decides to move (equation 14.1) it can move to the left or right with probability P_L or P_R, depending on the amounts of pheromone on the two branches, left and right. Equation (14.2) relates the probability of going left to the relative and absolute concentrations of pheromone on the left and right branches:

$$P_L = (k + C_L)^n / [(k + C_L)^n + (k + C_R)^n].$$
(14.2)

This is a simple function with two parameters for tuning to the behavioral and physiological characteristics of army ants. The parameter, n, specifies the degree of nonlinearity in the choice. A large value of n means that if one branch has slightly more pheromone than the other, the next ant that passes will have a high probability of choosing that branch. The parameter, k, specifies the attractiveness of the unmarked branch. The higher the value of k, the greater the marking *difference* needed to prevent an ant from choosing one path or the other by pure chance.

In the absence of trail-laying, the movement of the ants to the left and right is random, resulting in a simple binomial distribution of ants over the lattice. For example, if the ants were moving randomly outwards over the lattice their proportions across the second row of the lattice would be 1/4, 1/2, and 1/4. On row 3 of the lattice their proportions would be 1/8, 3/8, 3/8, 1/8.

In the simulations, parameter values of $n = 2$ and $k = 5$ were selected. These values are justified, in the absence of quantitative data for army ants, by fitting equation (14.2) to data from other ants. A detailed discussion of this type of trail-choice mechanism was presented in Chapter 13.

Prey Capture and Rate-of-Return

An ant that arrives at a node containing food items takes an item and returns toward the nest. The process of prey capture continues until all the food items at the nodes are exhausted. The simulations assume an initial distribution of food particles at time zero, before the swarm starts to raid, and that food is depleted only by the ants' foraging activities. The rate of army ant foraging in nature is very high compared to the rate of prey population growth or recovery in the foraging area, so we can neglect the dynamics of the prey populations except those caused by the army ants. Two parameters characterize the food distribution in the model: the fraction of nodes (x) containing prey, and the number of prey items (p) at a node. The product px corresponds to the mean density of prey per node. Since a node corresponds to 25 cm^2, the density/m^2 equals 400 px.) For simplicity, we assume that all nodes containing prey contain the same number of prey items. In the case of the general predator, *E. burchelli*, we assume that prey are distributed singly and homogeneously in the foraging area, so that ($p = 1$ and $x = 0.5$, meaning half the nodes are occupied with single prey items). In the case of a more specialized predator of social insects, such as *Eciton rapax* (see below), prey are highly clustered and therefore p is large and x is small. Estimates of prey density by Franks (1982) give a mean value around 30 prey/m^2. In the simulation we tested values of prey density ranging from 0 prey/m^2 to 1600 prey/m^2. This corresponds in the model to a range for px of between 0 to 4.

Dynamics of Pheromone Trails

Although there is good experimental evidence that *Eciton* army ants lay trail pheromones (see for example, Billen 1992), the chemical composition and physical properties of their pheromone trails are unknown. We can only speculate at this point on the dynamics of the pheromone trails. The concentration of pheromone on a trail reflects the relative rates of the competing processes of deposition by the ants and loss through evaporation, absorption and adsorption. In our model we assume that at each point a fixed fraction, f, of the pheromone at each point is lost per time step. The reciprocal, $1/f$, corresponds to the mean lifetime of the trail pheromone. Deposition per unit time is determined by the number of outgoing (N_o) and the numbers of incoming (N_i) ants at that point and by the amount of pheromone each ant lays (one for outgoing ants and q for incoming ants). The pheromone concentration is given by:

$$\frac{dC}{dt} = N_o + qN_i - fC, \tag{14.3}$$

where C is the concentration on a particular branch.

N_o and N_i, at each time step, are computed based on the number of ants at the neighboring nodes at the previous time step and the probabilities of the ants moving from one node to another. The mean lifetime of the trail pheromone is chosen to be 2 min, though the model gives similar results for values over a range of 2 to 10 minutes.

Crowding

Although the tendency of army ants to follow one another's trail pheromones should lead to large buildups of ants in particular areas, there is probably an upper limit to the amount of crowding that ants will sustain in the raid system. The ants will try to follow a stronger trail in preference to a weaker one, but if the stronger trail is already crowded—particularly by traffic flowing in the opposite direction—the ants may be forced to take the weaker trail. In the simulation we assume that a maximum number of ants can occupy a node. Should an ant decide to move ahead toward the right, but find the node overcrowded, the ant will move toward the left, and vice versa. Should both points ahead be full, the ant stays where it is. This assumption is based purely on physical constraints. Since each node in the lattice corresponds to a surface area of 25 cm^2, and we estimate that the area occupied by a single ant is slightly greater than 1 cm^2, then a node may contain a maximum of 25 ants.

Summary of the Model

At the level of individual ants the model works in the following way. An ant arriving in an empty zone (with no existing marks) chooses randomly the left or right branch. Having chosen a branch, the ant marks it and so modifies the next ant's probability of choosing the left or right branch. This autocatalytic process, previously described in Chapter 13, can lead rapidly to a preference for one branch over the other. This process, repeated at each point along the swarm's path, is the basis of how the swarm forms its trail system. The interplay between trail-laying and trail-following operates in both directions, i.e. between both ants leaving the bivouac and ants coming back to the bivouac after prey capture.

Results of the Model

Simulations with the model show that the distribution and abundance of prey is an important factor in determining the structure of the raid. Although these self-organizing swarm raids are dynamic, constantly growing and changing, they always maintain a particular pattern if they proceed through areas with similar prey distributions. Such patterns are clearly seen by the 500th step of the model, equivalent to about half an hour of real time. Thereafter, the shape of the structure remains basically the same, even as it grows.

Figure 14.7a shows a raiding pattern developing under conditions of pure exploration (with no food items available in the environment). Under these initial conditions, a diffuse front rapidly forms and advances with a single strong trail extending from the bivouac out to the front. Just beyond the front, the ground is relatively unmarked and the ants move slowly and randomly, thus accumulating and spreading out. The mean speed of the front for the first hour of raiding is approximately 10–15 m/h ($= 0.2 \text{ m/min}$), which is in agreement with empirical observations (Willis 1967). With a low prey density, the returning ants do not modify the swarm pattern seen in Figure 14.7a. With higher prey densities, however, the central trail splits into lateral trails which themselves branch out, giving a swarm pattern like the river delta (Figure 14.7b) characteristic of *E. burchelli*. As such a swarm advances, the older lateral trails are progressively abandoned while new ones are formed just behind the diffuse front. When the model is run with exactly the same parameter values but

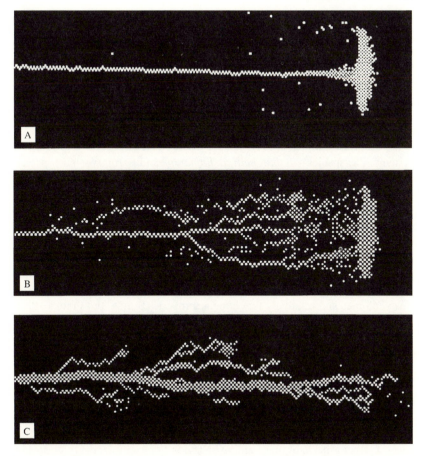

Figure 14.7 Three sets of foraging patterns developed by Monte Carlo simulations of the same model with three different food distributions: (a) without food; (b) each node has a 50 percent probability of containing one food item (as might be the case for a generalist predator like *E. burchelli*); (c) each node has a 1 percent probability of containing 400 food items, (as might be the case for an army ant like *E. rapax*, which is a specialist predator of other social insects). (From Deneubourg et al. 1989)

with prey in high-density clumps that are few and far between, the raid system consists of multiple subswarms (Figure 14.7c). This type of prey distribution and raiding pattern is typical of *E. rapax* (Figure 14.2), a close relative of *E. burchelli* that is more of a specialist predator of social insect colonies.

Influence Prey Distributions

An important prediction of the model is that changes in food distribution can generate different patterns of army ant raiding even if no change has occurred in the workers' behavioral and physiological parameters. The actors and their scripts are the same, but the collective play they perform can be very different when they are in a different theater. The model predicts that an army ant colony encountering different spatial distributions of prey will raid in different spatial patterns. This prediction leads to a novel interpretation of the different raid patterns seen in different species of army ants and suggests that a critical test of the model would be to determine whether raiding patterns change when prey distribution is experimentally manipulated.

In Peru, for example, three *Eciton* species produce large, above-ground raid systems. The species—*E. burchelli*, *E. hamatum*, and *E. rapax*—have species-characteristic raid patterns (Figure 14.2) as well as different prey preferences (Burton and Franks 1985). The model suggests that these species-specific patterns may result from differences in prey distribution for the three species rather than differences in the ants' behaviors and communication systems. *E. hamatum* attacks wasp colonies that are few and far between (Rettenmeyer 1963) but represent large packets of prey. *E. burchelli* eats both solitary arthropods (common but small packets of prey) and colonies of certain ant and wasp species (Franks 1982). *E. rapax* has an intermediate diet, attacking ant and wasp colonies and relatively few large arthropods (Burton and Franks 1985). Corresponding to each different diet, each species has a characteristic raiding pattern: *E. hamatum* has a dendritic raiding pattern, *E. burchelli* a large cohesive swarm raid, and *E. rapax* a raid system with many small swarm systems.

The prediction that these species-specific raiding patterns are in part generated by different species-specific prey distributions has potentially important evolutionary implications that we will discuss below. However, interspecific comparisons do not provide a strong test of the predicted importance of prey distributions because different army ant species may have significant differences in other aspects of their raiding behavior. For example, the three species have different colony sizes and worker running speeds and, possibly, different properties of their pheromone trails (Schneirla 1971; Burton and Franks 1985).

A much stronger test of the model's prediction that prey distributions are a major influence on the structure of army ant raids is to determine whether changing the prey distribution changes the raid structure in a single species. This is a strong test because one can assume that the behavior and physiology of the ants are the same regardless of the form of prey distribution.

Experimental Test

If indeed the prey distribution affects the structure of the raid, then it should be possible to manipulate an *E. burchelli* raid to split into subswarms simply by changing the prey distribution. Changing from a homogeneous distribution of prey to one with large, scattered clumps of prey should mean that return traffic is no longer initiated homogeneously across the swarm front but occurs intensively from a few dispersed sites. Thus, return traffic should be along a small number of divergent, heavily used and heavily marked columns that cause outbound traffic also to diverge. The swarm front should fragment into subswarms. This prediction of the model has been tested recently in the field (Franks et al. 1991).

Franks et al. performed their experiments, in which prey distributions in front of the raid systems of *E. burchelli* colonies were manipulated on Barro Colorado Island, Panama. Two methods were used to manipulate the food distribution encountered by *E. burchelli* swarm raids. First, since their prey are generally scattered, placing very large packets of immobilized arthropods ahead of the swarm should swamp the raid so that almost all prey would come from these packets during a considerable time after packets are discovered. Second, since their arthropod prey are mostly confined to leaf litter, raking away and redistributing the leaf litter to leave discrete bare patches that are relatively prey-free should make the prey distribution of the army ants much more heterogeneous.

Two large bags of prey were introduced in front of a swarm raid by using plastic mesh bags, each of which was filled with 100 dead crickets. The crickets were killed and preserved by freezing, and thawed shortly before being placed before the army ants. Each bag had the same fresh weight of crickets. In each experiment the total fresh weight of crickets introduced was between 50–60 g, or approximately one-fifth of a colony's total daily intake (Franks 1982).

The typical result of the experiment was that a large number of ants returned from each prey packet, presumably laying large amounts of trail pheromone. Thereafter, the original cohesive swarm front broke into smaller subswarms that diverged from the sites where prey had been introduced (Figure 14.8).

In the second set of experiments, a large garden rake was used to redistribute the leaf litter and the prey it contained in the swarm's path. In certain of these experiments, five patches of leaf litter were raked away to leave clear patches in a symmetrical distribution ahead of the swarm front (Figure 14.9). In this way, the prey were clumped into the deep piles of leaf litter that were raked up between the cleared patches. The large arthropod prey of these army ants usually hide under piles of leaf litter and avoid bare patches of soil where they are particularly conspicuous to birds that accompany army ant raids. These birds eat the large arthropods that army ants flush from the leaf litter (Willis 1967). Probably for this reason the prey remained in the deep piles of leaf litter into which they were raked.

On other occasions the leaf litter was raked away in a continuous band ahead of an army ant raid. These manipulations provided a control for the effects of the raking. If the continuous, prey-free band was narrower than the swarm raid, the sides of the swarm raid continued through the peripheral prey-rich areas and two lateral subswarms were produced. When the cleared prey-free area was broader than the swarm front, the swarm moved across it, without

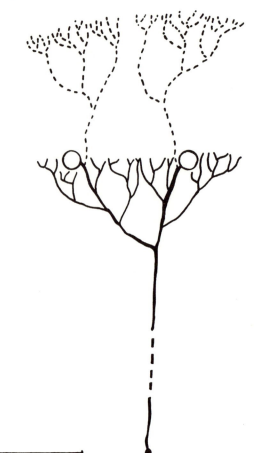

Figure 14.8 The progress of an *E. burchelli* swarm raid before (solid lines) and after (dotted lines) introduction of crickets (the two large circles). The scale bar represents 10 m. (From Franks et al. 1991)

breaking up. As the swarm crossed the bare patch of ground, however, there was momentarily little or no return traffic because there were no prey to be retrieved (Franks et al. 1991). Such a raid had a somewhat intermediate pattern between that shown in Figure 14.7a, for a raid in the absence of prey, and the normal pattern shown in Figures 14.2, and 14.7b. In addition, an *E. burchelli* raid was once observed crossing an extensive leaf-litter-free and prey-free area atop the nest of an immense colony of leaf-cutting ants. (Such ants are not preyed on by *E. burchelli*.) This swarm raid closely resembled the prey-free foraging pattern predicted by the model. The raid became narrower, with a

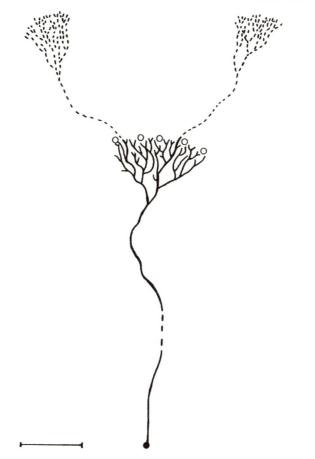

Figure 14.9 The progress of an *E. burchelli* swarm raid before (solid lines) and after (dotted lines) the clearance of five patches of leaf litter (represented by small circles). The scale bar represents 10 m. (From Franks et al. 1991)

reduced number of trails behind the swarm front, probably because there were no prey retrieved for a considerable period (Franks unpublished observations).

A Critique

The model shows rapid development of a stable pattern responsive to environmental factors such as differences in prey distribution. A drawback of the model is that the fit between its predicted pattern and the actual pattern observed in the field is only qualitative. There are some significant quantitative differences between the predicted patterns and those created by real army ants. For example, the anastamosing trails behind the swarm front are typically much longer in the model than in reality (see Figure 14.2 and Figure 14.3 and compare them to Figure 14.7b). Apparently the heavily used trails in the model last much longer than they do in nature. At least two possible explanations are that the pattern generated by the model is sensitive to the geometry of the substrate lattice, and in the model only ants with prey return to the bivouac, whereas in the field about half of the returning traffic consists of ants returning *without* prey (Franks et al. 1991). Ants returning without prey might begin their return at any point in the raid system and might have a large influence on the pattern of the raid, depending on how much trail pheromone they lay. Systematic investigation of these parameters in both the model and the field is likely to prove illuminating.

For simplicity, the model does not consider such factors as the running speeds of the different castes and their different speeds with and without prey. Also, the model assumes that all ants are identical.

It is important to recognize that changing the prey distribution in the model can switch the raiding patterns from one typical of *E. burchelli* to one resembling the pattern of *E. rapax* (Figure 14.2). However, the extremely dendritic raid pattern of *E. hamatum* has not been seen in the results of the modeling. This may be instructive. *E. hamatum* is a wasp specialist and climbs trees to discover the nests of such prey (Rettenmeyer 1963). Hence the raid pattern of *E. hamatum* occurs to a large extent over the natural, linear pathways provided by the tree branches it invades. In contrast, *E. burchelli* and *E. rapax* limit their raids much more to the two-dimensional surface of the ground. A different lattice may need to be used to represent the space that *E. hamatum* explores. The evident failure of the model to produce the patterns of *E. hamatum* draws attention to some important differences in the environment through which these colonies raid and to how the environment may structure their behavior.

Two other predictions of the model remain to be experimentally tested. One is that small army ant colonies cannot stage successful swarm raids. This prediction derives from the inability of small colonies to maintain a sufficiently high flux of ants into the raid. This might explain why large army ant colonies reproduce by binary fission, splitting into two parts (Franks 1985; Franks and Hölldobler 1987) each of which still has a large enough worker population so it can raid effectively (Deneubourg et al. 1989). If they started as small colonies, these army ants would not be able to use the same foraging system. Colony size and the associated sustainable rate of departure of ants from the bivouac is evidently a key parameter that natural selection tuned in this system.

The second untested prediction is that if an army ant raid hits simultaneously two large groups of prey, such as two large nests of social insects, it is likely to split into two subswarms (see Figures 14.8 and 14.9). In the field, *E. burchelli* attacks large social insect colonies in addition to relatively common and rather evenly distributed solitary invertebrates. The bigger and broader the *E. burchelli* raid, the more likely it is to hit two large social insect nests simultaneously. Large raids might then break into two parts under the influence of heavy return traffic from both large sources of prey. Intriguingly, Schneirla (1940) noted that *E. burchelli* swarm raids are widest at midday and it is at exactly this time of day that such large raids most often break into two separate swarm fronts. These natural events may occur because sometimes large raids encounter conditions similar to those in the prey redistribution experiments.

Evolutionary Implications

The simulation results support the hypothesis that differences in raiding patterns of *E. rapax*, *E. burchelli*, and *E. hamatum* may be based partly on their response to the foraging environment and not just differences in their communication systems. Of course, these results do not prove there are no differences in these communication systems. Indeed, such differences probably do exist. For example, it is highly likely that significant differences exist in the trail-laying and trail-following behavior of different *Eciton* species. Nevertheless, the model supports the suggestion that such differences may have occurred late in the evolution of the different raiding patterns.

The self-organization that appears to underlie the structure of swarm raids in army ants implies that large differences in this type of phenotypic expression may be based on small changes in the response of the organisms to their environment. For example, small changes in food preferences—including solitary arthropods in the diet in addition to social insects—even without changes in the ants' communication systems—evidently suffices to cause army ants to raid in a qualitatively different pattern.

Furthermore, phenotypic diversity can occur in these kinds of system even without prior genetic change. Imagine a primitive *Eciton* species, possibly one with a raid system like that of *E. rapax*, as a possible ancestor of *E. burchelli* and *E. hamatum*. If parts of this ancestral population migrated to areas where the only prey were in large, widely dispersed packets like social insect colonies, then it might start to raid qualitatively like *E. hamatum*, the specialist predator of neotropical social wasps, simply as a result of prey distribution. By contrast, if another sub-population moved into an area in which only large arthropods provided small but common packets of prey, then its colonies would presumably begin to raid like *E. burchelli*, the most polyphagous of neotropical army ants. It is further possible that these separate subpopulations might later change genetically as a result of selection pressures from specializing on social insects or general arthropods, and thence ignore other prey. If these now genetically distinct and reproductively isolated populations then became sympatric, they would have fundamentally different foraging patterns, thus limiting intraspecific competition between them. Such competition that did occur might not lead to exclusion but could cause selection pressure that reduces interspecific competition and further refines differences in their foraging patterns.

In summary, it seems that divergent raiding patterns among different species may be a by-product of their different ways of interacting with the environment and so may not be directly (explicitly) coded in their communication systems. For these reasons, the differences in communication systems that these species use in foraging may ultimately prove to be surprisingly small.

Self-Organization Is Not Always Adaptive

Another important finding of study is that the model shows aberrant, pathological patterns when unrealistic parameter values are assigned. Figure 14.10, for instance, shows the development of an aberrant structure when the mean pheromone lifetime is decreased from 2 min to 40 s. A result of this sharp decrease is that separate dense swarms follow in successive waves behind the initial swarm front. The reason for this aberrant raid pattern is that a short-lived trail pheromone disappears so rapidly behind the swarm front that the following ants find themselves in an area relatively free from trail pheromones. Thus they move slowly and are unable to catch up with their nestmates in the swarm front. This pattern is repeated, producing subswarms that follow one another at fairly regular intervals. This indicates that army ants would have an extremely inefficient raiding pattern, with subswarms repeatedly searching areas already depleted of prey, if they used a trail pheromone with a very short half-life.

It seems clear that changing a single parameter value to one beyond the relatively wide range of values for which the raid is stable could change what

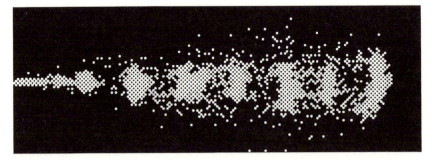

Figure 14.10 Simulation of a raid with the standard parameters for *E. burchelli*, except for short mean lifetime of trail pheromone (equal to 40 s). The raid pattern is pathological—multiple swarms following one another, repeatedly searching the same ground. Compare Figure with 14.7b, which shows the same simulation except that the mean lifetime of the pheromone is set at the standard value of 2 min.

is probably a highly efficient search pattern into an extremely inefficient one. We feel this is an important finding because it shows that self-organization does not automatically produce the patterns found in biological systems. This is contrary to the widely held belief that self-organization plays down the role of evolution by natural selection (see Chapter 7). The model actually demonstrates again that an adaptive biological pattern may only occur within a particular range of parameter values. This implies that natural selection is likely to have tuned the behavior of individual army ants to achieve these adaptive patterns.

Conclusion

Army ant swarm raids appear to have two major, probably highly adaptive attributes. First, they seem to be highly robust and insensitive to variation in the behavioral and physiological parameters of the ants. In other words, the pattern of raids is stable across a wide (but not unlimited) range of parameter values that determine the interaction among individual army ants. Second, the raid patterns are highly sensitive to an important external factor—the prey distribution. Robustness in the face of quantitative variation in individual rules of thumb used by ant workers suggests that the results of natural selection has moved these systems into a stable region in the parameter space. This is contrary to the view that natural selection favors systems at the "edge of chaos" (Kauffman 1991), where the system is highly sensitive to small changes in initial conditions or parameter values. This implies that it is the fundamental logic or algorithms that these army ants employ that leads to their robust foraging patterns. As we have seen, the logic of the mechanisms employed in self-organizing swarm raids seems to involve a relatively small set of rules. This set of rules results in a pattern that serves as an attractor of the system— a stable, robust pattern of army ant swarm raids.

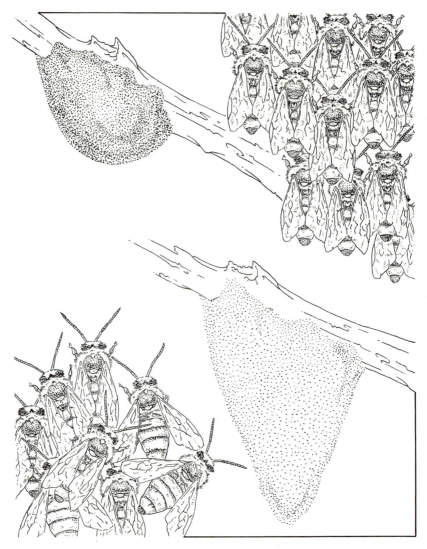

Figure 15.1 Thermoregulation by a swarm of honey bees. The upper portion shows the swarm under cold ambient temperatures: The bees are tightly clustered to prevent heat loss and may shiver to generate additional heat. The lower portion shows the cluster under warm ambient temperature: The individual bees are loosely clustered and the swarm structure is much more porous. (Illustration © Bill Ristine 1998)

15

Colony Thermoregulation in Honey Bees

"The ultimate adaptation to the physical environment is
control of the environment, and this is the level of
adaptation the social insects—social wasps, ants, social
bees, and termites—have repeatedly achieved with respect
to the thermal environment inside their nests."
—T. Seeley and B. Heinrich, *Insect Thermoregulation*

A Macroscopic View

Under a wide range of environmental conditions a cluster of honey bees
regulates its temperature (Figures 15.1, 15.2 and 15.6b) (Heinrich 1981a,b,
1985; Nagy and Stallone 1976; Southwick and Mugaas 1971). This group-level
thermoregulation occurs under two different circumstances. The first involves
swarms of bees. At certain times, a honey bee colony divides in a process called
reproductive swarming. A swarm consists of a queen and approximately half
of the workers from the parent colony. Together they depart from the hive and
form a cluster on the branch of a nearby tree. From this site scout bees search
for a new suitable nesting cavity in which to build comb and begin rearing
brood. The cluster may remain exposed for up to several days while the bees
decide which among several potential nesting sites will be selected. During
this interval the swarm precisely regulates its temperature (Figure 15.2). Over
an ambient temperature range of 1° to 25° C, the swarm's inner (core) temperature is generally maintained within a few degrees of 35° C. When the ambient
temperature is between 1° and 16° C, the outer surface (mantle) temperature
is maintained at 15° to 21° C. When ambient temperatures rise above 16° C,
the mantle temperature is not regulated but averages 2° to 3° C above ambient
temperature.

Swarming is a critical period in the life history of the colony and thermoregulation of the swarm cluster is a crucial adaptation for survival. The swarming
workers have a limited supply of honey stored in an outpouching of their gut
termed the *honey stomach*. This food is obtained prior to leaving the parent
hive when worker bees engorge themselves on the colony's honey stores. The
honey serves as fuel for the bees' metabolism that heats the swarm while it remains exposed to the environment. It also provides energy for the bees to fly to

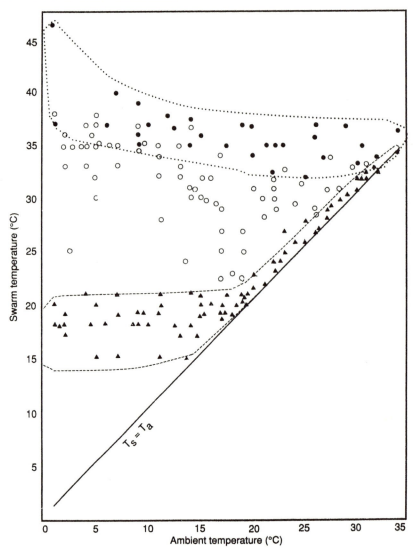

Figure 15.2 Core (circles) and mantle (triangles) temperatures of captive swarms in the laboratory as a function of ambient temperature. Open circles: swarms of 2,000–10,000 bees. Filled circles: swarms of 15,000–17,000 bees. (From Heinrich 1981b)

the new nest site, as well as an energy reserve for comb building and survival until foragers can gather sufficient nectar from the field.

Balancing a need for energy conservation is the requirement that the bees consume sufficient honey to stay warm enough to defend themselves against predators. In addition, the swarm needs to stay warm to assure that the queen is kept warm, since she is adapted to functioning in a warm, stable environment. Even when ambient temperatures are low, it appears that the bees are able to conserve energy by allowing some of the bees—those in the mantle—to remain relatively cool. This serves two purposes: to conserve energy by decreasing the honey requirements of the cooler, metabolically less active mantle bees, and to minimize heat loss from the swarm by maintaining a lower temperature difference between the cooler mantle and the air. The mantle bees need only be warm enough to move about and remain attached to the swarm. Heinrich (1981a) found that at air temperatures of less than 16° C mantle bees were unable to fly but could move about and raise their abdomens to expose their stingers to intruders. Their thoracic temperatures were regulated near 15° C. Even though mantle bees cannot immediately take off, they are able to warm themselves quickly by crawling into the interior of the swarm. Thus, the swarm as a whole conserves its limited honey supply by means of a mantle layer of cool bees that are less metabolically active than bees in the core.

Another instance of colony thermoregulation involves the winter cluster. Honey bees probably originated in the tropics but their range now extends to colder climates with harsh winters (Ruttner 1973). Rather than overwinter in a state of hibernation (diapause), as do most insects, honey bees remain relatively active throughout the winter. Their body temperatures are maintained above ambient temperature and they feed on honey the colony stores in large quantities.

Winter is a period of heavy mortality for honey bee colonies (Seeley 1978; Seeley 1985). To survive, the bees form a tight ball called a winter cluster. Over an ambient temperature range of −15° C to 10° C, the core temperature of the cluster is maintained between about 18° and 32°, and the mantle temperature between about 9° and 14° (Heinrich 1985; Simpson 1961; Southwick and Mugaas 1971).

The core and mantle temperatures of winter clusters are thus cooler than that of swarm clusters. Presumably, this is an adaptation for conserving the colony's limited store of honey. The bees on the surface of the swarm mantle pack tightly together, forming an insulative layer and they stay as cold as possible to conserve fuel. Winter bees enter a chill coma at 9°–12° C (Free and Spencer-Booth 1958), which is several degrees less than the chill coma temperature (15° C) of swarming bees (Heinrich 1981b). Bees on the mantle must maintain a body temperature above this level or they will fall from the cluster and die.

The precise thermoregulation of both the winter cluster and the swarm cluster obviously are collective adaptations to ensure colony survival over a wide range of environmental conditions. Such homeothermic behavior is well known among birds and mammals. For example, a chickadee or squirrel exposed to sub-zero winter temperatures has a number of adaptive features that enable it to stay warm. Morphological adaptations include a highly insulative layer of downy feathers or fur. An important physiological adaptation is a centralized thermostat within the brain. As arterial blood from throughout the body passes through the brain, its temperature is monitored by sensitive hypothalamic temperature receptors that act as a central thermostat. If the blood temperature falls below the thermostat's setpoint, the organism implements a series of physiological and behavioral adjustments aimed at correcting the imbalance (Brooks and Koizumi 1974). These include circulatory blood shifts that alter body surface temperatures to prevent heat loss, shivering—which generates heat by increasing metabolic rate—and rearrangements of pelage or plumage, which alter the insulative properties of the body surface (Southwick 1983).

Although the honey bee cluster has been likened to a "homeothermic superorganism" (Southwick 1983), a cluster of bees clearly lacks the kind of functional organization found in birds and mammals. A cluster of bees is not a single homeothermic organism with a brain and circulatory system. It is unlikely that the bee cluster harbors a mechanism analogous to a centralized thermostat. Furthermore, as we will discuss later, it is highly unlikely that bees gather information about the temperature from various places within the swarm and communicate this information across the swarm to other bees. Rather, the evidence suggests that thermoregulation arises through a mechanism based on the independent activities of bees responding to changes in the temperature of their immediate (local) environment. The goal of this chapter is to examine how the colony-level (macroscopic) process of thermoregulation is explained in terms of the activities of thousands of bees operating at the individual (microscopic) level, and to determine whether the process involves mechanisms based on self-organization.

A Microscopic View

The thermoregulatory responses of honey bee swarm clusters prompted Heinrich (1981b) to conduct a series of experiments examining the possible mechanisms of thermoregulation in swarm clusters. His simple yet elegant studies provide convincing evidence that swarm thermoregulation is not based on centralized control involving either leadership or the collection and dissemination of colony-wide information by a select group of bees. Instead, his conclusions accord with a mechanism based on self-organization in which each bee independently adjusts her behavior and physiology to regulate her own body temperature.

Centralized Control

Leadership by the queen is an obvious hypothesis for the coordination of activities within a colony of social insects. For example, the queen, through

Plate 1 Self-organized pattern formation in physical and chemical systems. (a) Wind-blown ripples on the surface of a sand dune. (b) Spiral waves produced by the Belousov-Zhabotinski chemical reaction. (c) Pattern of cracks produced by mud as it dries and shrinks along the shore of a pond. (d) Hexagonal pattern of Bénard convection cells created when a thin sheet of viscous oil is heated uniformly from below. A small amount of aluminum powder has been added to the oil to reveal the pattern of convection. (e) Polygonal pattern of paint cracks on a wooden surface. (f) Wrinkle pattern formed by a coat of varnish on a wooden surface. (a, © 1994 Bob Barber/ColorBytes; b, courtesy Stefan C. Müller; c–f, © 2000 Scott Camazine)

Plate 2 Animal coat patterns and insect coloration believed to involve self-organized pattern formation. (a) Zebra, *Equus grevii*. (b) Giraffe, *Giraffa* sp. (c) Tiger, *Felis tigris*. (d) Gila monster, *Heloderma suspectum*. (e) Rice paper or tree nymph butterfly, *Idea leuconoe*. (f) Locust borer beetle, *Megacyllene robiniae*. (All photos © 2000 Scott Camazine)

Plate 3 Self-organized pattern formation in plant and fungal systems. (a) Spiral phyllotaxis in the flower head of an oxeye daisy, *Chrysanthemum leucanthemum*. The number of interlocked clockwise and counterclockwise spirals of florets are two consecutive numbers in the Fibonnaci series. (b) A plasmodial slime mold (Myxomycota), *Physarum polycephalum*, growing on an agar plate. Compare the pattern of the plasmodial wandering stage as it searches for food with that of the advancing front of the army ant, *Eciton burchelli* (Figures 14.2 and 14.3). (c) Cross section through a head of red cabbage showing the interdigitating pattern of leaves. (d) Growth pattern of lichen on a flat stone. (e) A morel mushroom, *Morchella esculenta*. (f) Scanning electron microscope image of a Forsythia pollen grain (computer enhanced). (All photos © 2000 Scott Camazine)

Plate 4 Honey bees. (a) Queen bee surrounded by her retinue of workers, preparing to lay an egg. (b) Foraging worker, taking nectar from a flower. (c) Pollen forager, gathering pollen from a grass flower. (d) Waggle-dancing worker communicating to other workers the location of a rich nectar source. (e) Food-storer bee depositing processed nectar into a honey cell. (f) Worker bees marked with paint dabs and numbered tags allow researchers to follow the activities of individual bees and determine their behavioral stimulus-response "rules." (All photos © 2000 Scott Camazine)

Plate 5 Building behavior in ants and wasps. (a) Leaf cutter ants (*Atta* sp.) in their fungus garden. (© 2000 Scott Camazine) (b) *Leptothorax* ant nest housed between two panes of glass (40 mm × 40 mm) separated by a 1-mm-deep space. The nest is constructed of tiny grains of blue-dyed sand roughly 0.5 mm in diameter. (© 2000 Nigel R. Franks) (c) In the laboratory, the paper wasp, *Polistes dominulus*, will build its nest using fibers torn from paper of different colors. This provides a visual record of the progress of nest construction. This wasp has a ball of yellow fibers in her mandibles, and is about to add this to the existing structure. (© 2000 Guy Theraulaz) (d) Weaver ants (*Oecophylla* sp.). Workers use their larvae as "shuttles" to provide silk to sew together a nest of leaves. (e) Workers using their bodies to close the gap between leaves prior to sewing them together. (Weaver ant photos © 2000 Guy Theraulaz)

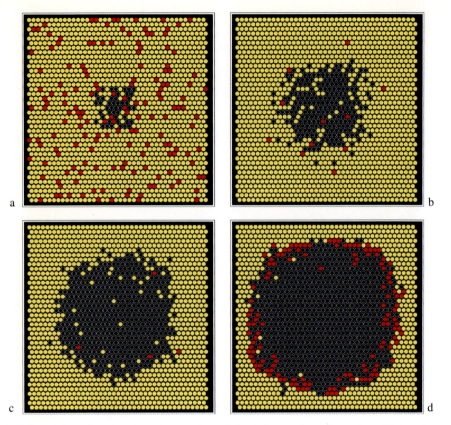

Plate 6 Simulation of pattern formation on the comb of a honey-bee colony. Brood (eggs, larvae, and pupae) are shown in black, honey in yellow, and pollen in red. In the simulation the queen starts her egg laying in the center of the frame. Initially, honey and pollen are scattered in the periphery, and a cluster of brood develops in the center (frame a). As the simulation continues, the brood area becomes more compact, and pollen gradually disappears from the periphery (frames b and c). The fully developed pattern shows a solid central region of brood, the periphery filled with honey, and the interface between these regions containing pollen (frame d). Software for simulating comb pattern formation can be downloaded at http://beelab.cas.psu.edu.

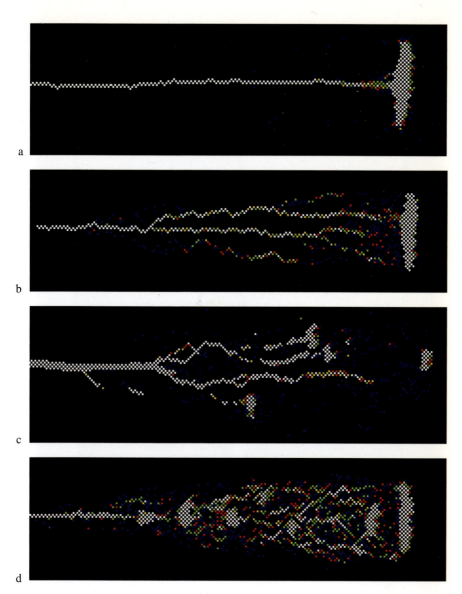

Plate 7 Monte Carlo simulations of army ant raid patterns. Color scale: white, yellow, green, red, and blue correspond to densities of ants from highest to lowest. The parameters of the simulations are identical to one another with the following exceptions: (a) No food on the substrate. (b) Each node with 50% probability of containing one food item, as might be the case for a generalist predator such as *Eciton burchelli*. (c) Each node with 1% probability of containing 400 food items, as might be the case for a predator such as *E. rapax* that specializes on other social insects. (d) Food distribution as in (b), but with a very short mean lifetime of the trail pheromone. The raid pattern is pathological, with multiple swarms following one another, repeatedly searching the same territory.

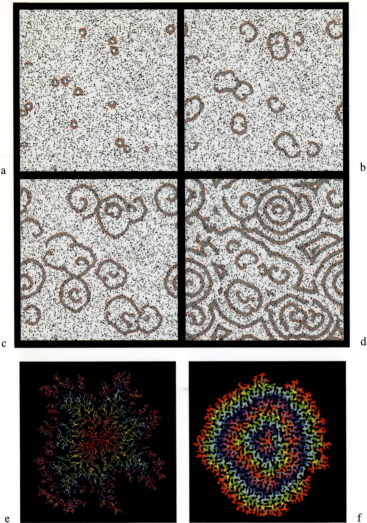

Plate 8 Simulations of self-organized processes using the programming language StarLogo, developed at the M.I.T. Media Laboratory (http://lcs.www.media.mit.edu/groups/el/projects/starlogo). (a)–(d) Simulation of spiral waves of cellular slime mold amoebae as they receive and relay cyclic AMP (cAMP). The simulation begins with a uniform distribution of amoebae seeded onto the substrate. (Empty sites are shown in black.) These starving amoebae are receptive (white) to cAMP signaling from nearby amoebae. Amoebae (red) that have sensed cAMP in the vicinity briefly emit cAMP. Then they enter a refractory state (gray) for a specified number of time steps. (e) Simulation of diffusion-limited aggregation (see Sander 1986,1987). In this process, particles diffuse towards the center of the field, collide with a neighbor and stick. (f) Another morphology develops when particles grow outward from the center by dividing at their leading, growing edge, as might be the case for a bacterial culture growing on an agar plate. (Simulations written by William Thies at MIT)

pheromonal communication, plays an crucial role in the cohesion of a swarm (reviewed by Free 1987). Heinrich (1981b), however, provided a simple, yet convincing demonstration that the presence of a queen is not necessary for swarm thermoregulation. He started with a swarm weighing 602 g that maintained the typical core and mantle temperatures of 35°–36° C and 19° C, respectively. He divided the swarm into two groups, one weighing 236 g and the other 366 g. Both groups thermoregulated like the parent cluster though only one group contained a queen.

Colony-Wide Information Gathering

Perhaps there are other, more complex mechanisms that do not involve the queen, but still involve a centralized control mechanism. As previously mentioned, vertebrates have a central thermostat where temperature information from throughout the body is gathered and monitored. In an analogous manner, Heinrich (1981b) wondered whether a special group of bees might move systematically through the swarm to gather temperature information that could be reported to a control center to where a thermoregulatory response is coordinated. This hypothesis was refuted by the simple means of creating a cluster of 8000 bees with part of the swarm segregated within a central gauze cylinder. The bees clustered within the gauze cylinder were unable to contact the swarm mantle, and the bees in the mantle could not reach the swarm core. Even so, the swarm thermoregulated normally, with a core temperature of 37° C and a mantle temperature of 20° C, at an ambient temperature of 4° C.

Using a different experimental design, Heinrich further tested the hypothesis that bees travel between the swarm core and the mantle to assess the ambient temperature or the temperature of the mantle bees. In a series of four experiments, he marked 50 bees on the surface of the swarm with spots of paint. Although some bees were disturbed by the marking procedure and moved into to center of the swarm, most of the remaining marked bees stayed in place, and remained visible in the mantle for the duration of the experiment, which lasted more than 24 hours. Heinrich also showed that at an ambient temperature of 3° C, individually marked bees tended to remain in relatively fixed positions for periods of over an hour. Clearly, no systematic movement of bees took place from the mantle to the core.

Even though swarms appear to thermoregulate without a significant exchange of bees between the core and the mantle, the possibility remained that information is transmitted acoustically throughout the cluster. If so, bees in the mantle could communicate to bees in the core. For example, mantle bees that became too cold might signal bees in the core to increase their heat production. Heinrich tested the hypothesis that thermoregulation is coordinated by colony-wide sound communication by performing a series of playback experiments. He reasoned that bees in the core may be of central importance in regulating the temperature throughout the swarm by producing heat in response to sounds produced by cold bees shivering on the mantle. To test this hypothesis, he allowed a swarm to cluster around a small speaker through which he could play taped recordings of bees. The swarm was kept at an ambient temperature of 20° C and its core temperature was 30° C. Temperatures were recorded continuously from the center of the swarm while two different recordings were played to the bees in the core: one was the sound of bees from the mantle of a swarm cluster at an ambient temperature of 3° C, and the other was the sound of bees from the core of the same swarm. No noticeable change in temperatures occured in response to any of the playback recordings. It is therefore unlikely that the bees in a swarm coordinate their thermoregulatory responses by means of sound communication.

Another possibility is that information is transmitted through the swarm chemically, not acoustically. Perhaps pheromones, carbon dioxide, or some other chemical stimulates bees in the core to produce heat when the ambient temperature falls. Heinrich disproved this hypothesis by exchanging gases from the core of one swarm with those of another. One swarm was kept in a controlled temperature room at 4° C, and maintained a core temperature of 26°–28° C. The other swarm was kept at an ambient temperature at 20° C, and its core temperature was 28°–30° C. Using a peristaltic pump, he alternately exchanged gases between the centers of the two swarms. In neither case did he observe any change in the core temperatures of the swarms.

Observations Support Self-Organization

Some activities in a honey bee colony involve sophisticated communication among colony members. The use of the waggle dance to communicate the location of a nectar source or the location of a potential nesting site is a classic example (Frisch 1967; Lindauer 1955). It is reasonable therefore to search for mechanisms of thermoregulation based on communication among colony members of global information. Thus, Heinrich (1981b) remarked "It would seem that communication could also be involved in coordinating activities between bees at the swarm core, where the highest temperatures are generated, and at the swarm mantle, the only place where bees experience the changes at the ambient thermal environment." Heinrich diligently looked for evidence of communication because the *presence* of communication is essential for centralized control (so that information can reach the controller). However, the presence of communication does not necessarily imply central control. The *absence* of communication, however, does contradict the idea of centralized control and provides support for self-organization. Heinrich's eventual conclusion was that communication among members of the cluster was *not* important for thermoregulation and that, "Many if not most of these responses can be understood, in part, in terms of the individual bees regulating their own body temperatures" (Heinrich 1985). By viewing the honey bee colony as a "homeothermic superorganism" (Southwick 1983), investigators may have been misled into looking for mechanisms directly analogous to a neural communication system linked to a centrally located thermostat.

In this chapter, we will see how a rather simple decentralized control mechanism, based upon self-organization, can account for cluster thermoregulation. What follows are some details of honey bee behavior and physiology involved in cluster thermoregulation. After that we will incorporate these observations into a model of thermoregulation based on self-organization.

We have already explained that a honey bee has a lower thermal limit for its body temperature below which it will fall from the cluster. A bee has several means of avoiding this chill coma. With a decrease in its body temperature, a bee's first response is to adjust its posture and position on the cluster. When the body temperature of the mantle bees becomes low, the bees move closer together and point their heads inward, as if trying to push headfirst into the cluster (Heinrich 1981b; Ribbands 1953; Southwick 1983). This has both local and global effects. Locally, it closes the spaces between the bees, creating a denser, thicker mantle layer (Figure 15.3). As the bees' heads and thoraces become tightly packed, the plumose thoracic hairs of adjacent bees become interlaced, providing insulation that may act like the down feathers of birds (Southwick 1983, 1985). In the headfirst-in-position, the bee's abdomen is pointing out of the swarm exposed to the cold, but low abdominal temperatures apparently are well tolerated as long as the bee's head and thorax are kept above the lower thermal limit. Honey bees appear to have a special mechanism for keeping their head and thorax warm, while at the same time preventing excessive heat loss to the environment; Their aorta is highly convoluted in the petiole region and functions as a countercurrent heat exchanger to minimize heat loss through the abdomen and keep the rest of the body warm (Heinrich 1980, 1987). If these behavioral and physiological adjustments are not sufficient to keep a mantle bee warm enough, she can also respond by shivering to raise her metabolic rate. However, since raising the metabolic rate incurs a significant cost in terms of honey consumption, shivering is probably used only as a last resort against falling into a chill coma.

From the view point of the individual bee, the local effect of these behavioral and metabolic changes is that the individual does not enter a coma and fall from the cluster. But collectively, these changes also have the important global effect of altering the temperature throughout the swarm. As bees on the mantle adjust their positions, they decrease the radius and porosity of the swarm. This severely restricts core cooling and causes a temperature rise in the center of the swarm. At high ambient temperatures, the heating of the core is not a problem. As shown in Figure 15.3, bees can open ventilation channels through the swarm. These passageways are important in allowing bees to dissipate excess heat by fanning their wings to increase heat loss by

5 °C   30 °C

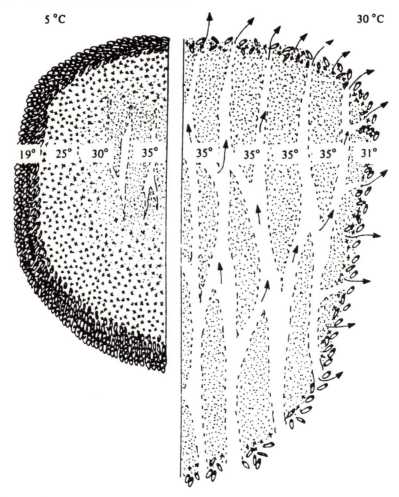

Figure 15.3 Diagrammatic illustration (Heinrich 1981b) summarizing thermoregulation of a swarm cluster at low ambient (left) and high ambient temperature (right), indicating positions of the bees, ventilation channels heat loss (arrows), areas of active (crosses) and resting (dots) metabolism, and approximate temperatures.

convection. At high ambient temperatures, Heinrich observed rapid temperature changes in the swarm interior. These temperature fluctuations were far more rapid than could be accounted for by passive cooling and suggested that heat was rising by convection through channels in the swarm (Heinrich 1981b). At low ambient temperatures, however, these ventilation channels disappear as individuals on the mantle huddle tightly together to keep warm (Figure 15.3).

According to Heinrich's (1981b) calculations, in large tightly clumped swarms, even at ambient temperatures below 5° C, the resting metabolic rate of bees in the core generates more than enough heat to raise the core temperature to 35° or more. Only for swarm clusters with less than 1000 bees (a very small cluster) will the bees have to expend additional energy above their resting metabolism in order to maintain a swarm core temperature of 35° C.

Heinrich's model of cluster thermoregulation, based on each individual acting independently according to its own direct perceptions, can explain a number of observations that might otherwise seem paradoxical. Individual bees appear to have a preferred temperature of about 35°–36° C (Heinrich 1981b). This is also the temperature maintained in the brood nest and in the core of most swarms. In a large swarm of 16,600 bees at an ambient temperature of 1° C, however, Heinrich measured excessively high core temperatures of 46° C (Heinrich 1985).

How can these anomalous observations be explained? If we invoke a thermoregulatory mechanism based on the coordination of temperature across the swarm, based on centralized control, this observation becomes difficult to explain. For example, if we assumed that the bees in the mantle receive instructions based on the core temperature and coordinate their activities in order to maintain a proper core temperature, then the high core temperature cannot be explained. Alternatively, if we assume that bees in the core receive instructions to increase their own metabolic rate in an effort to warm the mantle bees, then they are heating themselves up to near lethal temperatures. Either scenario is implausible, compared to the more likely explanation that each bee in the mantle or in the core acts independently to maintain her own temperature in the proper range.

Occasionally problems of temperature control arise, as in the case of large swarms at low temperatures. In this situation, the mantle bees—oblivious to circumstances in the core—are trying to keep themselves warm by huddling close together and shivering. As a result, the core overheats. Two processes contribute to this unavoidable overheating. First, core bees are unable to cool themselves because the cold mantle bees have closed the ventilation channels connected to the swarm surface. Second, as the core bees become warmer, their resting metabolic rate rises automatically with an increase in the local temperature (Kammer and Heinrich 1974).

This instance of core overheating is not an isolated occurrence. There is a tendency for the core temperature of swarms to rise as the ambient temperature declines, even though the mantle temperature stays the same (Figure 15.2) (Heinrich 1981b; Southwick and Mugaas 1971). This behavior is not unexpected when viewed in light of a mechanism based upon individuals independently responding to their local environment, not to instructions from a well-informed central supervisor.

Self-Organization Models of Thermoregulation

We began this chapter with a macroscopic look at the phenomenon of cluster thermoregulation, and followed with a microscopic look at the behaviors and physiology of individual bees within the cluster. We are now prepared to formulate a mathematical model that links those two levels of description.

Although Heinrich (1981a,b,c, 1985) did not formulate a mathematical model, his hypothesis that cluster thermoregulation arises through the independent activities of individual bees provides the basis for modeling the thermoregulatory mechanism. The goal of this section is to present a quantitative model of cluster thermoregulation based solely on the behavioral and physiological rules that the bees appear to be following. Earlier models were not based on this approach. One problem was that the models began with assumptions that, preferably, would be predictions generated by the model itself. For example, the models of Omholt and Lønvik (1986) and Omholt (1987) start with an explicitly specified core and mantle temperature, rather than having them arise from the model. Similarly Lemke and Lamprecht (1990) specify the metabolic heat output as an explicit function of the bees' position within the swarm. For this to be true, a bee would have to determine its position from the center of the swarm and adjust its metabolic rate accordingly. We believe that more recent models (Myerscough [1993] and Watmough and Camazine [1995]) are improvements on previous models because they better embody the actual mechanism of cluster thermoregulation used by the bees.

Biological and Physical Basis

In recognition of its priority, we will focus on Myerscough's (1993) model. As discussed below, the Watmough and Camazine model (1995) uses the same approach. It will be instructive for interested readers to compare the two models, especially to see how similar biological assumptions can be implemented in different ways and how small details in implementation can yield different results.

Although the models are implemented somewhat differently, they are based on the following biology:

1. Each bee uses information only about the temperature in her local environment.
2. Based on the temperature she experiences, a bee exhibits the following behaviors:

   (a) If she is too cold, she can move closer to her neighbors and also increase her heat output.
   (b) If she is too hot, she can move away from her neighbors.

3. The bee's heat output is based upon her metabolic rate. Below a thresh-old temperature she increases her metabolism above the resting level by shivering. Above this threshold temperature a bee remains at a resting metabolic rate, which is a function of ambient temperature.

The model uses a nonlinear heat diffusion equation with a nonlinear heat-source term that represents the bee's metabolic heat production. The equation is similar to equations for heat production and diffusion that model self-heating in combustible materials. Although heat is transferred through an actual clus-ter by both convection and conduction, with coefficients of heat transfer that increase rapidly at high temperatures, the model can partially account for rapid heat loss by convection. The model uses a single partial differential equation to model the transfer and production of heat as a function of time, t, and radial distance, r, from the cluster center:

$$c\frac{\partial T}{\partial t} = \frac{1}{r^2}\frac{\partial}{\partial r}\left(D(T)r^2\frac{\partial T}{\partial r}\right) + \rho(T)f(T), \tag{15.1}$$

where c is the heat capacity of the swarm, $D(T)$ is the thermal conductivity of the swarm, $\rho(T)$ is the density of bees per unit volume, and $f(T)$ is the metabolic output of a single bee at temperature T.

The model incorporates the following assumptions:

1. The swarm is modeled as a spherical cluster with only radial temperature variation.
2. The cluster contains a constant number of bees.
3. Each bee has a threshold temperature of $18°$ C, above which she re-mains at resting metabolism. Below $18°$ C, bees shiver, rapidly increas-ing their metabolic rate. The metabolic heat output per bee is shown in Figure 15.4.
4. The density of bees in the cluster is given as a function of temperature. Heinrich (1981a) measured bee density in a swarm as 7.7 bees/cm³ at $1°$–$5°$ C and 2 bees/cm³ at $30°$ C. The model assumes a simple linear relationship between bee density and temperature with a density of 8 bees/cm³ at $0°$ C and 0 bees/cm³ at $40°$ C. This mimics the movement of a bee within the cluster in response to the local temperature she expe-riences. As the temperature drops to $0°$ C, the bees cluster together until they are packed as tightly as possible. As the temperature rises the bees move apart until the integrity of the cluster is lost and the swarm breaks apart (0 density at $40°$ C).
5. The thermal conductivity of the swarm governs how heat flows within the swarm. Heinrich's (1981a) data on the passive cooling of clusters of dead bees provides an estimate of the thermal conductivity of a uniformly packed cluster of bees. The thermal conductivity was about

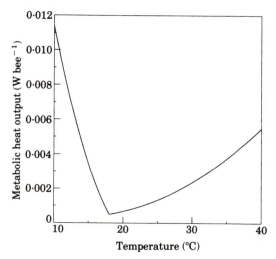

Figure 15.4 Metabolic heat output per bee as a function of temperature for Myerscough's (1993) model.

0.001 W/cm/°C. Of course, in a live swarm, the thermal conductivity varies with temperature because of behavioral adjustment by the bees. At low temperatures the bees crowd tightly together, approximating the properties of a dead cluster of bees. However, at high temperatures the bees will move about and open channels within the swarm allowing heat to be transported largely by convection. Myerscough approximated this behavior with a simple function: $D(T) = D(0)[\rho(0)/\rho(T)]^2$ where $D(T)$ is the thermal conductivity as a function of ambient temperature, $D(0)$ is the minimal thermal conductivity of 0.001 W/cm/°C based on Heinrich's data, $\rho(T)$ is the density of the swarm as a function of ambient temperature, and $\rho(0)$ is 8 bees/cm^3 as mentioned above. Since the density of the swarm becomes 0 at $T = 40°$ C (the temperature at which the swarm is assumed to break up), this function assumes that the thermal conductivity begins at the low value of 0.001 W/cm/°C at 0° C and rapidly rises to infinity at a temperature of 40° C. Although this function is not based on direct experimental evidence, it undoubtedly captures the essential features of the relationship between thermal conductivity and ambient temperature within the swarm.

6. At the edge of the swarm, where $r = $ R, the flux of heat into the surrounding air is proportional to the difference between ambient temperature T_a and the mantle temperature T(R):

$$-D(T)\left.\frac{\partial T}{\partial r}\right|_{r=R} = L(T)\big(T(R) - T_a\big), \tag{15.2}$$

where $L(T)$ is the coefficient of heat transfer between the cluster of bees and the air. It is also a function of temperature. Heinrich (1981a) determined the coefficient of heat transfer between a cluster of dead bees and the surrounding air to be 0.5 mW/cm^2/°C. A reasonable model for $L(T)$ is $L(T) = L(0)[\rho(0)/\rho(T)]^2$ based upon the same reasoning given for the form of $D(T)$: At low ambient temperatures, the swarm behaves like a cluster of dead bees. At high ambient temperatures the open channels in the swarm promote rapid loss of hot air from the swarm. $L(0)$ is taken to be 0.0005 W/cm^2/°C.

7. The heat capacity of the swarm has not been experimentally determined. However, Myerscough estimated the value as 1 Joule°C/cm^3 based on the known heat capacity of water and consideration of the density of bees in the swarm.

Results of Myerscough's Model

As shown in Figure 15.5a, the core temperatures of the model swarms vary between 18° C and 35° C. At a given ambient temperature smaller clusters exhibit lower core temperatures. Mantle temperatures for the different sized swarms are shown in Figure 15.5b. For various ambient temperatures, the temperature profiles of a cluster of 16,000 bees are shown in Figure 15.6, along with profiles observed by Heinrich (1985, figure 1).

How well do these results compare with observations? Core temperatures for the model swarms were within a realistic range of cluster temperatures, although core temperatures below 27° C are more characteristic of winter clusters than swarm clusters (compare Figures 15.5 and 15.7). The model swarms, however, failed to show the observed trend for core temperatures to rise as the ambient temperature falls. Heinrich (1981b) and Southwick and Mugaas (1971) noted that as ambient temperatures fell below 7°–8° C, temperatures on the cluster mantle remained independent of air temperature, while temperatures in the core often *increased*. In addition, the model swarms showed unusually large warm core regions (Figure 15.6). These large warm cores may also account for the excessively high average metabolic rates of swarms at high ambient temperatures.

In defense of this model, however, it should be noted that the lack of close agreement with certain experimental results probably results from gaps in our knowledge of several of the model's parameters. These include bee density, thermal conductivity, and the heat transfer coefficient, all as a function of the ambient temperature. Lacking this information, Myerscough's goal was not to produce results that correspond precisely with observations, but to determine whether a reasonable hypothesis is that cluster thermoregulation arises through the bees' independent behavioral and physiological responses to the local ambient temperatures they perceive. It would undoubtedly have been possible to

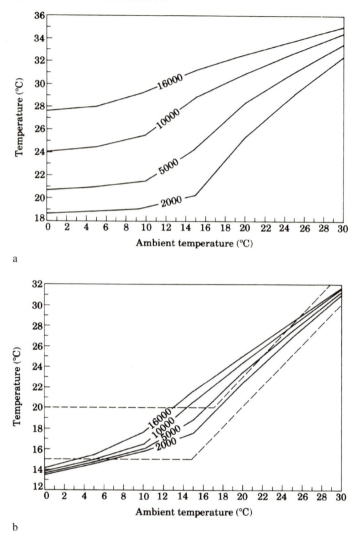

Figure 15.5 Results of Myerscough's (1993) model for four different swarms consisting of 2,000 to 16,000 bees: (a) core temperatures plotted against the ambient temperature, and (b) mantle temperatures plotted against the ambient temperature, where the dashed curve is a plot of observed mantle temperatures measured by Heinrich (1981a).

produce better correspondence with observations by choosing different functions for density, thermal conductivity, and heat transfer equally consistent with experimental evidence. But such an exercise in "tweaking" parameters would have served little value. The importance of the model is that it demonstrates

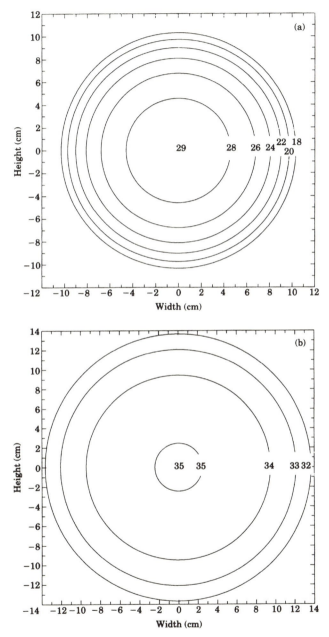

a

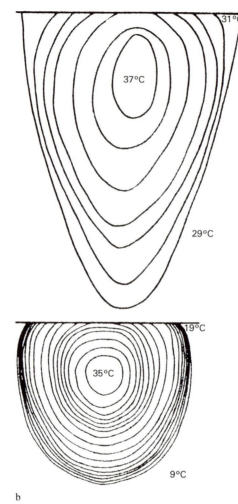

31°C

37°C

29°C

19°C

35°C

9°C

b

Figure 15.6 Temperature profiles of a model swarm (Myerscough 1993) compared to the temperature profiles of an actual swarm measured by Heinrich (1981a,b): (a) the contour plot of the temperature profile of a model swarm with 16,000 bees at ambient temperatures of 10° C and 30° C. (b) measured profiles in a swarm of 16,600 bees at ambient temperatures of 9° and 29° C.

that cluster thermoregulation can be achieved, at least in theory, through an entirely decentralized mechanism in which individual bees follow simple rules of thumb based on local temperature cues. Furthermore, these rules have not been chosen arbitrarily but are based on extensive experimental observations and thus are likely to be the rules that bees actually use.

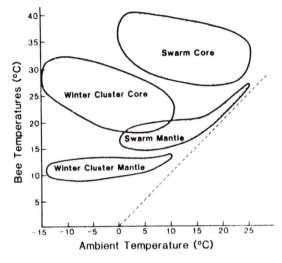

Figure 15.7 Swarm and cluster temperatures as functions of air temperature (from Heinrich 1985). The regions in the diagram exclude the extremes but enclose at least 80 percent of the observed temperatures. Swarm data were derived from Heinrich (1981b), winter cluster cores from Southwick and Mugaas (1971), and winter cluster mantle from Simpson (1961).

Comparison with the Model of Watmough and Camazine

Watmough and Camazine (1995) have developed a model of cluster thermoregulation similar to Myerscough's. The major difference is that the density of the bees is not given explicitly as a function of temperature. Instead, the model is taken one stage further by replacing the assumed temperature dependent density profile of the bees with a second partial differential equation for honey bee density that models the motion of the individual bees in a manner suggested by Heinrich's (1981b) experiments. The equation generates the density of bees in terms of an individual bee's behavioral response to the local temperature it experiences. When the local temperature is below a specified value, a bee moves inward. When a bee is too warm it moves outward. Counterbalancing the thermotactic response is a factor expressing random motions of a bee that are uncorrelated with temperature gradients.

The model consists of two partial differential equations, one that describes heat transfer and production as a function of time, t, and radial distance, r, from the cluster center, while the other relates the density of bees as a function of t, r, and temperature, T:

$$c \frac{\partial T}{\partial t} = \frac{1}{r^2} \frac{\partial}{\partial r} \left(r^2 \lambda(\rho) \frac{\partial T}{\partial r} \right) + \rho f(T) \tag{15.3}$$

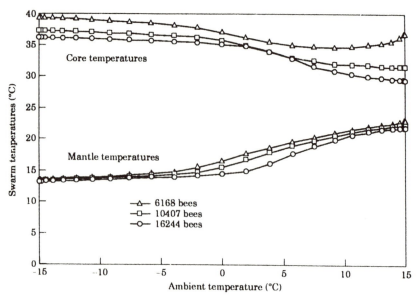

Figure 15.8 Numerical computations of the core and mantle temperatures of different sized model swarms are plotted as a function of ambient temperature. Note that the core temperature rises as the ambient temperature falls. (From Watmough and Camazine 1995, figure 6)

$$\frac{\partial \rho}{\partial t} = \frac{1}{r^2}\frac{\partial}{\partial r}\left(r^2\mu(\rho)\frac{\partial \rho}{\partial r}\right) - \frac{1}{r^2}\frac{\partial}{\partial r}\left(r^2\chi(T)\rho\frac{\partial T}{\partial r}\right), \tag{15.4}$$

where c is the heat capacity of the swarm, $\lambda(\rho)$ is the heat conduction coefficient at a bee density, ρ, $f(T)$ is the metabolic output of a single bee at temperature, T, $\mu(\rho)$ is the motility coefficient for the random movement of the bees, $\chi(T)$ is the thermotaxis coefficient at temperature T. The similarity between equation (15.1) and (15.3) should be apparent. The density of bees, which is an explicit function of temperature $\rho(T)$ in (15.1) has been replaced by density, ρ, in equation (15.3), and equation (15.4) expresses density in terms of bee movements. The main results of the model are shown in Figure 15.8 (Watmough and Camazine 1995, figure 6), which shows the core and mantle temperatures of different sized swarms as a function of ambient temperature, and in Figure 15.9 (Watmough and Camazine 1995, figure 8) which shows temperature and density profiles for a swarm of 13,109 bees at different ambient temperatures. The core temperatures for the model swarms are well within the observed range of cluster temperatures. The model swarms also exhibit the trend that has been observed for core temperatures to rise as ambient temperature falls, and for bees forming a thick, dense mantle at low temperatures.

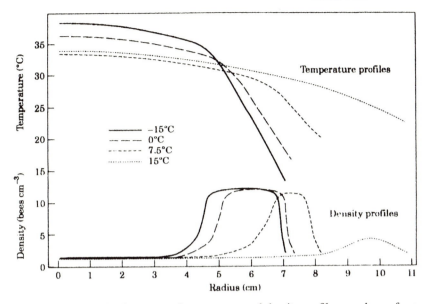

Figure 15.9 Numerically computed temperature and density profiles are shown for a model swarm of 13,109 bees at different ambient temperatures. As the temperature decreases, the thickness of the mantle increases. Note again that, as in Figure 15.8, the core temperature increases with decreasing ambient temperature. (From Watmough and Camazine 1995, figure 8)

Comments on the Model

What purpose do these two models serve? Unlike a model that is merely descriptive, such as the diagrammatic, qualitative model shown in Figure 15.3, a mathematical model makes quantitative predictions that can be experimentally tested. Validated models also allow investigators to perform certain simulated experiments that may be difficult, or even impossible, to perform. For example, Heinrich (1985, p. 400) remarked, "Obviously both the contraction of the cluster as well as increased heat production could increase the temperature of the core, but the two have never been adequately separated." To understand the mechanism of thermoregulation in cold ambient temperatures, it would be useful to know whether an elevated core temperature arises primarily through the prevention of heat loss (as bees cluster tightly together) or through an increase in heat production (as the bees shiver). Unfortunately these two behaviors are tightly coupled. When bees become cold they both cluster *and* increase their metabolic rate. We know of no way experimentally to uncouple these behaviors. A model, however, can easily separate these two variables and assess their relative importance in maintaining adequate core temperatures.

Finally, we can ask what these models tell us about mechanisms based on self-organization? Watmough's and Camazine's model, as well as Myerscough's, examined the hypothesis that cluster thermoregulation could be explained by a mechanism based entirely on the independent, individual behavioral and physiological activities of bees, each responding to local temperature. As illustrated in Figure 2.4, each bee in the cluster utilizes a pathway of information that flows from the local environment (work in progress in the form of the warming swarm) to the individual bee itself. In this system, there is no need for direct communication among the bees because each bee is able to use an indirect means of interaction through the medium of the common temperature environment of the swarm.

These models support a regulatory process for cluster temperature based on self-organization. The most important result of the model is the demonstration that temperature profiles and average metabolic outputs per bee can be obtained that are qualitatively—and often quantitatively—similar to those observed in actual honey bee clusters. Cluster thermoregulation does not appear to require central coordination within the swarm.

Southwick and Mugaas (1971) and Southwick (1982) viewed the thermoregulatory mechanisms of a honey bee colony as qualitatively and quantitatively similar to those of active homeothermic birds and mammals. They proposed that the cluster was thermoregulated to maintain a constant mantle temperature at low ambient temperature. This hypothesis—that a centralized control mechanism regulates a particular temperature within a region of the cluster—appears to be invalid. Paradoxically, cluster thermoregulation does not appear to be a process in which bees actively regulate the temperature of the *cluster per se*. Instead it seems that each bee simply attempts to regulate her own body temperature, with the result that the cluster as a whole generally stays within an acceptable temperature range.

Box 15.1 Heinrich's Experimental Techniques

In this chapter, we described two experiments by Heinrich (1981b) that failed to support the hypothesis of centralized control mechanisms for swarm thermoregulation. In one experiment, Heinrich found that bees need not move from the core to the mantle, or vice versa, for the swarm to thermoregulate, thus refuting the hypothesis that thermoregulation requires the collection and monitoring of information throughout the swarm. Figure 15.10 adapted (from Heinrich 1981b, figure 19) shows the temperature of a swarm of 8000 bees that was partially produced inside a gauze cylinder to prevent bees from moving between the core and the mantle.

Box 15.1 continued

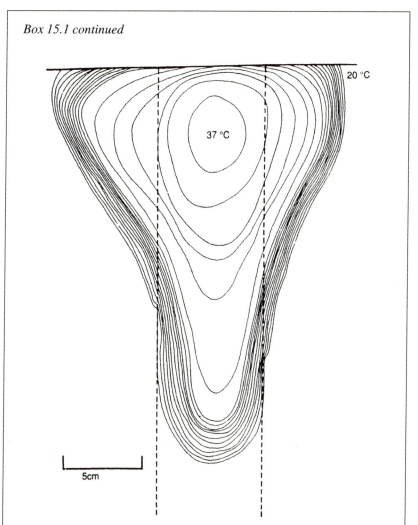

Figure 15.10 The contour plot exhibits isotherms (1° C) in a swarm of 8,000 bees that is partially built in a gauze cylinder attached to the plexiglass foothold of the swarm. Ambient temperature = 4° C.

In another experiment, Heinrich tested the hypothesis that information is chemically communicated between the core and the mantle by means of pheromones, carbon dioxide levels or other volatile substances. Figure 15.11 shows the experimental apparatus that Heinrich devised to enable gases from one swarm (kept at a low ambient temperature) to be

Box 15.1 continued

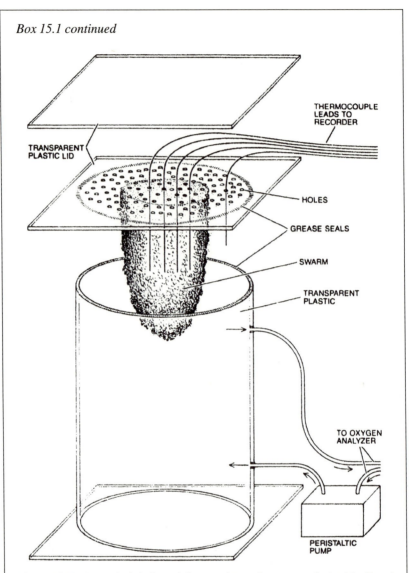

THERMOCOUPLE
LEADS TO
RECORDER

TRANSPARENT
PLASTIC LID

HOLES

GREASE SEALS

SWARM

TRANSPARENT
PLASTIC

TO OXYGEN
ANALYZER

PERISTALTIC
PUMP

Figure 15.11 An exploded view of the experimental apparatus devised by Bernd Heinrich to measure the temperature and metabolic rate of captive swarms.

pumped into the closed environment of another swarm (kept at a higher ambient temperature). With this apparatus a swarm can be maintained in the laboratory while its metabolic output and temperature in various

Box 15.1 continued

locations can be measured without disturbing the bees. The holes in the lower lid serve as a foothold for the bees, and as a source of ventilation. During periods when the metabolic rate of the swarm is measured, the solid lid is sealed firmly over the ventilating lid and air is circulated from the closed system to an oxygen analyzer.

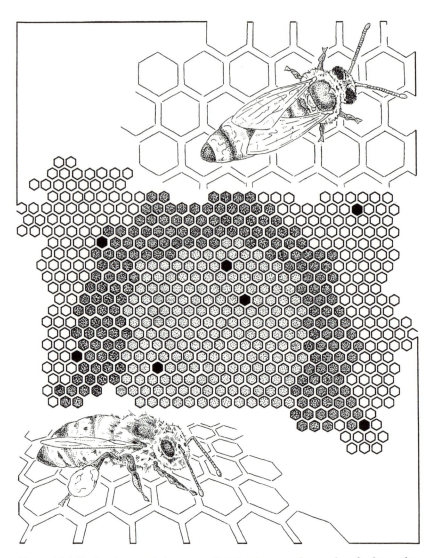

Figure 16.1 In the characteristic pattern that develops on the combs of a honey bee colony, the central portion of the figure represents an area of comb near the center of the colony. The brood area (lightly stippled) is surrounded by a band of pollen (darkly stippled) and a peripheral region of honey (unstippled). Occasional, empty cells (dark) are scattered across the comb. In the upper right is the queen and in the lower left is a forager returning from the field with a pollen load. (Illustration © Bill Ristine 1998)

16

Comb Patterns in Honey Bee Colonies

Honey and pollen are stored above and alongside the
brood nest.
> —T.D. Seeley and R. Morse, *Insectes Sociaux*

Pollen is usually placed in cells next to the brood nest,
where it is easily accessible to nurse bees.
> —M. Winston, *The Biology of the Honey Bee*

In earlier chapters, we dealt with self-organized patterns built *of* organisms themselves, such as a school of fish moving in a coordinated manner, a group of fireflies flashing in unison, or a multitude of raiding army ants. In this and several of the next chapters we examine self-organized structures built by social insects. The challenge here is to understand how a colony-level (global) structure emerges from the activities of many individuals, each working more or less independently. In the building of these large biological superstructures, it is difficult to imagine that any individual in the group possesses a detailed blueprint or plan for the structure it is building. The structures are orders of magnitude larger than a single individual and their construction may span many individual lifetimes. It seems far more likely that each worker has only a local spatial and temporal perspective of the structure to which it contributes. Nonetheless, the overall construction proceeds in an orderly manner, as if some omniscient architect were carefully overseeing and guiding the process.

This chapter describes how a regular pattern on the comb of honey bee colonies emerges through a self-organizing process. First, we present a description of the pattern and its biological significance. We then discuss alternative hypotheses for how the pattern might develop. Following that we present observations and experimental results describing the biological basis of the pattern development, and show how this pattern-formation process is most consistent with a mechanism based on self-organization. Along the way, three approaches to modeling the pattern are presented and we discuss advantages and drawbacks of using a particular modeling approach. Discussions are based on experiments and models presented in Camazine (1991) and Camazine et al. (1990).

The Colony-Level Pattern

The typical honey bee colony comprises approximately 25,000 female worker bees, a few thousand male drones, and a single queen. In addition to the adult bees there is immature brood, consisting of developing eggs, larvae and pupae, as well as a variable amount of accumulated honey and pollen stored within the hive in a series of parallel wax combs subdivided into approximately 100,000 hexagonal cells. A characteristic well-organized arrangement of brood, honey, and pollen develops on the comb, consisting of three distinct, concentric regions—a central brood area, a surrounding band of pollen, and a large peripheral region of honey (Figures 16.1 and a). In times of low pollen availability the region between the brood and honey may be empty rather than filled with pollen (Figure b).

The Adaptive Significance of the Comb Pattern

The highly organized pattern of the comb suggests that it may be an adaptive structure. Indeed, a compact brood area may help to ensure a precisely regulated incubation temperature for the brood. As discussed in the previous chapter, the temperature within the colony is carefully maintained at about 33°–34°C. In addition, a compact brood nest may facilitate efficient egg laying by the queen and efficient brood care by the nurse bees. The central location of the brood may also allow it to be better defended against predators and protected from environmental vagaries. The location of the pollen, in a band adjacent to the brood area, where it is readily accessible to nurse bees, may promote efficient feeding of the nearby larvae.

The pattern on the combs is both well-organized and consistent throughout the season. This feature may also be adaptive. Each day honey and pollen from tens of thousands of foraging trips are deposited in the cells, stores of honey and pollen cells are continually consumed, hundreds of eggs are laid, mature adult bees emerge from their cells, and diseased brood is removed. Despite the constant turnover in which cells are often refilled with something different, the stability of the pattern is maintained, a feature that may contribute to colony efficiency.

Of course, one should avoid circular reasoning in arguing the adaptive significance of the comb pattern. The existence of a regular concentric pattern does not mean that the pattern is adaptive, nor does it mean that one pattern is more efficient than others. One can imagine other arrangements that may be equally adaptive, such as small islands of brood each surrounded by a ring of pollen, making it easier to feed nutritious pollen to the developing brood. One needs to keep in mind that such alternative patterns may not have arisen either because they were less adaptive and thus were not selected, or because of historical quirks during evolution.

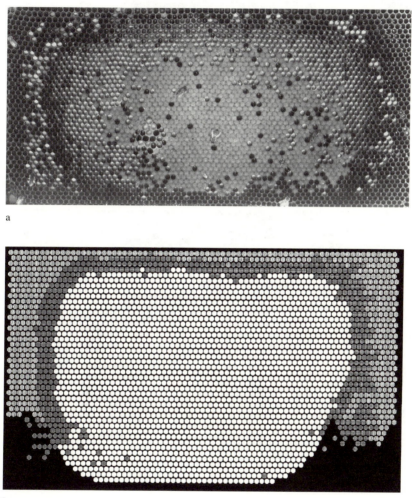

Figure 16.2 Eggs, pollen, and honey are deposited in a characteristic pattern within the bee colony: (a) photograph of a frame showing the characteristic pattern of centrally located brood, a band of pollen, and a peripheral region of honey; (b) diagram of a similar frame (white circles = brood, darkly stippled cells = pollen, lightly stippled cells = honey, and black cells = empty cells). (c) In this photograph of a frame taken from a hive during early August, little pollen was available; there is a band of empty cells in the region previously containing pollen. A similar frame is shown in (d). Symbols are as in (b).

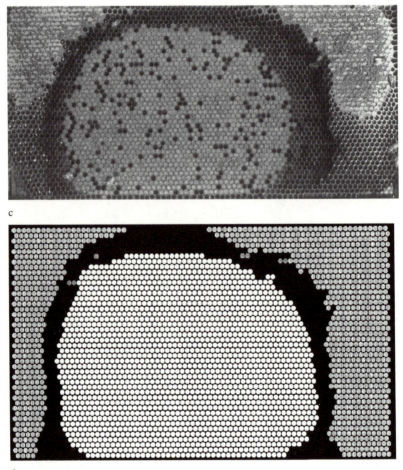

c

d

Alternative Hypotheses of Pattern Formation

Although these arguments do not provide proof that the concentric pattern is adaptive, the consistency of the pattern and its presumed adaptive significance raise the important question of what mechanisms account for its origin and maintenance. How does this well-organized colony-level pattern emerge from the activities of thousands of bees? We suggest two alternative hypotheses.

The conventional theory is the blueprint or template hypothesis, which suggests that honey, pollen, and eggs are deposited by the bees in particular locations on the comb. Many authors have described the characteristic pattern on the comb, and one might be tempted to infer from these descriptions that the mechanism involves deposition in specified locations known to the bees. For example, Winston (1987) states, "Pollen is usually placed in cells next to the brood nest, where it is easily accessible to nurse bees." Seeley and Morse (1976) note that, "Honey and pollen are stored above and alongside the brood nest." In describing the feral colony, Seeley (1983) states, "Honey is stored in the upper region, pollen is packed in a narrow band directly below, and brood is reared in the lowermost portion of each comb." Perhaps the bees have some means of knowing where on the comb to deposit eggs, pollen, and honey. One wonders whether the pattern develops simply as a consequence of the bees filling in the cells of the comb according to an arrangement latent in the blueprint, much as a child creates a pattern when it fills in regions of a picture in a coloring book.

Can this type of mechanism explain the orderly pattern of brood, pollen, and honey on the combs? If so, this raises the question of the nature of the presumed blueprints or templates that the bees use. Conceivably, a temperature gradient from the center of the comb to the periphery could act as a template indicating where eggs, pollen, or honey are to be placed. Alternatively, there may be an innate blueprint, with each bee knowing that the central portion of the comb is reserved for brood and that pollen and honey are to be placed more peripherally. Whatever the precise mechanism, it would be the external blueprint or template that imposes the pattern, rather than some form of internal dynamic interaction.

If the bees filled in the comb by means of some innate knowledge of where to deposit food or eggs, or if they used a template inherent in the comb, this would directly contradict a mechanism based on self-organization. Although, in principle, the blueprint hypothesis could account for the pattern formation, it is contradicted by certain observations that were the initial stimulus for this study. In particular, the blueprint hypothesis is inconsistent with the observation that pollen and honey are often deposited throughout the comb (Figures 16.3a, 16.4). This becomes particularly obvious when the developing bees mature and become adults, vacating their pupal cells within the central brood area, or when an initially empty comb is placed in the colony. In both situations, the central, empty cells are frequently filled with pollen or honey and not reserved for brood. Such observations have probably been ignored, as being either the aberrant behavior of a few bees, or errors that occur when a high rate of honey or pollen deposition reduces the available number of storage cells.

These observations suggested that a different, more subtle mechanism might account for comb-pattern formation. In place of the blueprint hypothesis it seems possible that a mechanism based on self-organization might provide an explanation for how pattern emerges spontaneously as a dynamic process involving local interactions and a limited overview of the developing pattern. Taking this approach, we analyze lower level individual components of the pattern-formation process—egg laying, pollen and honey deposition, honey and pollen consumption, and the emergence of mature bees from their pupal cells—in order to understand how the higher level pattern arises. In the self-

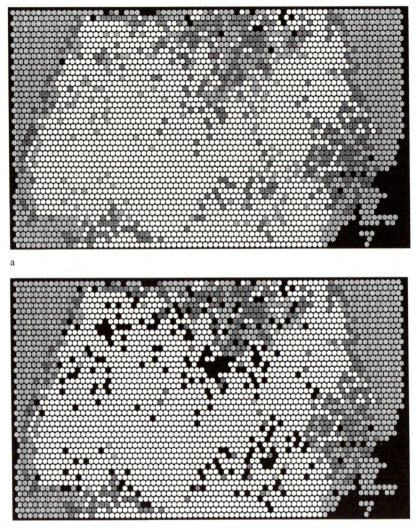

a

b

Figure 16.3 Two tracings of the same frame at different times showing preferential emptying of honey and pollen from cells nearby brood: (a) in the evening at the end of a day of good foraging, and (b) the following morning, before foraging. Symbols are the same as in Figure 16.2b.

organization hypothesis, there is no need to invoke a blueprint that specifies locations for brood, pollen, and honey, since the dynamic relationships among the component processes of deposition and removal are sufficient to organize

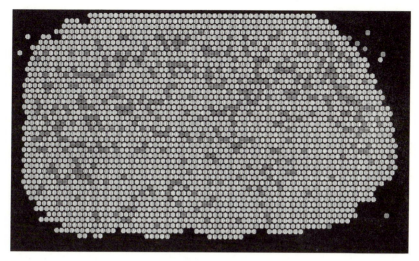

Figure 16.4 Tracing of a frame showing the pattern of deposition of honey and pollen on an empty frame. An empty frame was placed in a hive at the beginning of the day, before foraging, and removed at the end of the day. Symbols are the same as in Figure 16.2b.

a pattern on the combs. At first the self-organization hypothesis may appear counterintuitive because a global spatial pattern emerges automatically from processes based solely on local cues that lack long-range spatial information. Nonetheless, we hope to convince readers that such a mechanism is more consistent with the available evidence than are other explanations.

Biological Basis of the Pattern Formation

A theory of pattern formation should explain not only the final concentric pattern of brood, pollen, and honey, but also the development of the pattern from its initial stages. A parsimonious explanation should also account for both the initial and mature patterns by means of the same process.

In this section, we describe the biological basis of pattern formation by considering each process that affects the deposition and removal of brood, pollen, and honey. For each substance—brood, pollen, and honey—we describe how the behavioral rules of individual bees contribute to the spatial pattern on the comb, and we estimate the rate of deposition or removal.

To increase the didactic value of this chapter, we present the biological observations and experimental data in the chronological sequence in which they were originally acquired, saving a final set of data for later in the chapter. The reason for this will become clear as we go along.

Behavior of Individual Bees

Most of the colony's activities involve the combs, either for the storage of honey and pollen, or for brood rearing. A particular cell may be used for either honey or pollen storage, or for brood rearing, but at any particular moment the cell is used for only one of the three. We now consider data regarding the rate and spatial pattern of the input and removal of brood, honey, and pollen from the cells.

Egg-Laying by the Queen

The egg-laying rate of a queen honey bee has been reported in many studies and is generally between 1000 and 2000 eggs/day, or about 1 egg/min (Brünnich 1923; Nolan 1925; Bodenheimer 1937). An additional estimate of the egg-laying rate in observation hives was obtained by following the queens of two hives for eight periods of 30–90 min each for a total of 350 min (Camazine 1991). The average egg-laying rate was 0.71 ± 0.3 eggs/min (mean $\pm$ s.d.; range 0.4 to 1.1 eggs/min), essentially the same as that in full-sized colonies. Values at the lower end of the range corresponded to times when the hive had few available empty cells, while greater values were obtained when an empty frame was placed in the hive.

In addition to providing data on the rate of oviposition, these observations also revealed the pattern of egg laying. Figure 16.5 is a typical record made by tracing the path of the queen as she laid. It shows the locations of forty-three consecutive ovipositions (indicated by numbered cells) made over a one-hour period, 9:00 A.M. to 10:00 A.M. The empty frame had been placed in the colony the previous evening. Prior to the morning observation period, eighty-two eggs were laid on one side of the frame (indicated by white circles). As shown, these initial ovipositions were not placed contiguously but were widely scattered. Although the subsequent (numbered) ovipositions filled many of the gaps, the queen's pattern of egg laying was nonetheless unsystematic. The queen often zigzagged over the frame, doubling back to a spot where she just oviposited. Rather than exhaustively searching for all available cells in an area, the queen successively returned to an area to fill in empty cells.

This raises the question of how the queen selects cells. Each cell is carefully inspected prior to oviposition, but not every empty cell is accepted. The queen apparently takes into account the presence of nearby brood, since her ovipositions are restricted to cells relatively close to other cells which contain brood. Figure 16.6 (derived from the tracing of Figure 16.5 and from similar data) demonstrates this tendency for eggs to be laid near one another. Of 143 ovipositions, the mean distance to the nearest egg (i.e., the number of intervening nonbrood cells) was 0.7 ± 1.0 (mean $\pm$ s.d.). For 95 percent of all ovipositions, there were less than three intervening nonbrood cells.

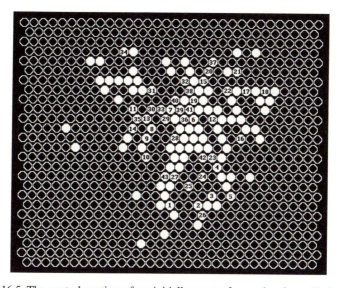

Figure 16.5 The central portion of an initially empty frame showing cells in which the queen oviposited. The empty frame was placed in the colony in the evening. The queen was observed for 1 h (9:00 A.M. to 10:00 A.M.). The numbers indicate the 43 consecutive ovipositions observed during this period. Unnumbered white circles are 82 ovipositions made before the period of observation.

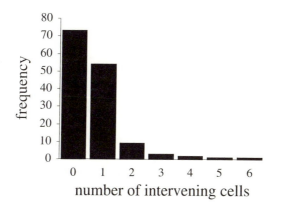

Figure 16.6 Frequency histogram demonstrating the queen's tendency to lay eggs near other brood cells. The bars represent the frequency of egg deposition at specific distances from the nearest brood cell. Values indicate the number of intervening nonbrood cells for each of the 143 ovipositions.

Once an egg is laid in a cell, it remains there for approximately twenty-one days, the developmental period of the worker bee. As far as we know, eggs and larvae are not moved from one cell to another. Thus, we can summarize the pattern of brood deposition by describing the queen's route while ovipositing. She moves rather unsystematically, zigzagging across the comb and laying eggs, on average, within a few cell lengths of another brood-containing cell. How the compact brood area develops from this pattern of movement will become more apparent after examining the dynamics of honey and pollen cells.

Honey and Pollen Foraging

We now consider the dynamics of honey and pollen within the colony. During the foraging season a typical colony collects 60 kg of honey. Based on a weight of 40 mg per nectar load, workers require about 3 million foraging trips a season for nectar collection. A typical pollen load weighs 15 mg, and foragers collect about 20 kg of pollen each season. This amounts to approximately 1.3 million foraging trips (Seeley 1985).

Approximately twenty-five nectar loads are required to fill a cell with honey, based on an average sugar concentration of 40 percent for nectar, 80 percent for honey, and a weight of 40 mg per nectar load (Seeley 1985). About fifteen pollen loads to are necessary fill a pollen cell, based on a load size of 15 mg and an average weight of 0.23 g per pollen cell (Camazine et al. 1990).

Counterbalancing this food collection is food consumption. A reasonable estimate for the amount of honey consumed during the foraging season is 35 kg, or about 60 percent of the collection. In contrast, a colony collects about 20 kg of pollen through the season but consumes nearly all of that, overwintering with about 1 kg. Since pollen input and consumption are nearly equal, pollen often does not accumulate in the colony to a significant extent. However, wide fluctuations occur regularly in pollen foraging as pollen-producing plants come into bloom and result in periods of large pollen input that rapidly fills available cells. In one study, the ratio of incoming pollen foragers to nectar foragers, averaged over the day for many days, was 0.26 ± 0.19 (mean $\pm$ s.d.) with a range of 0.06 to 0.83 (calculated from Visscher and Seeley 1982).

With some idea of the rate at which honey and pollen are brought into the colony and consumed, we can now examine *where* on the comb the bees put this food. Pollen foragers returning from the field select a cell in which to deposit their pollen loads. By contrast, honey foragers regurgitate their nectar to the food storer bees which then select a cell in which to unload.

To document the pattern of honey and pollen deposition, an empty frame was placed in an observation hive in the morning (prior to foraging) and removed at the end of the day. The location of each cell containing honey or pollen was recorded by photographing the frame. This was repeated several times during a period when both nectar and pollen were being collected. Figure 16.4 is a typical record. Pollen and honey can be seen throughout the frame. Cells near the edge of the frames received little nectar and pollen, and cells in the upper portion of some frames were used in preference to cells lower on the frame. Nonetheless, there was no indication that the bees selected particular regions for pollen and honey, or that a portion of the frame was reserved for brood, as would be predicted by the blueprint hypothesis. The predominance of nectar on the frame reflects the relatively greater rate at which the bees were bringing in nectar during this period.

One might argue that although bees deposited pollen and honey randomly in cells on an empty frame, perhaps the pattern of deposition would be different on a frame containing brood. The central brood area potentially could provide bees with a cue enabling them to deposit nectar and pollen in specific locations. This is not the case, however. The original observation that prompted the self-organization hypothesis was that pollen and honey were regularly, but unexpectedly, deposited within the brood area. This can be observed both in observation hives and in full-sized colonies. An example is shown in Figure 16.3, a tracing of a frame from an observation hive. Although much of the pollen is confined to a band at the edge of the brood area, and most of the honey is on the periphery, much pollen and honey are also seen scattered within the brood area.

Self-Organization Model

At this point in the study of pattern formation it appeared that we might have the essential components of the system and understand their interactions sufficiently to formulate a model of how the concentric pattern of brood, pollen, and honey developed on the comb. Referring back to Figure 6.1, we see that a global phenomenon—comb pattern formation—was identified, and that experiments and observations were made to identify the lower-level processes involved as well as their interactions. Quantitative data were obtained on the rates and pattern of deposition and removal of brood, pollen, and honey. If all the biological details have been identified and adequately quantified, it should be relatively easy to formulate a model that serves as the hypothesized mechanism of pattern formation. The first such model developed was a Monte Carlo simulation.

Since so much is going on simultaneously in a honey bee colony, it is difficult to fathom the complex dynamics of deposition and removal of brood, pollen, and honey, and determine the effects of various parameter values without the aid of a computer model. Thousands of foraging trips for nectar and pollen occur each day and hundreds of eggs are deposited by the queen. At the same time, thousands of bees enter cells to consume honey and pollen. We had no way of knowing intuitively, without some form of mathematical model or computer simulation, whether we had all the pieces of the puzzle.

To begin, the biological details described in the previous section were incorporated into a simulation of bees filling one side of a standard Langstroth frame (approximately 75 by 45 cells, or a total of 3375 cells). We hoped that the simulation would allow us to determine whether the behavioral rules and parameter values described in the preceding section were sufficient to generate the observed pattern, and to determine which components of the model were necessary for the pattern formation. Based on details gained from observations and experiments in the field and from reports in the literature, we incorporated the following data in the initial simulation:

1. To simulate the probability distribution of ovipositions shown in Figure 16.5, the queen starts from the center of the frame and lays an egg in any empty cell that is less than 4 cells to the next nearest brood cell.
2. The maximum egg-laying rate was set at 1 egg/min, 24 h/day.
3. Only one egg is laid per cell.
4. After 21 days the brood cell is vacated.
5. Honey and pollen are deposited in randomly selected cells that are empty or partially filled with the same substance.
6. Honey and pollen are deposited in cells during daylight hours, 12 h/day.
7. The average capacity of a honey or pollen cell is 20 loads.
8. Honey and pollen are removed from randomly selected cells.
9. The ratio of honey removal to input is 0.6.
10. The average ratio of pollen removal to input is 0.95.
11. The average ratio of pollen input to honey input is 0.2. This ratio is varied over the course of a simulation to match the natural variation in pollen availability.

During the initial stages of developing the simulation, we incorporated as much biological detail as possible in the model. This approach was taken since we had no a priori idea which features were crucial for generating pattern. We will see later, when we present the cellular automaton version of the model, that fewer details are sufficient to capture the essence of the pattern formation process.

The parameter values specifying the egg-laying rate and rates of honey and pollen deposition and removal were incorporated in the Monte Carlo simulation as follows: During each minute of the simulation a maximum deposition and removal rate was specified. Only during the initial stages of the simulation were these maximal rates achieved, since the frame begins to fill and it becomes more difficult (at times impossible) to find cells in which to deposit honey and pollen or in which to oviposit. When this occurs, the maximum number of depositions or removals is forfeited, thus simulating the normal feedback that occurs in colonies.

When all eleven features described above were incorporated into a simulation, the simulation unexpectedly failed to show the characteristic pattern

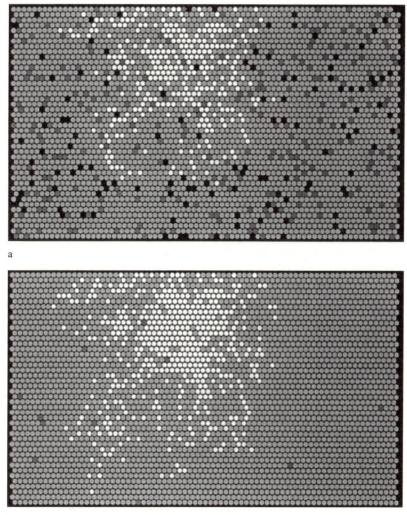

a

b

Figure 16.7 Computer[1] simulation of pattern formation on the comb. (a) day 2; (b) day 11. Symbols are the same as in Figure 16.2b. Parameter values: ratio of honey removal to honey input $= 0.59$; ratio of pollen removal to pollen input $= 0.99$; average ratio of pollen input to honey input $= 0.21$ (range, 0.03 to 0.56), and no preferential removal of honey and pollen nearby brood (preferential removal factor $= 1$).

(Figure 16.7). The compact brood area did not develop and pollen was scattered haphazardly on the frame. At this stage, we either had to demonstrate that the model was incorrectly implemented or that some crucial biological detail was omitted or incorrect. After careful examination of the program code, we felt that the simulation was working correctly, but an incorrect assumption about the biology had been made. Of the eleven components of the model listed above, the eighth was incorrect. Since specific data were lacking, it had been assumed that pollen and honey were removed randomly over the comb, as had been shown to be with honey and pollen deposition. Although it seemed a reasonable assumption at the time, it was determined experimentally to be incorrect.

A year later, during the next summer season, experiments were performed specifically to address the question of how honey and pollen were removed from the cells. The results revealed that honey and pollen are not removed randomly across the surface of the comb. In striking contrast to the essentially random pattern of *deposition*, the *removal* of both honey and pollen occurs preferentially from cells near brood. The pattern of removal was demonstrated in two sets of experiments. In the first, we documented the overall dynamics across the comb. In the second, we focused on individual cells and the behavior of the bees that visit them.

Experiment 1. The contents (brood, pollen, or honey) of all cells on one side of a frame was documented by a tracing made on a pane of glass laid against the glass wall of the observation hive. After a specified time interval, we noted which cells had been emptied in the intervening period. By choosing a period during which there was little or no foraging (overnight in two cases and during a cloudy day with little foraging in a third case), the changes on the frame largely reflected the removal of honey and pollen, rather than the complicating combination of deposition and removal.

Results are shown in Figures 16.3 and 16.8. Figure 16.3a and b are tracings from an experiment showing the overnight emptying of honey and pollen cells. Here, all of the emptied pollen and honey cells were within two cells or less from a cell containing brood; no pollen or honey cells further from brood were emptied. Similar, but less striking, results were found in other replicates of the experiment. Figure 16.8, based on the data from three experiments, illustrates how brood proximity affects the probability that a honey or pollen cell will be emptied. For each of the three trials we divided the honey and pollen cells on the frame into eleven categories indicating the distance of the cell to the nearest brood cell. (The zero category represented a honey or pollen cell adjacent to brood, and category ten included all cells ten cell-lengths or more from the nearest brood cell.) We then calculated the percentage of honey or pollen cells in each category that was emptied in the intervening period. The graph indicates that a cell containing honey or pollen is more likely to be emptied if it is near a brood.

Experiment 2. This set of experiments was based on the following reasoning: If honey and pollen are removed at a greater rate from cells near the brood compared with more peripheral cells, then the bees should visit these cells

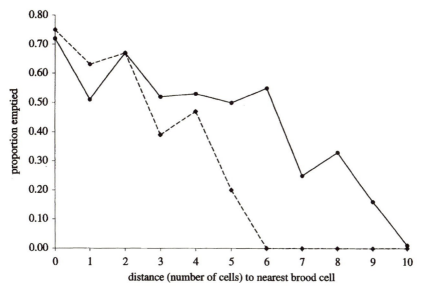

Figure 16.8 The proportion of pollen (dotted line) and honey (solid line) cells emptied as a function of proximity to brood.

more frequently or spend more time in them than in the more peripheral cells. This was confirmed by comparing the occupancy rate of pollen cells located in or adjacent to the brood area versus peripheral cells. For each of seven trials, ten pollen cells were randomly chosen from within or adjacent to the brood area and another ten cells were randomly chosen more peripherally, at least three cells from the nearest brood cell. At eight minute intervals over a two-hour period, each cell was observed to record the presence or absence of a bee. In addition, throughout the observation period all cells were watched for bees depositing pollen in any cells.

No bees were observed to make pollen deposits; therefore the occupancies are assumed to reflect visits in which bees were removing (presumably consuming) pollen from the cell, rather than packing pollen shortly after a deposit. For each trial the occupancy (the percentage of the observations in which a cell was occupied by a bee) for the cells near brood was greater than the occupancy of peripheral cells (Camazine 1991). For all seven trials, the mean occupancy was significantly greater for the central versus peripheral cells ($p = 0.001$). For five of the seven trials, the cells were examined the next day to determine whether the increased occupation actually reflected removal of pollen. All fifty of the peripheral cells still contained pollen, but only 40 percent of the fifty central cells had pollen.

The most likely explanation for the preferential removal of pollen and honey near brood is that food is consumed largely in the brood area. The brood area holds a high density of young bees occupied with the task of feeding the larvae. These nurse bees restrict most of their activity to the brood area (Seeley 1982), and may therefore account for the preferential removal. An interesting feature of the preferential removal of honey and pollen near the brood is that it probably arises automatically. One need not assume any sophisticated, genetically programmed behavior on the part of the bees. It may simply arise as nurse bees working near the brood search randomly for food to feed the developing larvae. A random walk centered in the brood area will automatically result in nearby cells of pollen and honey being emptied at a greater rate than more peripheral cells.

From the second experiment, we are able to estimate the relative rate at which pollen is removed from a central versus a peripheral cell, a parameter that we term the *preferential removal factor*. For each trial we took the ratio of the occupancy of central versus peripheral cells. The mean of these ratios (9.9) is the preferential removal factor for pollen used in the simulations below. Lacking an experimental estimate for the preferential removal of honey, we have used the value obtained for pollen cells. This is justified on the basis of Figure 16.8, which shows that the probability distributions for cell emptying as a function of distance to the nearest brood cell are similar for both honey and pollen cells. Although, for the honey cells, there is less of a difference between the central and peripheral cells, there is clearly preferential removal of honey that is near the brood.

Having determined that pollen and honey are removed preferentially from cells nearby brood, we can now correct point 8 above to read as follows:

> Honey and pollen are removed preferentially from cells near the brood. The observed magnitude of this effect is such that a pollen cell completely surrounded by brood is emptied about ten times as quickly as a cell with no brood neighbors.

Once this detail is incorporated in the simulation, the characteristic concentric pattern readily appears. Starting from an empty frame, the pattern develops gradually. Initially (Figure 16.9a), pollen and honey are found throughout the frame as bees deposit their loads randomly on the empty frame. At the same time the queen wanders over the frame from her central starting point. Similar to what is seen in Figure 16.5 (an actual tracing of the queen's oviposition), the center of the frame begins to fill with eggs, but a compact brood area has not yet developed. In fact, many of the cells interspersed among the eggs contain honey and pollen. This is the early, disorganized stage of the pattern formation process.

Several simulation days later (Figure 16.9b) the characteristic well-organized pattern formed. The central brood area is now compact. Honey and pollen on

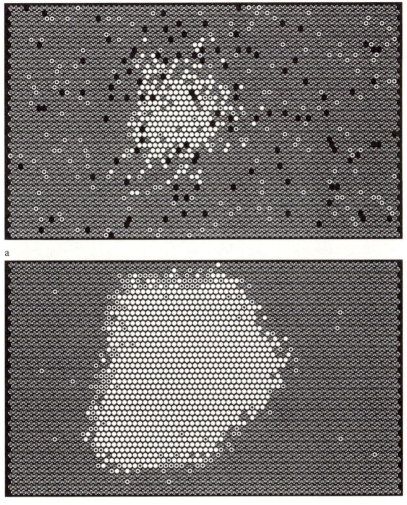

a

b

Figure 16.9 Computer simulation[1] of pattern formation on the comb: (a) day 1; (b) day 22. Symbols: open cells = brood; black cells = empty; open cells with black dot = pollen; darkly stippled cells = honey. Parameter values: ratio of honey removal to honey input = 0.59; ratio of pollen removal to pollen input = 0.99; average ratio of pollen input to honey input = 0.21 (range, 0.03 to 0.56); preferential removal factor = 10.

the periphery have segregated into a peripheral region, almost entirely honey, and a band of pollen adjacent to the brood area. How has this transformation occurred?

Observing the progression of the simulation reveals the interacting processes that contribute to the pattern formation. First, the central compact brood area results, in part, from the queen's attempts to lay eggs near one another. Even though pollen and honey are deposited within the brood area at the same time, their preferential consumption near the brood does not permit their accumulation within the brood area. Thus, the brood area is continually freed of honey and pollen and filled with eggs, enhancing its compact structure. Even though the queen appears to be rather inefficient by continually retracing her steps, we see that it is necessary for her to return repeatedly to an area in order to discover new empty cells which are gradually cleared of pollen and honey. It would only be efficient to search for oviposition sites exhaustively and thoroughly in one area if that area were reserved for eggs. Since many of the cells nearby brood are often filled with honey and pollen, a better search strategy may be to quickly and cursorily pass through an area, and later return when more cells are emptied.

A second process explains the segregation of honey and pollen on the periphery. Since both are deposited randomly, initially pollen as well as honey appears on the periphery. However, recall that in a typical colony, an average 95% of the collected pollen is consumed. With the normal fluctuations in pollen availability, much of the time there is a daily net loss of pollen. Thus, pollen is consumed on the periphery at nearly the same rate as it is deposited. Since honey is brought into the colony at a much greater rate than pollen, cells on the periphery that have been emptied of their pollen are therefore more likely to be replaced with honey. Gradually any pollen deposited on the periphery is removed, leaving this region almost entirely honey. Where, then, is pollen stored?

Eventually the only place available for storage of pollen is in the band of cells adjacent to the brood. Once a cell in the brood area is occupied by an egg, it is reserved as a brood cell for the next twenty-one days of honey-bee development. But in the interface zone between the brood and the peripheral stores of honey, the preferential removal of honey and pollen continually provides a region where cells are being emptied at a relatively high rate. These cells with a high turnover rate are available for pollen. The only other region where empty cells regularly become available are those cells in the brood area from which fully developed bees emerge. Bees readily put both honey and pollen in this area, but it is rapidly removed (Figure 16.3).

An interesting feature of this process of pattern formation is that an interface region emerges automatically. This zone develops without the bees specifically maintaining a region for pollen storage; nevertheless it functions as if it were an area reserved for pollen. When much pollen is available, this region quickly fills up. When the pollen input decreases, the band empties slowly.

Another, more speculative feature of the system is the possibility that the rate of egg laying and pollen collection may be self-regulated. The preferential removal of pollen and honey as a function of nearby brood suggests positive feedback. As the amount of brood on the comb increases, the rate of honey and pollen removal nearby would also increase, providing more empty cells in which the queen can lay. With enough positive feedback, a band of excess empty cells might develop surrounding the brood, thus allowing the queen to lay eggs at her physiological maximum rate. If the width of this empty zone is indeed a function of the amount of nearby brood, then the size of the pollen storage area may be automatically regulated in proportion to pollen requirements. As the amount of brood increases, consumption of honey and pollen near the brood would probably increase as well, thus increasing available pollen storage space in parallel with the nutritional needs of the developing larvae.

For those who would like to observe the simulation in progress, or would like to explore the behavior of the model under different parameter conditions, a Macintosh compatible version of the Pascal simulation can be downloaded at our self-organization website: http://beelab.cas.psu.edu.

Other Modeling Approaches

In contrast to other case studies presented in the book, three different modeling approaches have been used to analyze this pattern formation process. This was done in part to address the question of whether one approach or another might yield information not available using other approaches. Another reason for using three different approaches was to determine which would be the easiest to implement, and which would be the easiest to understand. The answers to these questions obviously vary from individual to individual depending upon one's background and expertise. After finishing this chapter, and perhaps referring to some of the original literature for more detail, consider the following assessment. Conceptually, the Monte Carlo simulation is easiest to understand. As much biology as possible was directly incorporated into the model. Unfortunately, this approach requires a great deal of programming that is time-consuming and generates lengthy program code that is difficult to make error-free.

The cellular automaton model is the easiest to implement, but is more difficult to relate to the biology of the system than the Monte Carlo simulation. The differential equation model is likely to be preferred by those who have the expertise to deal with the mathematics. Its advantage is that it allows one to determine more precisely the conditions under which the pattern develops and is perhaps the most concise (and "cleanest") formulation of the pattern formation hypothesis.

We now present these other modeling approaches and discuss some of the relative merits of each approach.

A Cellular Automaton Model

The richness of detail incorporated into the Monte Carlo simulation makes it difficult to get an intuitive feel for the behavior of the system, and to know which of its features are essential for pattern formation. For example, does the length of the foraging day affect the pattern? Would the pattern develop if the bees were to forage twenty-four hours a day? Does the particular pattern of movement of the queen affect the comb pattern? The problem with a realistic simulation—one that faithfully captures much of the detail of a system—is that, in the extreme case, the simulation becomes nearly as complex as the biological process itself, making it difficult to understand which features are necessary and sufficient for the pattern to develop. For this reason, a simpler and more abstract model, such as a cellular automaton model, can be useful.

In the case of the comb pattern system, a cellular automaton model is ideal because the comb naturally consists of a two-dimensional hexagonal lattice of cells. Furthermore, since each cell on the comb contains a single substance—brood, pollen, or honey—each may be assigned a value based on its contents. In the simplest case, we can consider four alternative states for a cell: brood, honey, pollen, or empty. A more complicated model incorporating the proper time scale could subdivide the brood into twenty-one states corresponding to its twenty-one days of development (three days in the egg stage, seven days in the larval stage, and eleven days in the pupal stage). Similarly, rather than considering only full cells of pollen or honey, we can subdivide the pollen and honey states into a number of substates corresponding to partially filled cells.

All that is needed to complete the model is to identify a set of transition rules that specify the probabilities that a particular cell type becomes transformed to another cell type. These transition rules depend, in part, on the content of nearby cells. We must therefore specify the relevant neighborhood of cells which affect the cell in question. For example, we know that the queen rarely lays an egg in cells that are further than three or four cells from another brood cell. We also know that pollen and honey cells completely surrounded by brood are emptied at a greater rate than cells surrounded by honey. The transitions from empty cell to brood cell or to pollen cell and then to empty cell depend on the content of neighborhood cells. The radius of this neighborhood is small, amounting to just a few cell lengths. In the model, we consider the neighborhood to consist of all cells within a radius of four cell-lengths. Although the choice of four cell-lengths is somewhat arbitrary, this value can be changed in the simulation. Certainly, a small value, less than ten cell-widths, would be reasonable.

We can now describe a set of rules specifying the probability of transition from one cell state to another. These rules were developed to incorporate the known biology, as previously discussed, and applied to empty cells, honey and pollen cells, and brood cells.

Empty Cells

1. An empty cell can receive an egg, a load of pollen or a load of honey. It may also remain empty. If a threshold number of neighboring cells contain brood, then the cell will get an egg with a certain probability, P_b, a function of the queen's rate of egg laying and her ability to discover all available empty cells. If the cell does not receive an egg, there is a certain probability it will receive either honey (P_h) or pollen (P_p), or remain empty (P_e). The probabilities P_h, P_p, and P_e, do not depend on the states of the neighboring cells but are only a function of the colony's foraging rate. For example, during a period when ample nectar is available and the bees are actively foraging, P_h may be close to 1, while P_p and P_e will be close to zero.

Honey and Pollen Cells

2. The contents of honey and pollen cells are incremented in units of a single load, corresponding to the amount of honey or pollen a bee brings in from the field. For simplicity, honey or pollen is also removed from a cell in decrements of a single load. The capacity of a cell is taken to be twenty loads.

3. Honey or pollen can be added to an empty cell or a cell already partially full of the same substance, or added to randomly chosen cells without reference to the contents of nearby cells. The relative amounts of honey and pollen added to cells is determined by the foraging rates for these substances. The model includes fluctuations in the relative foraging rates for honey and pollen corresponding to variations in the availability of flowering plants during the season.

4. Honey is removed from a cell with a probability determined by the number of neighboring brood cells and by a baseline rate of removal that is unaffected by nearby cells. For honey, this background removal is small compared to the large amounts of honey stored for winter use. Only about 60 percent of collected honey is consumed during the foraging season.

5. The probability that pollen is removed from a cell also depends on the two factors listed for honey. Removal probability depends on the number of nearby brood cells, as well as on the baseline removal rate. However, the baseline removal rate of pollen is much higher than that for honey since 95 percent of the pollen collected during the season is consumed.

Brood Cells

6. With each time increment the age of each brood cell is increased by one day. A brood cell becomes empty after twenty-one days as the adult bee emerges from its cell.

These simple rules capture the essence of the comb pattern formation process. They have been implemented as a cellular automaton model using the programming language StarLogo. The simulation can be downloaded at our self-organization website: http://beelab.cas.psu.edu. The StarLogo application can be downloaded from the StarLogo site at http://www.media.mit.edu/starlogo.

As shown in the simulation (Figure 16.10), the pattern develops in much the same way as in the Monte Carlo simulation. The advantage of this model is that it compensates for the loss of realism by offering a better conceptual understanding of how the pattern forms. Although many parameters are involved in the cellular automaton model, it is easier to understand the key features of pattern formation from this model than from the simulation.

A systematic analysis reveals that the characteristic pattern is insensitive to the particular parameter values chosen, as long as the following relationships are maintained: The queen deposits her eggs near other brood cells; the influx of honey is greater than the influx of pollen; the removal of honey and pollen is greater near brood, the brood-development time is relatively long with respect to other processes occurring on the comb. The choice of the size of the neighborhood, the particular values of honey or pollen input and removal, or egg laying rate are not important, affecting only the rate at which the pattern develops but not its general form. What this means is that the characteristic concentric pattern of brood, pollen, and honey emerges under a wide variety of conditions, essentially always present in nature.

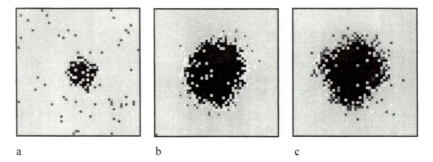

a   b   c

Figure 16.10 Results of a cellular automaton[1] model of comb-pattern formation using the programming language StarLogo. (a) day 3 of the simulation; (b) day 7; (c) day 9.

A Differential Equation Model

The proposed model comprises four differential equations incorporating what is known about behavior of the queen bee and the collection and consumption of pollen and honey. A detailed presentation of the model can be found in Camazine et al. (1990). Three of the model's equations govern the rate of change in the amount of honey (H), pollen (P), and brood (B) in the cells. The fourth equation reflects the movements of the egg-laying queen (Q). A honey bee colony has discrete numbers of individuals and cells, but inasmuch as the numbers are quite large it is reasonable to model the process of pattern formation as continuous and monotonic in time and space. Each model equation is first presented in plain language so that each term is clearly related to observed honey bee behavior.

1. Rate of change of honey concentration = honey input − background consumption by adult bees − consumption by nurse bees for feeding of nearby brood:

$$\frac{\partial H}{\partial t} = a_1 s(H)\theta_1(P)\theta_2(B) - a_2\left(\frac{H}{k_1 + H}\right)[1 + c(B)]. \qquad (16.1)$$

2. Rate of change of pollen concentration = pollen input − background consumption by adult bees − consumption by nurse bees for feeding of nearby brood:

$$\frac{\partial P}{\partial t} = a_3 s(P)\theta_1(H)\theta_2(B) - a_4\left(\frac{P}{k_1 + P}\right)[1 + c(B)]. \qquad (16.2)$$

3. Rate of change of brood concentration = egg input by queen:

$$\frac{\partial B}{\partial t} = a_5\theta_1(H)\theta_1(P)(1 - B)Q. \qquad (16.3)$$

4. The movement of the queen = a diffusion process, or simple random walk:

$$\frac{\partial Q}{\partial t} = \frac{1}{r}\frac{\partial}{\partial r}\left[\left(\frac{Dk_2 r}{k_2 + H + P}\right)\frac{\partial Q}{\partial r}\right]. \qquad (16.4)$$

H, P, and B are, respectively, the honey, pollen, and brood concentration at time, t (days), and r is the distance (number of cell widths) from the center of the comb or from the cell where the queen lays the first egg. The role and form of the various functions $s(H)$, $\theta_1(P)$, $\theta_2(B)$, $c(B)$, and the parameters a_1–a_4, k_1, k_2, and D are described below.

We consider the radially symmetric, two-dimensional pattern formation on a single comb in the center of the colony. A brief explanation of the terms in the equations is given below. A complete derivation is presented in the appendix of Camazine et al. (1990).

Equations (16.1), (16.2): Input and Consumption of Honey and Pollen

The rate of change is described with three terms. The first term is the rate of input by foragers, the second represents the background consumption by the adult bees, and the third represents consumption by brood. Each term involves a parameter indicating the maximum rate of input or removal, and factors which modify these rates and depend on the local levels of honey, pollen, and brood.

In the honey input term, $a_1 s(H) \theta_1(P) \theta_2(B)$, a_1 is the maximum rate of honey input. The other three factors are functions of the current levels of honey, pollen, and brood. They modify the input rate in a region according to local levels of honey, pollen, and brood. Since a cell contains only one substance (honey, pollen, or brood), the presence of pollen or brood in a cell strongly inhibits honey input. However, honey input to a cell is not inhibited by the presence of honey in the cell until it is full. The typical forms of these inhibitory functions are described in detail in the appendix of Camazine et al. (1990). They are decreasing functions of H, P, and B and exhibit threshold-like behavior. The pollen input term $a_3 s(P) \theta_1(H) \theta_2(B)$ is of the same form as that for honey, so that a_3 is the maximum rate of pollen input.

There are two honey removal terms, $a_2[H/(k_1 + H)]$ and $c(B)a_2[H/(k_1 + H)]$, reflecting a background rate of consumption by the adult bees across the entire comb and the additional consumption nearby brood, respectively. The parameter a_2 is the maximum rate of honey consumption. The factor $[H/(k_1 + H)]$, where k_1 is much smaller than 1, assures that as long as there is honey in a cell the bees remove it at a uniform rate. The function $c(B)$ in the second removal term is directly proportional to the number of nearby brood cells and thus reflects the increased honey consumption that occurs in the presence of brood.

The two removal terms in the pollen equation have the same form as those in the honey equation. The parameter a_4 is the maximum rate of pollen consumption, and the function $c(B)$ influences the rate at which pollen is consumed in the presence of nearby brood.

Equations (16.3) and (16.4): Input of Brood

The brood equation consists solely of an input term. The model considers only the first twenty-one days, the length of the developmental period of worker bees; the emergence of adult bees from the comb need not be considered. To model the spatial characteristics of egg input, an equation for the queen's movement describes the probability, Q, that the queen will be a specified distance from the center of the comb as a function of time. The queen's movement is modeled by a process analogous to a simple random walk, or a diffusion process. Initially the queen lays eggs in the center of the comb. As time progresses the queen moves outward, inhibited only by the presence of honey or pollen. This ensures that she is likely to stay within the brood area. The rate of input of eggs is determined by the product of a_5 and the queen's maximum physiological rate of egg-laying per unit time, the value of Q at that location, and the inhibition factors $\theta_1(P)$, $\theta_1(H)$, and $(1 - B)$. The brood level is normalized to 1, so $(1 - B)$ is the fraction of cells in a neighborhood without eggs. These factors, similar to those in the honey and pollen equations, modify the input of brood (eggs) according to the local presence of honey, pollen, or brood.

Parameter Estimation

It is crucial to obtain biologically realistic estimates of the parameters. The key parameters of the model are the rates of honey and pollen input and removal, and the queen's maximum egg-laying rate.

A cell filled with honey weighs approximately 0.5 g. A worker cell has a hexagonal cross section with a width of 5.25 mm and an area of 23.87 mm². Given a cell depth of 15 mm, the volume of such a cell is approximately 360 μL. The specific gravity of honey is approximately 1.4 (Morse and Hooper 1985), so a honey-filled cell weighs 0.5 g. During a foraging season, assumed to be five months in duration, a typical colony collects 60 kg of honey, giving an average rate of $(60,000 \text{ g}/ 150 \text{ d}) \times (1 \text{ cell full}/ 0.5 \text{ g}) = 800$ cells full per day. A reasonable estimate for the amount of honey consumed during that period is 35 kg, or slightly more than half the input (Seeley 1985). Thus the honey consumption is approximately $35/60 \times 800 = 467$ cells full per day.

The specific gravity of honey differs from that of pollen, and this must be taken into account when calculating the corresponding input and consumption of pollen. A pollen-filled cell weighs about 0.23 g (the mean weight calculated from 22 full pollen cells; Camazine, personal observation). A typical colony collects about 20 kg of pollen during the season (Seeley 1985), so the average input by pollen foragers is approximately $(20,000 \text{ g}/150 \text{ d}) \times (1 \text{ cell full}/ 0.23 \text{ g}) = 580$ cells full per day.

The pollen consumption term is estimated from the total colony pollen consumption during the foraging season, which is about 95 percent of its input (Seeley 1985), or approximately 551 cells full per day. Since pollen input and consumption are nearly equal, pollen does not significantly accumulate in the colony. As pointed out previously, pollen input fluctuates widely since the periodic blooming of pollen-producing plants leads to periods in which net pollen input is relatively large and the comb rapidly fills with pollen. In one study, for example, the average ratio of incoming pollen foragers to honey foragers was 0.26 ± 0.19 (mean $\pm$ s.d.) with a range of 0.06 to 0.83 (calculated from Visscher and Seeley 1982). In the computations presented below the pollen input was varied while the average pollen input was fixed at 580 cells-full per day.

With these estimates of honey and pollen input and consumption, we can now estimate the parameters a_1–a_4, measured in cells-full per cell per day. Our simulations used a comb area of approximately 10,000 cells. A typical colony in the middle of the season would have relatively few available empty cells, so that if an empty frame were placed in the hive a large portion of the colony's input would go into that empty frame. For simplicity we assume the total colony input goes into these 10,000 empty cells. Thus, dividing the above input estimates by 10,000 gives $a_1 = 0.08$ and $a_3 = 0.058$. Note that the choice of 10,000 cells only affects the rate at which the comb is filled, and is therefore arbitrary. This is discussed in more detail in the appendix of Camazine et al. (1990). Using these values results in filling the frame in about seven to ten days, in agreement with experimental observations (Camazine 1990). The *ratios* of the input and removal parameters, not their absolute values, are important for the behavior of the model.

The estimation of the removal parameters is more complicated. As noted previously, pollen or honey near the brood is removed at a rate about ten times greater than on the periphery. Since experimental observations show that pollen accumulates mostly in regions close to the brood, it follows that most of the pollen is being removed not at the rate a_4 but at a rate intermediate between a_4 and $10a_4$. For convenience we assume that the pollen is removed at the rate $5a_4$. Since pollen removal is approximately 95 percent of the average input, a_3, we estimate a_4 to be $0.95a_3/5 = 0.011$. However, honey appears mostly on the periphery of the frame; thus only a small fraction of the total removal occurs at the higher rate and it is unnecessary to include the factor of 5 in the estimation of a_2. Recalling that the honey-removal rate is 35/60 the input rate, $a_2 = 35/60 \times 0.08 = 0.047$. The parameter k_1 cannot be estimated directly from biological data and is discussed in the appendix of Camazine et al. (1990).

The egg-input parameter, a_5, is the maximum physiological egg-laying rate of the queen in the absence of honey, pollen, or brood. The presence of food decreases the queen's ability to lay at the maximum rate (since the queen must search for empty cells). The rate of egg-laying varies considerably, depending on various factors, including the queen's age, the season, available food supplies, and crowding. However, maximum daily egg-laying rates of between 1000 and 2000 have been reported (Bodenheimer 1937; Brunnich 1923; Nolan 1925). We estimate, therefore, that the maximum egg-input is 1500 eggs per day. The parameters k_2 and D have no direct biological interpretation. They are discussed in the appendix of Camazine et al. (1990).

Results of the Model

Results obtained by numerical solution of the model equations are in good quantitative agreement with the actual pattern-development process observed in honey bee colonies. Furthermore the emergence of the characteristic pattern is not very sensitive to the particular parameter values chosen, provided they are in the range observed in nature. Figure 16.11 shows computed results with parameter values indicated in the figure legends. All the parameter values were kept constant during the simulation except for the pollen-input rate, which was increased for two days out of ten, simulating the natural fluctuations in pollen availability. On days six and seven, the pollen input rate was 0.174, and on the remaining eight days the level was 0.029, giving the proper average pollen input-rate of 0.058 described above.

Figure 16.11a shows the pattern in its early disorganized phase, a day (in the simulation) after an empty frame was placed in the hive and the queen began to lay in the center of the frame. At this time, low levels of pollen and honey are found across the entire frame, except in the immediate vicinity of the brood. Figure 16.11b and 16.11c show the pattern on days five and seven, respectively, by which time the brood area has expanded radially and the periphery is dominated by honey. The concentric pattern is now apparent. Throughout much of the season the pollen-input rate is low and only a small amount of pollen accumulates in the interface between brood and honey. Nonetheless, the interface persists as a band of relatively empty cells (Figure 16.11a and 16.11b).

What Have We Learned?

It may be useful to pause briefly and ask two few critical questions: Have we proved that the comb pattern is self-organized? Where do we go from here?

Regarding the biology of pattern formation, we would like to be able to say that we understand how the pattern develops. We have described eleven features that are important for pattern formation. But are there any features that we omitted, and are any of the features that we included superfluous? Are we confident that we know which bees play a role in the process? Are we confident that we understand what information these bees gather and how they act on that information?

Based on the observations and experiments we described, and using the models as tools to test our assumptions, we do have some degree of confidence that the eleven features of the process are *sufficient* for pattern development. Based on our definition of self-organization, we can also conclude that the system as we presented it is self-organized—a colony-level pattern emerges from interactions among individuals at the lower level of the

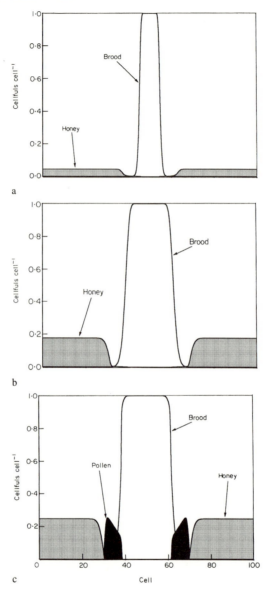

Figure 16.11 Computed solutions of the differential equation model. The graphs show the concentration profiles of brood, pollen, and honey along a cross section of a radially-symmetric frame 100 cells in diameter. The parameter values: $a_1 = 0.08$, $a_2 = 0.047$, $a_3 = 0.029$ (on days 1–5), $a_3 = 0.0174$ (on days 6, 7), $a_4 = 0.011$ (in units of cells full/cell/day); $a_5 = 1500$ cells full/day; $D = (\text{cell lengths})^2/(\text{day})$, $k_1 = k_2 = 0.01$, $k_3 = 1$. (a) day 1, (b) day 5, (c) day 7. In (a) and (b), the pollen profile is too small to be seen on the graph.

system. Furthermore, the behavioral rules of the workers are executed using only local information concerning the content of nearby comb cells. We have no evidence that any bee knows about the overall pattern developing on the comb.

However, two key questions remain: Do the bees really behave as described in the simulation? What alternative hypotheses, if any, are consistent with the data presented? We believe that the answer to the first question is, "yes." We have generated our hypotheses and built the model using a bottom-up approach. In so doing, we have only included in the model those features derived from direct observation and experimentation. To the extent that we believe those eleven biological features are valid, we should have confidence that the bees generate the pattern as described in the simulation. Nonetheless, much stronger evidence in support of the self-organization hypothesis might be obtained by manipulating or perturbing the comb-pattern experimentally, and then observing whether the bees' responses are consistent with the self-organization hypothesis.

Perturbations of the System

One type of experiment that has been done is a series of "transplants" inspired by techniques used in developmental biology. In embryonic induction, for example, notochord mesoderm induces nearby ectoderm to develop into nerve tissue (Gilbert 1994). By analogy, what would happen if one transplanted a small block of cells containing honey, pollen, or brood from one area of the comb to another? Consider the following experiment performed in the field (Camazine, unpublished data). Blocks of a hundred comb cells containing honey were cut from comb in one part of the colony and carefully implanted in the center of the brood area. The edge of the block was sealed to the adjacent comb with melted beeswax. Every other day the comb frame was removed and the contents of the block and nearby cells were recorded. After several days, the bees emptied the honey in the block of cells and the queen began to lay eggs in the empty cells. This is precisely the result predicted by the self-organization hypothesis. Based on the differential removal of honey near brood cells (point 8, p. 325), we would have predicted that that this comb would eventually be emptied, found by the queen, and replaced with eggs. Furthermore, the same experiment was simulated, and yielded the same result.

Many similar experiments (and their corresponding simulations) were done using blocks of pollen and brood. In each case, the results of the experimental manipulations were those predicted by the self-organization hypothesis and corresponded to the results of the simulation.

An interesting feature of these experiments, incidentally, is that the pattern formation process appears quite robust in the face of perturbations. There are several situations in a bee colony when the comb pattern is suddenly disrupted. For example, the continual emergence of adult bees results in irregular gaps in the compact brood nest, as does disease, such as fungal infection which kills larvae that the bees then remove. Predators may also gain access to the colony and disrupt an area of comb. The sporadic blossoming of forage plants results in large influxes of pollen and nectar that are often stored within the brood area in cells vacated by bees that completed metamorphosis. These perturbations of the pattern are self-repairing, with the pattern quickly returning to its previously ordered state. In self-organization, pattern-formation processes are generally robust; the pattern is an attractor of the system (Kauffman 1993). This robust self-repair is a common feature of self-organization and may contribute to efficient colony functioning. Indeed, to repair a defect in the pattern one need not invoke a special repair process requiring a new or special set of individual behavioral rules. The same self-organized process that built the initial pattern can operate to repair the pattern.

Returning to the question of whether alternative hypotheses could account for the comb pattern: We know, for example, that temperature gradients exist within the colony—perhaps bees use this gradient as a template or guide to deposit their honey, pollen, and eggs in particular regions of the comb. If a clearly defined temperature gradient existed across the comb and were used by the bees it would provide just the kind of global blueprint that directly contradicts the self-organization hypothesis. An experiment that has not yet been performed would be to alter the temperature gradient across the comb and determine whether the bees alter the comb pattern. An altered patern would be evidence for an alternative hypothesis of pattern formation.

Where Do We Go from Here?

After sufficient testing of the self-organization hypothesis and its alternatives, we expect to be reasonably convinced that the comb pattern develops as we have described it in this chapter. If that is the case, we can ask, "Where do we go from here? What does the model tell us? How can we use the model as a tool?"

Once it has been demonstrated that a simulation generates the observed pattern, and once the answers to the questions above provide further support for the self-organization hypothesis, we should have sufficient confidence in the model to use it as a tool to increase our understanding of the system. We can begin by exploring a range of parameter values and initial conditions. This can help us determine which of the many features initially incorporated into the model are crucial for generating the observed pattern. Explorations of this sort reveal that the model is quite robust. Pattern formation does not depend on a particular, narrow range of parameter values or initial conditions. The characteristic concentric pattern forms consistently when the following conditions hold: Eggs are deposited near previously laid eggs; the preferential removal factor is greater than unity; and the ratios of pollen and honey deposition-to-consumption correspond to those typically found in nature.

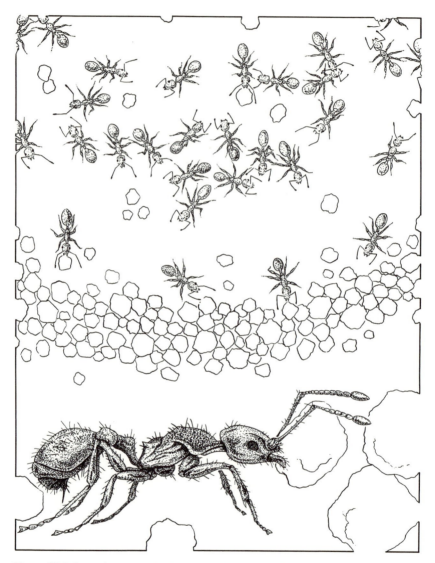

Figure 17.1 *Leptothorax* ants building a wall to surround the central cluster of brood and the queen. (Illustration © Bill Ristine 1998)

17

Wall Building by Ants

Stone walls do not a prison make
            —Richard Lovelace, "To Althea, From Prison"

Introduction

The *Leptothorax* ant colony and its fortifications are so small that together
they could fit into the empty case of a wrist watch. The advantage of studying
self-organization in *Leptothorax* ants is that they provide a simple and experimentally tractable structure: that of a social fortress in miniature.

Structure of a *Leptothorax* Nest

The fortress nests of the tiny ant, *Leptothorax albipennis*, formerly known
as *L. tuberointerruptus* (Orledge 1998), occur within the tiny crevices of rocks
that have been fractured by weathering. This species can be found nesting on
rocky hillsides in several coastal areas in Southern England. In other parts of
Europe similar rock crevices may be utilized by these and many other species
of leptothoracine ants. It seems that horizontal crevices are preferred. The col-
onized crack may be only 2 or 3 cm wide and long, with a gap between the
roof and floor of the crevice of only 1 or 2 mm.

L. albipennis colonies generally contain a single queen, up to 500 workers
and, in total, a similar number of brood: eggs, larvae, and pupae. Each worker
is only 2 to 3 mm long. Careful observation in the field often reveals that each
colony, tightly packed into its rock crevice, has arranged a C-shaped wall of
debris to surround its cluster of brood and workers. Between the wall and the
cluster, there is an empty zone (a corridor). The wall is pierced by usually only
one passageway, about 1 mm across, through which the foragers leave and
enter the nest (Figure 17.1) (Franks et al. 1992).

This tiny C-shaped nest wall is the structure that we will consider in this
chapter. This wall must be one of the least elaborate nests constructed by any
social insect.

The Adaptive Function of the Wall

The habit of *L. albipennis* to colonize crevices in rocks means that they do
not have to construct a roof or a floor to their nest. It seems likely that the

C-shaped wall they construct with its single narrow entrance is a defensive structure against physical factors and biological enemies. Colonies overwinter in these rock crevices and it is possible that the nest wall helps prevent water from entering the nest and drowning its inhabitants. It is also possible that the wall excludes drafts and prevents dry air currents from desiccating the ants' vulnerable brood. The nest wall may also help the colony defend itself against biological enemies. Larger colonies of the same species might try to invade the crevice of a smaller neighbor to take over a good nest site. *Leptothorax* colonies are vulnerable not only to intraspecific enemies but also to socially parasitic ants such as *Chalypoxenus* and *Epimyrma* (Buschinger 1986, 1989) that may try to invade their society. Social parasitism is probably more common in the genus *Leptothorax* than any other group of ants (Hölldobler and Wilson 1990). Certain species that keep slaves attack *Leptothorax* nests to capture worker pupae, which the slave-makers take to their own nests to be raised as useful slaves. Alternatively, in the case of inquiline social parasites and some colony-founding slave-makers, newly mated parasitic queens may try to invade a *Leptothorax* host colony where they kill and usurp the position of the existing queen so that they can commandeer a complete workforce and nest site. In all these cases physical factors and biological enemies can effectively destroy the colony. Enslaved or parasitized workers have no fitness—biologically they are the living dead. The C-shaped wall, no matter how simple it may appear, especially if it is breached by only a single narrow entrance that is easy to defend, can be a lifesaving biological structure.

How the Nest Wall Is Constructed

To understand how the nest wall is constructed, we need to explain the following observed features of wall building in *Leptothorax*:

1. The efficient construction of new nests in new cavities of a novel geometry. Given a nest with some complete cardboard walls in a pattern that the ants could never have encountered over their lifetimes or indeed over their evolutionary history, the ants only do enough building to close up the open walls (Figure 17.2d).
2. The ants reach a consensus on where the single nest should be situated when many different possibilities for the position of such a nest are simultaneously available.
3. Ants build a nest cavity proportional to the size of the population.
4. The ants build a smaller nest of appropriate size if some of their nestmates are removed before building begins (see later).

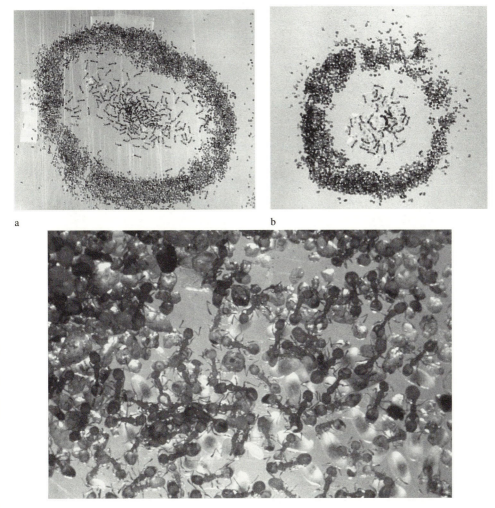

a

b

c

Figure 17.2 (a) A *Leptothorax albipennis* nest in the laboratory. The ants were given a 1-mm-deep cavity between 40 mm × 40 mm glass plates in which they built a dense wall from sieved sand (see text). Each grain of sand is roughly 0.5 mm in diameter and has been dyed blue. The worker ants, each approximately 2.5 mm in length, are densely clustered around the central queen, the largest individual present, and the white brood. There is a corridor between the dense brood/adult ant cluster and the inside of the wall. The single entrance to the nest is at "7 o'clock." (b) A nest built under similar circumstances to Figure 1a, but housing a much smaller colony. Note that the density of material is uneven in the wall and the ants have yet to select just one entrance way through the thinner part of the wall from the several prototype entrances present between "12 o'clock and 3 o'clock." (c) Individually marked *Leptothorax albipennis*

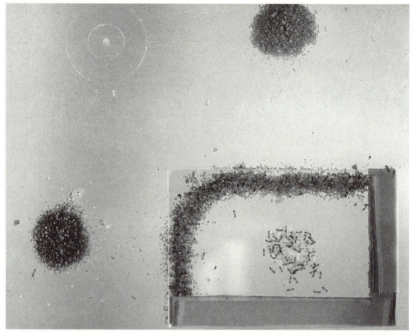

d

(*Figure 17.2 continued*)   workers in close proximity to part of the nest wall they have constructed. (d) The ants have been given a nest cavity with pre-existing L-shaped cardboard walls. They have also been given two piles of building material. Notice how they have used the respective closest pile of sand to build the long and short walls of the nest (as shown by the light and dark shades of sand). Also notice the greater width of the corridor between the tight cluster of adults and the wall of the nest compared to the corridor in (a) and (b). ((a), (b), (d) © N.R. Franks; (c) © James King-Holmes)

Dynamics of Building Activity

Observations of nest building in the field are virtually impossible since the nest is encased above and below in solid rock. However, *Leptothorax* ants are easy to culture in the laboratory. A suitable artificial crevice can be formed from two microscope slides held 1mm apart by a cardboard pillar at each corner. When a colony collected from the field is provided with such an artificial nest crevice housed in a petri-dish, it will readily emigrate into the gap between the slides. Building material can be provided in the form of grains of carborundum, sand, or even fragments of glass. Under such circumstances, a colony will encircle itself within a few hours behind a dense cohesive wall with a single narrow entrance. Since the ants only build a wall and neither a roof nor a floor, nests between microscope slides can be easily illuminated from above and below so they can be observed with a microscope or video camera. All members of the colony can be observed and individually marked with paint (Sendova-Franks and Franks 1993,1994), and the walls they build can be recorded on film or video tape or automatically recorded using digital image analysis (Franks et al. 1992; Stickland and Franks 1994). When a *Leptothorax albipennis* colony emigrates into a new nest crevice in the laboratory, the workers bring with them all their brood. The brood are carefully arranged into concentric rings, with the eggs in the middle, progressively larger larvae in bands around these, and the pupae and prepupae arranged so they are in an intermediate position between the two largest larval stages (Franks and Sendova-Franks 1992). After the brood are sorted, the adult ants, with the queen at the center over the egg pile, form a rather tight cluster over and around the brood. Only when these formations of brood and adults are complete does a small minority of the workers start building.

The organization of worker activity into two phases, moving brood and building walls, can be easily understood as a by-product of competition between two types of carrying activity and a higher preference among workers for hauling brood rather than building material. Unladen workers can pick up a brood item or a stone, but given a choice they are more likely to pick up the brood than the stone. Workers are first employed carrying brood, and only when this task is almost done (when most of the brood is carried and dropped at the right place) are they secondarily available to pick up and transport stones. This script is sufficient to produce a delay between the tasks in the absence of any direct inhibition produced by the brood on stone-carrying behavior.

In this chapter, we will concentrate on building behavior. A colony is capable of building a new set of walls in about twenty-four hours. The 'global' rate of stone deposition ranges from 0.003 to 0.025 granules/s (see Figure 17.3 and Franks et al. 1992).

Nest Size Regulation

One result of experimental analysis is that there can be a rather precise relationship between the population size of the colony and the area enclosed by the nest wall (Franks et al. 1992). The ants appear to build themselves a nest of just the right size. Figure 17.3 shows that each adult worker has about 5 mm^2 of floor area in the nest. This relationship between the size of the colony and its nest holds true whether the ants have to clear a cavity of scattered building materials or bring new building materials into the nest from the outside. In the following account we will examine nest building in nest cavities that are initially free from debris.

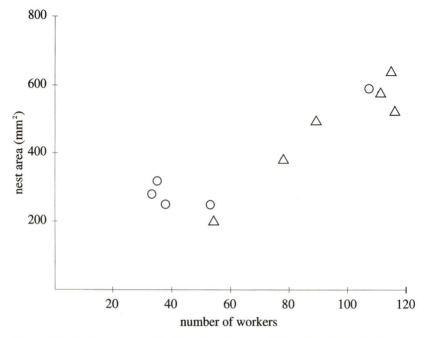

Figure 17.3 Nest area varies directly with colony-population size (from Franks et al. 1992). Circles represent nests made in empty cavities that constrained the area in which the ants place the building materials. Triangles represent cavities that were originally full of material which the ants had to clear to make their living space. In both cases the ants allocated approximately 5 mm^2 for each adult ant in the colony.

Response to Changes in Population Size

One of the most attractive features of social insect colonies, from an experimental viewpoint, is the ease and speed with which a colony can be taken apart by the investigator and put together again by the insects themselves. An *L. albipennis* colony can build a new nest, have that nest taken apart by an investigator, and respond by emigrating to a new nest and building a new set of nest walls all within a few days. The disruptive process can also be modified so that between one nest and the next a fraction of the worker population is temporarily taken from the colony and housed separately. In such experiments, the remaining part of the colony builds a nest in proportion to the existing population size and current requirements. When the removed members of the colony are returned they typically squeeze into the existing small nest. Intriguingly, approximately 50 percent of the colony can be removed and returned in this way without initiating any rebuilding (Figure 17.4a). But if approximately 75 percent of the workers are initially removed and returned along with their

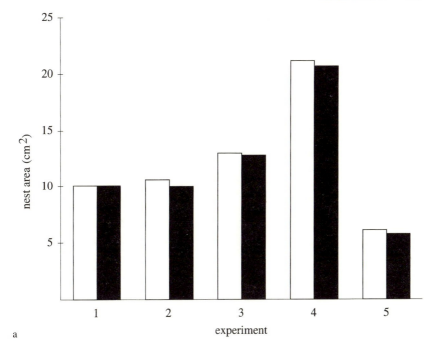

a

Figure 17.4 Two sets of experiments that further reveal the control of nest size in *Leptothorax albipennis*: In (a), approximately half of the adult ant population in each of four colonies was removed and later returned after building by the first quarter of the colony was completed (from Franks and Deneubourg 1997); in (b), approximately three-quarters of the adult ants in each nest were removed and later reunited after the building by the remaining quarter of the colony was completed (Franks and Deneubourg 1997). The white bars show the size of nests constructed by each of the reduced colonies. The black bars show the final size of the nest after reunification.

nestmates, the colony unleashes an explosion of new building activity (Figure 17.4b) (see Franks and Deneubourg 1997 for further details).

The sequence of events just described is enigmatic. Why do the ants first build a nest of just the right size if they will also temporarily tolerate substantial overcrowding as their population grows? Why does the addition of just a few extra workers trigger a wave of nest destruction and rebuilding? We will suggest one way in which this mystery may be resolved later in this chapter when we consider a model of building behavior for these leptothoracines.

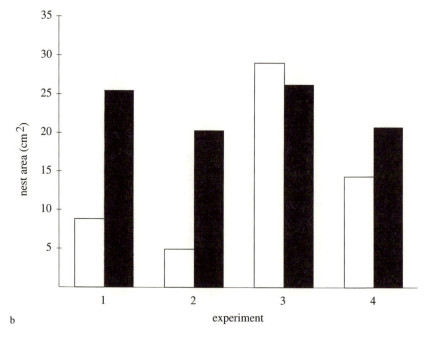

b   experiment

Individual Building Behavior

The building process has the following stages.

1. Observations of colonies with individually marked workers showed that only a small fraction of the workers collect building materials. These workers depart from the tight cluster of their nestmates inside the nest, collect a single grain of building material, which can be larger than their entire heads, and return with it to the nest. The first few operations of this kind often have a characteristic pattern.
2. The ant with its stone walks back into the nest and drops the stone within a distance of one or two of its own body lengths from the cluster of its nestmates.
3. Ants carrying stones tend to release their stone after they make direct contact with a cluster of nestmates or with other previously deposited stones. The latter stimuli grow in importance as more and more stones are deposited. At this stage the ants seem to use the stone they are carrying as a battering ram and only release it after they sense the resistance of other stationary stones. Such ants actively bulldoze their stone into others (Franks et al. 1992).
4. Ants that retrieve building material from outside the nest rarely if ever pick up a stone that they have dropped inside the nest.

5. Ants that remain in the nest, particularly those on the outside of the tight cluster of workers around the brood frequently pick up stones that are close to them and bulldoze them outwards again, toward the periphery of the nest.

These simple behaviors have been quantified in some detail. We will consider first the behaviors that initially deposit stones and then behaviors that relocate such material.

Deposition of Material

Qualitative and quantitative records have been made of the behavior of individual workers by videotaping the building process and analyzing the ants' movements. In order to get accurate unbiased records its is often necessary to play such videotapes both forwards and backwards many times, often at reduced speeds, to examine each event of interest. Here are the major findings from this analysis (see Franks and Deneubourg 1997 for further details).

The initial location of each of 300 dropped stones was recorded in terms of the stone's proximity to other items in the nest. Most of the stones (70 percent) were dropped near, or pushed against, other stones; however, stationary nestmates were also used in the first stages of building as cues for dropping stones (9 percent). Spontaneous dropping of stones was also recorded (that is, dropping with no known stimulus) (21 percent). This study was conducted during the first stage of building, when less than 20 percent of the final number of stones had been carried into the new nest cavity. For the same set of observations investigators recorded the elapsed time from when an ant entered the nest with a stone to when it deposited the stone. The elapsed time was recorded for each of the 300 dropped stones. Figure 17.5 shows that these times have a simple exponential distribution and that the average duration of stone-carrying is 21 s. The corresponding times to drop a stone near an 'obstacle,' such as a stationary stone or ant, spontaneously or are not significantly different, so in this analysis these time periods have been combined.

Investigators examined the pathway taken by each ant after it enters the nest with a stone and recorded the number of stationary stones the ant passed that were within one antenna's length from the ant. In Figure 17.6 the natural logarithm of the number of ants still carrying a stone is plotted against the number of stones the ants encountered. The approximately linear nature of the relationship suggests that each ant carrying a stone has a fixed probability of between 0.3 and 0.5 of releasing its stone on encountering a stationary stone or ant.

The above findings are consistent with one another. External workers seem to make an approximately random walk within the nest until they encounter nestmates or stationary stones. The presence of a stationary stone was the major cue indicating that another stone should be deposited. Since in the first stages of nest construction stones may be encountered fairly haphazardly, and

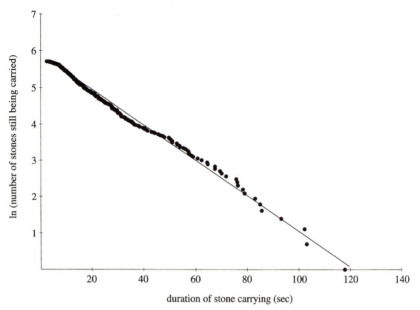

Figure 17.5 Three hundred ants were observed while carrying stones into their nest cavities (from Franks and Deneubourg 1997). Investigators recorded the time, in seconds, that each stone was carried from the moment the laden ant entered the nest cavity to the moment it dropped the stone. The natural log of stones still being carried is plotted as a function of carrying time within the nest cavity. The relationship can be thought of as a 'survivorship curve' for moving stones. A line fitted by the least squares method indicates that the relationship is best described by: the natural logarithm of stones being carried $= 5.86 - 0.048t$ (seconds) ($r^2 = 0.994$), where t is time, in seconds, and r is the correlation coefficient. Note that a negative relationship between the variables is inevitable in such a survivorship curve—hence a high r^2 value is to be expected. However the extremely high value observed and the extreme linearity of the plotted relationship suggest that the probability of an ant dropping a stone is constant per unit time. That is, there is an exponential decay in carrying times. The mean time that a stone is carried is $(1/0.048) = 21$ s.

since there is a constant probability of deposition when a stone is encountered, this explains why the probability per unit time of dropping a stone is approximately constant. This constant probability accounts for the exponential distribution of carrying times (Franks and Deneubourg 1997).

   In the vast majority of cases internal workers who moved stones pushed them outward from the center of the nest toward the periphery. In fifty-one independent observations, 72 percent of the ants moved the stone outward, 12 percent pushed the stone inward, and 16 percent pushed the stone laterally. Internal workers also tend to bulldoze the stones they displace into one another.

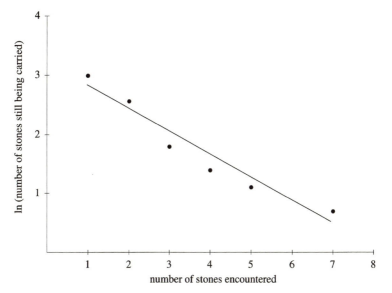

Figure 17.6 Fifty independent observations of ants carrying stones into the nest (from Franks and Deneubourg 1997) indicated a constant probability of 0.3–0.5 that a laden ant will drop its load when it encounters a stone. Video tapes of such behavior were played at slow speed and the number of stones each laden ant passed by, within one antenna's length, before dropping its stone were recorded. The relationship is best described by the natural logarithm of the number of stones still being carried = 3.20 − 0.395 (number of stones encountered) ($r^2 = 0.92$).

To determine the probability that an internal worker will pick up a stone, the videotape of the building process was played forward until an internal builder was observed to move a stone. At that point the video tape was played backward and investigators recorded the number of ants passing within one antenna's length of the stone before it was moved. This period of observation terminated when the stone was seen on the reversing tape to be moved previously. In this way, the number of ants moving past a stone in the time between it was moved and left and moved again could be recorded. These observations were made in areas of active building. Figure 17.7 shows that the probability of a stone being removed is independent of the time it has remained in that location. Figure 17.8 shows that ants have a constant probability of picking up a stone that they encounter independently of the number of other ants that have previously encountered that stone. (Constant probabilities of picking up stones also can be deduced from the reasonably high level of linearity in the relationships shown in Figures 17.7 and 17.8.) Hence, Figures 17.7 and 17.8 show that stones in the same area are essentially the same in terms of the chance they will be picked up.

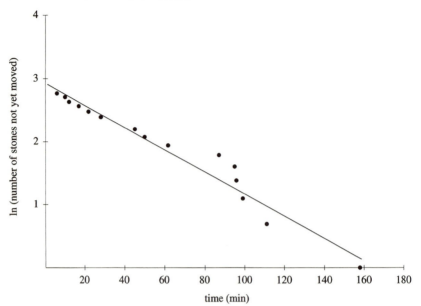

Figure 17.7 Twenty independent observations were made of more-or-less isolated stones lying within a nest, in what was to become the corridor (from Franks and Deneubourg 1997). The graph shows the natural logarithm of the number of stones not yet moved against time in minutes and indicates that the mean life time of a stationary stone in this part of the nest is (1/0.0171 minutes), i.e. almost exactly 1 h. The graph can be thought of as a "survivorship curve" for stationary stones. The relationship is such that the natural logarithm of stones not yet moved $= 2.91 - 0.0171t$ (min) ($r^2 = 0.95$).

This does not imply that the *location* of a particular stone is unimportant in terms of the likelihood that it will be moved. The more ants that pass a stone, the more likely that the stone will be moved in any given moment. In simple terms, it appears that the stones that most frequently "get in the way" of the ants inside the nest are the ones most often shifted outwards. Stones deeply buried in the wall can only be passed by near their exposed face, whereas stones out in the open can be passed by on all sides. Workers relocating material generally move stones outwards, and the stones that are most exposed to ant traffic are the most likely to be moved. Hence stones are only likely to remain stationary for long when they have been incorporated appropriately into a wall. Again, although this behavior is simple, this is not a trivial observation. We will also assume, on the basis of the data presented in Figures 17.5–17.8, that the activity level of each worker is independent of the number of workers involved in the entire building activity (there is no 'group effect'; see Chapter 18).

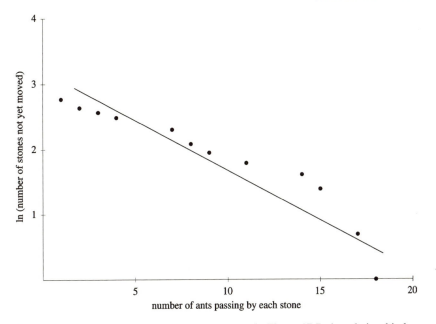

Figure 17.8 Using the same set of observations as in Figure 17.7, the relationship between stones not yet moved and the number of ants passing each such stone within one antenna's length is best described as the natural log. of stones not yet moved = 3.06 − 0.132 (number of ants passing by) ($r^2 = 0.88$). The probability of an unladen ant picking up such a stone in the nest is therefore relatively constant at about 0.1 (from Franks and Deneubourg 1997).

Observations suggest that these *Leptothorax* builders do not use either pheromone trails or cement pheromones as termites do to coordinate their building activity (see Chapter 18). Pheromone trails are used only in special circumstances by *Leptothorax* workers (but see Aron et al. 1986, 1988; Maschwitz et al. 1986; Mallon and Franks 2000). The large stones are not porous and not manipulated in the mandibles of these ants as are the soil particles that termites use and that are mixed with cement pheromones.

In general, these results show that the building workers inside the nest and those coming into it with new building materials seem to have rather simple behavior patterns and almost no direct communication with one another.

The observation that both internal and external builders tend to bulldoze stones into other stones is very important. As more stones accumulate in the nest, more "active sites" become available into which other stones can be pushed. In a sense, through the ants behavior stones "attract" other stones and this is an important source of positive feedback in the building process that speeds up building work. Such bulldozing also explains how the wall comes to be so tightly packed and densely consolidated (see Figure 17.2a).

Evidence for a Template

There is good evidence that *Leptothorax* workers use a template partly to organize their building work. Just as a *Macrotermes* queen and the building pheromone she produces act, as a template for locating the wall of her royal chamber (see Chapter 18), the cluster of adult *Leptothorax* workers and brood in the new nest may produce pheromone signals or cues that act as a template to determine where building material should be deposited.

It is also well established that numerous ants species, including *Leptothorax*, mark their nest ground and environment and that this marking affects the movements or "walk" of the individuals (see Aron et al. 1986; Maschwitz et al. 1986; Hölldobler and Wilson 1978, 1986; Hölldobler and Wilson 1990, pp. 286–291). Simply as a by-product of their centralized organizations, a circular gradient of marks can be created where the density of marks decreases with distance from the nest center.

When a *Leptothorax* colony emigrates to a new nest from an old one that had complete cardboard walls, and the colony is given building material, some ants return to the old nest site after it has been cleared of brood and place some building material around the area that the brood had occupied (A. B. Sendova-Franks, personal communication). This suggests the existence of a long term chemical template that might have a role in building. We suspect that a marking effect is probably important in many ant species.

In winter, the general activity of these colonies decreases. This occurs for two strongly related reasons: The metabolism of adult ants and their brood falls with decreasing temperature, and as a consequence of less demand from the brood for food the activity of the adult ants falls even further. Thus everything that determines the effective size of the template is reduced. We have observed that in the late autumn, as winter approaches, the ants tend to remake their nests with a smaller size in relation to the adult worker population compared to nests at the height of the summer.

Leptothorax ants exhibit building behavior of a kind that is widespread in ants, namely the active exploitation of existing heterogeneities in the building environment. When ants cluster their brood after an emigration, they are stimulated to put the brood near existing crevices. In our experimental design, ants given a crevice with two existing cardboard walls in an L-shape tend to place their brood close to the internal corner of this structure (see Figure 17.2d). Also, a difference is found between the size of corridors in such nests and the corridors in nests constructed in open areas with no cardboard walls, yet with a similar population of ants. The width of the corridor is approximately two times greater in with L-shaped walls than in a nest with round walls (Figure 17.2b and d). This suggests that the presence of the cardboard walls (which required no additional building) resulted in approximately twice as much traffic occuring in the area under construction, and that this traffic directly or indirectly contributes to the size of the template.

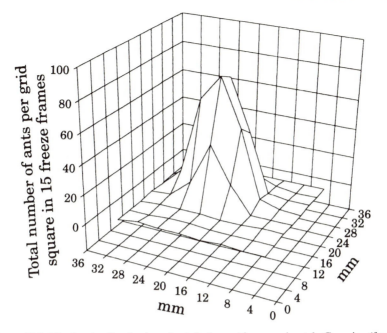

Figure 17.9 The density distribution of ants in the nest is approximately Gaussian (from Franks and Deneubourg 1997). Data for three different colonies were obtained, each showing the ants in the nest during the building process. The data graphed are derived from the pooling of data from fifteen separate freeze frames evenly distributed over the period in which most building work occurred.

This chemical template can be reinforced by a mechanical one. The cluster of adult workers around the carefully sorted brood cluster may also serve as a mechanical template that determines where the nest wall should be built. Indeed, the behavior of the first building workers who make contact with the cluster of their nestmates and then seem to 'pace out' a relatively short distance before depositing their building material strongly suggests that the cluster itself acts as a template. The density and location of adult worker ants in nests just before or during the first stages of building have been recorded. Figure 17.9, shows that such ants have a roughly Gaussian density distribution (bell-shaped curve). The workers distribute themselves at a fairly high and homogeneous density over the brood, but away from the brood cluster their density declines in all directions. Altogether, the workers in the nest form a cluster that has a fairly high degree of radial symmetry.

As in termite building, however, the use of templates in building does not explain all aspects of the building process. In these leptothoracine ants, how could a template explain the production of a single nest entrance?

The Basic Model for Wall Formation

In the previous section we reviewed a large body of information on individual worker behavior during wall-building by certain *Leptothorax* colonies. In this section, we will develop a model and simulation based on those empirical findings. The model is an approach to determining whether a self-organization mechanism can account for the pattern of the nest wall with an entrance. The simulation is based on the way that ants pick up and drop stones under the control of an amplifying mechanism and the template of the cluster of workers and their brood. Figure 17.10 summarizes the behavioral rules that can lead to the formation of the nest structure.

In the model, the space over which the *Leptothorax* ants build is simplified and represented by a square lattice (Figure 17.10d). The distance between two nodes in the lattice is assumed to be 1 mm. The probability that an unladen ant will pick up a stone is a function, $P(r)$, of the distance, r, to the cluster of workers and their brood. The center of the nest is at the center of the lattice in the simulations. The reference distance (r) is measured from this center. For simplicity we do not explicitly consider the physical influence of the worker-plus-brood cluster (see Figure 17.10).

An unladen ant (U) meets a stone, S, and can pick it up and become a laden ant, L. The frequency with which stones are picked up in a particular zone is related to $P(r)$. At any moment, a laden ant can drop its stone with a probability given by $D(r)Q$. We assume that a maximum number of stones can occupy a node. An ant that decides to drop its stone but finds the node is overcrowded moves toward another node. This assumption is based purely on physical constraints. Each node in the lattice corresponds to a surface area of 1 mm². The area occupied by a single stone is slightly greater than 0.2 mm² and the stones are around 0.5 mm in diameter. A node may contain a maximum of 5 stones; $Q = 0$ if the number of stones at a node is equal to the maximum number of stones that can be placed in such an area, and $Q = 1$ if all the space in such an area is free. The quantities D and P are influenced by the template. If the ants are in an area where building is "authorized" by the template, D is large and P is small; whereas if the ants are in an area not authorized for building by the template, D is small and P is large. The function for dropping a stone, $D(r)$, is maximum, and the function for picking up a stone, $P(r)$, is minimum, at $r = r_0$, the center of the nest. This depends on a particular density of stimuli (both mechanical and pheromonal) derived from the physical presence of brood and adult ants, as well as the chemical signals and cues these provide (see Figure 17.10). A larger population generates a larger r_D. In this simulation $D(r)$ is:

$$D(r) = \frac{D_M}{1 + \tau(r - r_0)^2} \tag{17.1a}$$

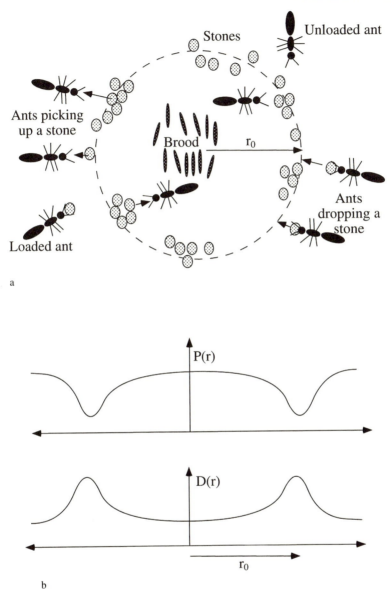

a

b

Figure 17.10 The diagram in (a) shows some of the key behaviors involved in building by *Leptothorax*. A cluster of brood is surrounded by, in turn, nest workers, building materials, and external workers. The probability that workers may deposit $D(r)$ or pick up stones $P(r)$ depends on the distance to the brood and to the center of the nest (b), and on the density of stones in the neighborhood of the ant (c and d) (see text). The insects walk on a square lattice and the behavior of each ant is affected by the configuration of stones directly beneath the ant's head and in the four nearest lattice points (see text).

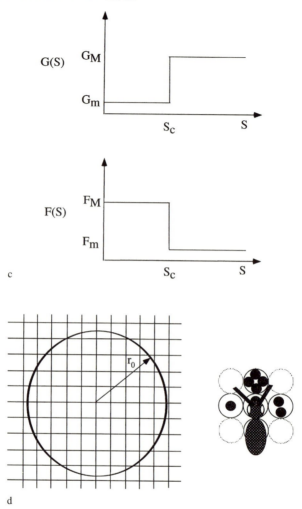

c

d

where τ is a measure of the wall thickness and D_M is the maximum value of the dropping function, $D(r)$. A large or small value of τ corresponds to a wall with, respectively, a small or large width.

The probability of picking up a stone, $P(r)$, is:

$$P(r) = P_M \left(1 - \frac{1}{1 + \tau (r - r_0)^2} \right). \tag{17.1b}$$

The first version of the model assumed that only the template, equations (17.1a) and (17.1b) affected the builders' behavior. It did not consider any amplification in the probability of picking up or dropping stones as a result of the presence of other stones. To account for the amplification effect we replace $D(r)$ by $D(r)G(S)$ so that the probability for dropping becomes:

$$D(r)G(S)Q, \qquad (17.2)$$

where $G(S)$ expresses the finding that ants drop stones mainly when they collide with obstacles. The following all-or-nothing function is a simple form for $G(S)$: if S is small and less than S_c, $G(S)$ is small and equal to G_m. If S is large and greater than or equal to S_c, $G(S)$ is large and equal to G_M (see Figure 17.10). So,

$$G(S) = G_m \quad \text{if } S < S_c; \quad G(S) = G_M \quad \text{if } S \geq S_c; \quad G_m < G_M \quad (17.3)$$

where S are the stones within the ant's radius of perception. In this simulation, the radius of perception covered the node occupied by the ant and the first four neighboring sites (see Figure 17.10). Equation (17.3) means that as the number of stones in a zone increases, the probability of dropping a stone increases, in agreement with the observed biology. Similarly the probability of picking up a stone is:

$$P(r)F(S). \qquad (17.4)$$

To reduce the number of parameters we assume that for $F(S)$, the inequality, (17.3), is reversed (see Figure 17.10):

$$F(S) = F_m \quad \text{if } S \geq S_c; \quad F(S) = F_M \quad \text{if } S < S_c; \quad F_m < F_M. \quad (17.5)$$

This means that as the number of stones in a zone increases, the probability of picking up a stone decreases. We neglect loading and unloading times as they have a relatively short duration.

Unladen ants are essentially ants moving outward from inside the nest. This is so because of the observations given above; and because ants must leave the nest for foraging and to look for building material. The movement of ants can be complex, and in this version of the model we simplified this greatly. At each time step, each examined individual may pick up a stone (if a stone is present at its current node). If the ant is successful, a new position is randomly selected and the probability of dropping is computed. If the ant does not drop its stone, a new random position is examined, and so on. Thus each stone has, in effect, a specified probability of moving from one node to another at each time step. This approximation is justified by the difference in the time scale between individual building activities and the global dynamics of building.

Two limitations may apply to this model. First, the movement of the ants is represented by a very crude approximation. A better spatial model would introduce the exact walk of ants between the different zones and would examine the relationships among the movements of the workers and their building materials. Second, the model does not take into account certain sources of negative feedback. For example, it does not consider negative feedback that might inhibit wall-building activities when the wall is finished. In the model, the ants never stop building. The model ants will continue to try to find more stones even when the wall is finished. However, our purpose here is to capture only the important properties of this system, and focus on them to further our understanding (see below). The local description we have employed is sufficient for these goals.

Role of the Parameters

From the data analysis presented in Figure 17.6 there was a roughly 40 percent chance that a laden ant dropped its stone when it collided with a stationary stone or a stationary ant in an area of active building. This serves as an estimate for the maximum value in the model for $D_M G_M$. The probability that an ant picked up a stone when it encountered one was 0.1 (see Figure 17.8). This serves as an estimate for the maximum value in the model for $P_M G_M$.

The area on which ants moved was between 4,000 mm^2 to 10,000 mm^2 when ants had to go a little way outside the nest to pick up stones. The total number of stones offered to the ants in the experiments was between 400 and 1,000. In the simulations, the area was 6,400 mm^2 and 500 or 1,000 stones were offered to some 10–60 workers.

Amplification Effect

With the model we can separately examine the roles of both the template and amplification mechanisms. In the absence of a template effect, $D(r)$ and $P(r)$ would be constant and the location, r, would not affect the probability of picking up or dropping stones. At each time step, stones are picked up and dropped. $F(S)$ and $G(S)$ are, respectively, the probability of picking up and dropping stones. The fewer the stones at a particular site, the greater the probability that an ant will pick one up. Conversely, the fewer the stones at a particular site the lower the probability that an ant will drop the stone it is carrying. As a result we observe classical clustering of the stones. This is a by-product of the amplification inherent in stones "attracting" other stones and stones "inhibiting" one another's removal from a site.

In the simulation, at time zero a number of stones are placed at random at nodes in the network. When a small but loose cluster appears by chance, it encourages the addition of more stones to the growing cluster, thus further increasing its attractiveness. This positive feedback mechanism causes clusters

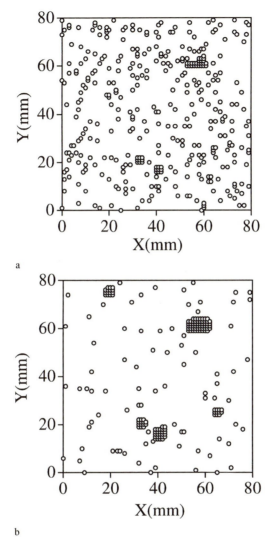

Figure 17.11 Distribution of stones with an amplification but not a template effect is shown at 2 h (a) and 4 h (b). At time zero, 500 stones were placed at random in the network. $D(r) = 0.5$; $P(r) = 0.35$; $G_m = F_m = 0.01$; $G_M = F_M = 0.55$; $S_c = 4$; maximum number of stones per node $= 3$; number of ants: 30.

to grow. Through the ants' action, clusters tend to absorb both isolated stones and clusters smaller than themselves. Figure 17.11 shows how randomly distributed stones are rapidly grouped into small clusters, which, after a longer period, merge into a smaller number of larger clusters.

The Template Effect

Figure 17.12a shows the results of a simulation in which only the template rules were applied. There was no interaction between the stones and the workers, so that $P(S)$ and $F(S)$ were constant—the number of stones did not affect the probability of picking up or dropping a stone. The function for dropping stones exhibited a maximum, and the function for picking up stones a corresponding minimum, at r_0. In other words, the stones were distributed on a circle of radius, r_0.

At time zero in the simulation, a number of stones were placed at random in the network. For a population of a few tens of workers, after a few hours, the stones were neatly clustered on a circle of radius r_0. This distribution of material is a by-product of the template rules that determine the building behavior of the ants, and of the form of the functions $P(r)$ and $D(r)$, leading to a maximum density of stones at a distance r_0. Given this script, the stones are quite regularly distributed on a circle of radius r_0 and no pile of stones is observed inside this wall.

Both $P(r)$ and $D(r)$ are directly related to local building activity and to the characteristics of the colony and are influenced by physical and chemical cues emanating from adult ants and their brood. Such cues decrease from the colony center toward the periphery. A change of colony composition leads to modification of the spatial distribution of such cues and in turn to modification of the probability of dropping and picking up material. For a particular point

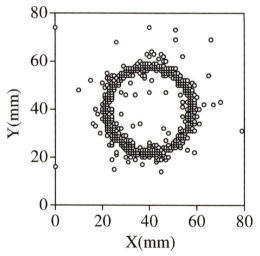

a

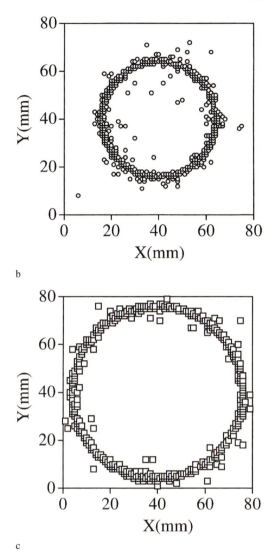

b

c

Figure 17.12 Distribution of stones with a template effect but no amplification effect at $t = 48$ h is shown in (a). At time zero, 1000 stones were placed at random in the network. $G(S) = F(S) = 0.25$; $D_M = 0.5$; $P_M = 0.35$; $\tau = 0.025$; $r_0 = 18$ mm; maximum number of stones per node $= 3$; number of ants $= 30$. In (b), the distribution of stones with a template effect is shown at $t = 240$ h. At $t = 0$, the stones were randomly (homogeneously) distributed. As in Figure 17.11a, between $t = 0$ and $t = 48$ h, $r_0 = 18$ mm. Between $t = 48$ h and $t = 240$ h, r_0 is increased to 24 mm. Other parameters were the same as in (a). Note that the radius of the wall increases. (c) Distribution of the stones with a template effect at $t = 240$ h. At $t = 0$, the stones are randomly

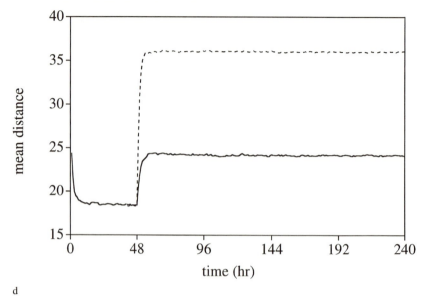

d

(*Figure 17.12 continued*)   distributed (homogeneous). As in (a), between $t = 0$ h and $t = 48$ h, $r_0 = 18$ mm. Using the same initial parameters, between $t = 48$ h and $t = 240$ h r_0 was increased to 36 mm. In (d) the mean distance of the stones to the center of the nest is displayed as a function of time. Up to 48 h, $r_0 = 18$ mm; at 48 h through 240 h, $r_0 = 24$ mm; at 48 h through 200 h, $r_0 = 36$ mm.

initially far from the center of the nest, $D(r)$ is small. As the ant population grows, $D(r)$ in this zone also grows initially. But, as the ant population grows beyond a certain threshold, $D(r)$ decreases. The pattern of change for $P(r)$ in the same zone over the same time is the exact opposite: It begins large, decreases, and then increases. So if the colony grows, the value r_0 increases and is shifted towards the periphery. In natural conditions, the rate of change of $D(r)$ and $P(r)$ is related to population growth in the colony and is low compared to the time scale of building. So we can consider that the wall will always reach a stationary value before there is a change in $D(r)$ and $P(r)$. In the laboratory, however, it is possible to induce abrupt changes in the size of colony populations. The simulation model based on this template script shows how nest size can respond flexibly to changes in colony size in a way that is independent of the history of the system.

Figure 17.12b and c shows this flexibility in response to increased size of an ant-colony population under laboratory conditions. After a period of building (e.g., 48h) with $r_0 = 18$ mm, the value r_0 was modified (e.g., 24 or 36 mm). In this case, the old wall ($r_0 = 18$ mm) was destroyed and a new wall appeared at

$r_0 = 24$ or 36 mm. In the stationary state, the new wall was identical to a wall built when r_0 was kept constant at, for example, 24 mm or 36 mm throughout the simulation. This flexibility is observed for any value of parameters, the rate of 'reorganization of the material' being the main difference observed in comparisons between different sets of parameters.

Figure 17.12d shows how the mean distance of the stones to the nest center increases with time when r_0 is modified. Initially, $r_0 = 18$ mm. After 48 h, we simulated the introduction of different populations, replacing $r_0 = 18$ mm with $r_0 = 25$ mm or 36 mm. The figure shows that for both modifications the mean position of the stones changes and reaches a mean value of 24 mm or 36 mm, respectively.

Amplification and Template Effects Combined

The behavior of a more complete model takes into account both of the key processes. When we imposed an initial value of r_0, the stones were mainly on the circle of radius r_0, as in the previous case (under the influence of the template alone). When amplification was added a number of clusters of stones occur in the wall with entrances between them (Figure 17.13a; see also Figure 17.2b and Figure 17.16). The explanation for this pattern is simple. The template effect, $D(r)$ and $P(r)$, concentrates the stones around r_0, and the mutual attraction of the stones (related to $D(S)$ and $P(S)$) favors cluster formation within this zone. When a small but loose cluster appears by chance in the building zone around r_0, this fluctuation promotes the addition of other stones to itself, thus increasing the cluster's attractiveness even further. This process, combined with competition between these clusters, leads to the formation of alternating entrances and dense stone clusters around the wall.

Probably, however, certain other processes related to the foraging traffic are also involved in door formation. On both outbound and inbound journeys the foragers may encounter stones and try to evade them and to search for other paths. This searching behavior may also be associated with the picking up of stones. Ants that encounter stones try to push the stones out of their path. If the number of stones is large in a particular zone, Z, compared to the quantity of stones in other areas, the proportion of the traffic in Z will be relatively small. If entrance formation occurs elsewhere than in Z the corresponding traffic in Z will decrease. Thus entrance-formation is extremely unlikely to occur in Z (for a description of such dynamics see Franks & Deneubourg 1997).

When a colony is removed from its original nest and about half of its adult workers is taken away, the remaining population builds a nest of an appropriately small size. When such rebuilding has been completed and the remaining half of the colony is returned, all the ants squeeze into the small nest and in most cases no new building occurs (see Figure 17.4a). Hence a colony can be housed in two different nest sizes corresponding to two different solutions to the building problem. However, when this experiment is repeated, but approximately 75 percent of the original population is initially removed and returned, an explosion of new building activity may occur. In most cases the nest with the remaining 25 percent is rapidly demolished and rebuilt so that the wall is shifted toward the periphery, thus yielding an appropriate amount of internal area to the nest in proportion to its complete population. It is noteworthy, however, that great variability occurs in the rebuilding response when 75 percent

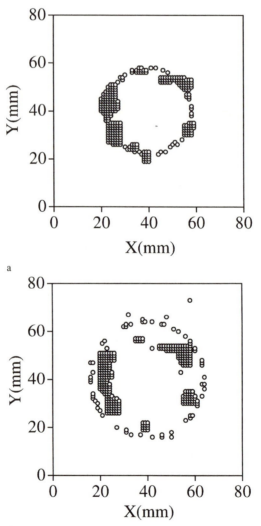

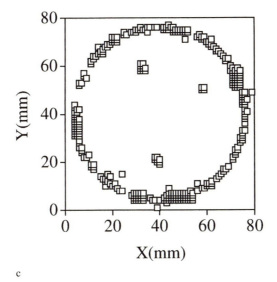

c

Figure 17.13 (a) The distribution of stones with both an amplification effect and a template effect is shown at $t = 48$ h. At $t = 0$, 1,000 stones were placed at random (homogeneously) in the network. $D_M = 0.5$; $P_M = 0.35$; $\tau = 0.025$; $r_0 = 18$ mm; $G_m = F_m = 0.01$; $G_M = F_M = 0.55$; $S_c = 6$; maximum number of stones per node = 3; number of ants: 30. (b) The distribution of stones with both an amplification effect and a template effect is shown at $t = 240$ h. At $t = 0$, the stones were randomly distributed. Between $t = 0$ and $t = 48$ h, $r_0 = 18$ mm. Between $t = 48$ and $t = 240$, r_0 was increased to 24 mm. Other parameters were the same as in (a). (c) The distribution of stones with both an amplification effect and a template effect is shown at $t = 240$ h. At $t = 0$, 1,000 stones were placed at random in the network. (*continued*)

of the worker population is returned to a nest built by 25 percent of the colony (see Figure 17.4b). This variability might appear to show that the response is typically noisy. Much of the variability might also be explained, however, by the existence of multiple stationary states in such systems.

The synergy between amplification and the template reproduces these experimental results. Multiple stationary states occurred when we simulated the experiment with reintroduction of the missing 50 percent of the colony. Figure 17.13a shows the pattern resulting from a computer simulation with $r_0 = 18$ mm, which corresponds to the result after 48 h of real time. After 48 h, $r_0 = 18$ mm is replaced by $r_0 = 24$ mm. This procedure is used to simulate the reintroduction of the missing 50 percent of the colony; the ratio between the two radii is about 1.4, so the ratio of the area of the two corresponding circles is about 2. Under these circumstances, the wall is maintained close to $r_0 = 18$, and thus does not move. But if at time $t = 0$ we impose $r_0 = 24$ and keep it

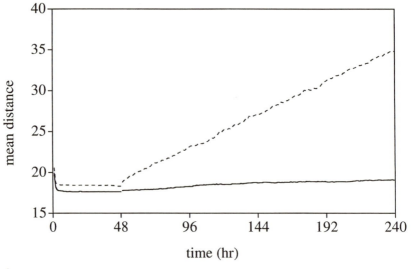

d

(*Figure 17.13 continued*)   Between $t = 0$ and $t = 48$ h, $r_0 = 18$ mm. Between $t = 48$ h and $t = 240$ h, r_0 was increased to 36 mm. Other parameters were the same as in (a). (d) The mean distance of stones from the center of the nest is displayed as a function of time, with three different boundary conditions: 0 through 48 h, $r_0 = 18$ mm; 48 h through 148 h, $r_0 = 24$ mm; 48 h through 148 h, $r_0 = 36$ mm. (Compare with Figure 17.12d.)

constant, the wall is built at this distance from the center of the nest. For the same value of $r_0 = 24$ mm, two different solutions are observed.

This result is a classical example of multiple stationary states. When modification of the respective r_0 is small, the dynamics of the system will be in a regime with multiple stationary states. When modification of the respective r_0 is much greater, as is likely to occur when the colony population is increased from 25 percent to 100 percent, the system falls outside the limits of this regime and can lead to a complete rebuilding of the walls further from the center of the nest. After 48 h, we simulated the introduction of a large population (the remaining 75 percent of the population), replacing $r_0 = 18$ mm by $r_0 = 36$ mm (the ratio of the area of the two corresponding circles is 4). In this case, the template effect strongly modified the probability of picking up and dropping stones and the wall was relocated to the new value of $r_0 = 36$ mm. Figure 17.13c shows the pattern after 240 hours of simulation. In this case a complete rebuilding of the walls further from the center of the nest can occur.

Figure 17.13d summarizes these results and shows the mean distance between the stones and the cluster of brood and workers for two different simulations in which initially $r_0 = 18$ mm. After 48 h, we simulated the introduction of more ants by replacing $r_0 = 18$ mm with $r_0 = 24$ or 36 mm. The figure shows

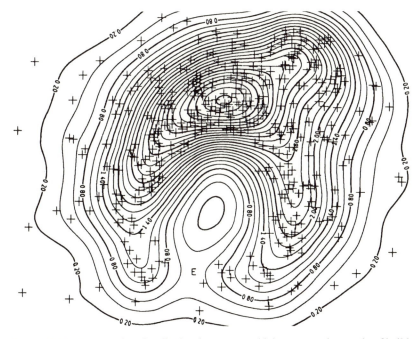

Figure 17.14 Contour plot of walls showing greatest thickness near the supply of building materials. The walls taper away in a 'pincer movement' towards the nest entrance, which is diametrically opposite the thickest part of the wall.

that the mean position of the stones was not modified for a small modification in r_0, but was changes for a larger one. These dynamics should be compared with those in the simulation that only considered the influence of the template (see Figure 17.12d). In the latter case, when r_0 was modified the walls always moved. This shows that such multiple stationary states are the by-product of an interplay between the amplification behavior, ($D(S)$ and $F(S)$), and the template dynamics. Such multiple stationary states are, therefore, clearly predicted by the model and the experimental evidence validates the prediction.

Initial Distribution of the Stones

Observations show the strong influence of the original distribution of stones on the final pattern. Placing a large quantity of material in one area tends to result in an exceptionally thick wall near that area and that progressively the walls are constructed away from this site in a pincer movement (Figure 17.14). The result is that the wall becomes progressively thinner away from the original building site and the nest entrance tends to occur diametrically opposite the

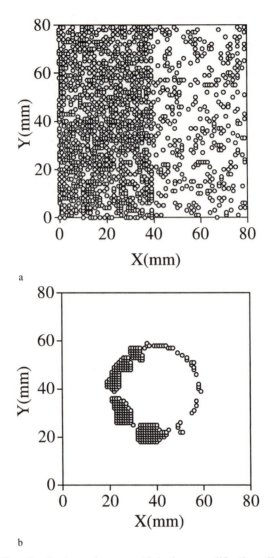

Figure 17.15 Two distributions of stones with both an amplification effect and a template effect at $t = 48$ h, with initially inhomogeneous conditions. (a) At $t = 0$, the left-hand half of the setup contained 800 stones (mean density of 0.25 stones/mm^2) and the right part contained 200 stones (mean density of 0.06 stones/mm^2). Within each half, the stones were homogeneously distributed. $D_M = 0.5$; $P_M = 0.35$; $\tau = 0.025$; $r_0 = 18$ mm; $G_m = F_m = 0.01$; $G_M = F_M = 0.55$; $S_c = 6$; maximum number of stones per node = 3; total number of stones = 1000: number of ants = 30. Two runs of the simulation (b) and (c) show the distribution of the stones at $t = 48$ h.

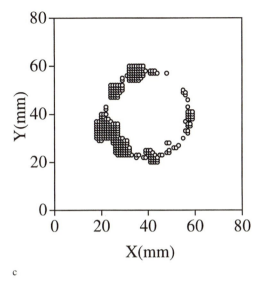

c

original source of the building materials. The simulation easily produces this effect (see Figure 17.15a, b, and c). The thickest part of the wall is constructed in the zone with the highest initial density of stones.

Discussion

Many structures and artifacts produced by animals are circular or hemispherical (e.g., see von Frisch 1975; Hansell 1984). A frequent by-product of digging in ants and other fossorial animals is excavated material distributed symmetrically around the nest hole. In such situations, the symmetry is easy to understand. Ants carrying soil particles may all issue from a common point (the nest hole) and move toward the periphery where they drop their soil pellet (Wehner 1970). In the case of a solitary builder, the animal exhibits behavior that induces a "fidelity of a part of his body" to a particular point. In this case, the animal's body behaves like a template or living compass (see the description of nest building in the weaver bird, in Chapter 4). In some instances, however, the structures produced have no initial reference point or are far larger than the builder. In these cases, the animals' initial behavior create a "center" that becomes a reference point. This is well illustrated by the geometric web in certain araneid spiders (Foelix 1982).

Three important questions arise from these brief descriptions. First, what is the reference point from which the structure is organized and what are the cues for orienting the structure? The reference point is not always evident to the observer nor is it obvious in many cases how the animal or animals select

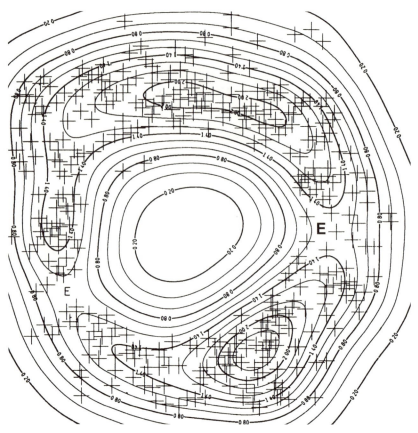

Figure 17.16 Contour plot from a real nest with two diametrically opposed entrances (E).

a specific reference point. Second, what determines the size of the structure? How is its size regulated to accommodate a variable population? Third, how can a circular structure be breached in one or more places to allow contact with the outside world. How, for example, can cues from traffic-flow in and out of the structure be used to create the appropriate size and distribution of such passageways?

Our simulations of *Leptothorax* constructions offer some answers to these questions. The *Leptothorax* nest is characterized by a cluster of brood surrounded by workers that seem to serve as a reference point for the building activity. By using this colony population as a reference-template, the ants are able to build a nest of the appropriate size for their current needs. In addition, entrance formation seems to be a direct by-product of amplifications that govern the piling behavior and of the movements of the foragers.

Brood clustering is common in ants (Franks and Sendova-Franks 1992, and references therein), and in the case of a *Leptothorax* nest emigration it is the first activity that occurs in the new site. Clustering clearly involves positive feedback: Larger clusters grow more and more quickly. Moreover, this behavior can amplify the attractiveness of favorable nest sites and help to locate the entire colony in only one of several possibly suitable sites (see Chapter 4, Franks and Sendova-Franks 1992; Deneubourg et al. 1990c). For *Leptothorax*, therefore, the question of a reference point for nest construction has been resolved. Indeed the problem of nest size regulation has also been resolved through this activity. Size is regulated through rapid clustering and sorting of the brood, surrounding this cluster with the colony's population of workers, and then using this organized grouping as a template to determine nest size.

The model we have developed in this chapter illustrates a means of automatically regulating nest size. When the colony population grows, the template grows, both in terms of the mechanical presence of the ants and their brood and any pheromone concentration produced by these members of the society. The point where this "concentration" is at an optimum for building moves toward the periphery. In such a model, the colony can react as the population increases. The growth of the template acts as the driving force for nest growth. In addition, the interplay of this template with amplification phenomena inherent in the building process is a source of the different dynamics of size regulation.

The *Leptothorax* colony is characterized by highly dynamic changes in population size (e.g., the brood develops into a new generation of workers at a particular time during the year). The findings of our model suggest that the colony can react automatically and instantaneously as the population increases or it can delay reaction so as to build less often and avoid unnecessarily exposing the nest population to dangers. The automatic and instantaneous reaction occurs when the template is dominant. The delayed reaction occurs when the mutual attraction of the building materials (mediated, of course, by the behavior of the ants) is dominant, leading to a delayed response of new building activity (Deneubourg and Franks 1995). This situation in which there exist multi-stationary states would prevent frequent episodes of explosive building and would also result in a large enough new structure that allows the colony to grow considerably before its safety is jeopardized again with another building phase. Using relatively simple building procedures, the colony appears able to anticipate and plan for the future.

The apparent simplicity of individual behavior in *Leptothorax* is shown in two ways. First, it has not been necessary to invoke group effects (see below as well as Chapter 18 and Bonabeau et al. 1998a, b). Second, as we will discuss presently, there is no need to posit the existence of two castes—internal and external workers.

Our observations suggest, for example, that the probability of an individual picking up a stone is relatively constant even in areas of different traffic flow.

This implies that a high density of ants does not change the internal state of each of them. In other words, stones in high traffic-flow areas are moved more quickly because more ants are passing by, with each ant exhibiting a constant probability of picking up a stone. There is no need to suggest a group effect whereby ants at high density stimulate one another to increase their individual rates of stone-moving behavior.

Simplicity of behavior is also suggested by the observation that each ant may follow the same basic qualitative rules that are modified, if at all, purely quantitatively according to the ant's current context. The relative simplicity of this nest building system is illustrated by considering one possible alternative form of self-organization in *Leptothorax* building.

A possible form of organization in *Leptothorax* that, in principle, could have been naturally selected is construction through competition between castes. Wall location might result from competitive activity between external workers and internal workers. The first retrieve material from foraging trips, the second carry material from the neighborhood of the brood towards the periphery. Consider, as before, that the building material can be picked up or put down. If the rule for picking up and dropping is the same as in the model described above, the only essential difference would be at the level of the pattern of movement of the ants (i.e., relating to the paths typically described by external and internal workers). When this alternative scenario was modeled (Deneubourg and Franks, unpublished), the key result was that the wall location did not depend on the absolute population but on the relative proportions of the two castes. Though this scheme might initially appear to be a simple and plausible alternative, it is actually much more complex. Not only must there be hardwired differences between internal or external workers, or hardwired switches between the two castes' behavior, but the creation of a suitable nest size would require the presence of precise, self-adjusting mechanisms to set up critical caste ratios. We suspect, however, that natural selection generally favors the simplest available mechanisms and, in this particular case, we observe that the mechanism almost certainly is simple context-dependent behavior, not competing castes.

It is even possible to mark individual leptothoracine ants (Sendova-Franks and Franks 1993, 1994, 1995a, b, c) and to examine their personal contributions to building activity. In the future this should enable us to look at the role of individual variation and possible specialization in biological structure formation.

One of the important challenges in evolutionary biology is to show plausible pathways by which sophisticated systems can evolve. The creation of simple walls and entrance passageways by *Leptothorax* suggests how termites could have made similar rudimentary but adaptive structures before additional, refining communication systems evolved. This is important because it shows how simple structures can evolve first and then how these can give rise to selection pressures for pathways of communication and information flow that could be used to refine those structures. The flux of foraging traffic in and out of the *Leptothorax* nest, coupled with, in effect, the mutual attraction between building stones (as seen in wall formation) leads to the creation of a nest entrance.

In the case of *Leptothorax*, we suspect that trail pheromones do not play a role in the formation of entrances. This contrasts markedly with the formation of passageways in and out of the royal chamber in termite nests and with the formation of termite foraging galleries both of which involve the use of cement pheromones mixed with the building material (Chapter 18). Perhaps, one of the reasons for the difference in these systems is that *Leptothorax* building is essentially two dimensional, whereas termites use trail pheromones as a template for the creation of three-dimensional arches. The importance of this study of rudimentary building by *Leptothorax* is that it shows how doors and piles of material can be formed using mostly mechanical interactions. Similar mechanisms may have been used by termites in the early stages of evolution of their building, but now involve more sophisticated systems for chemically labeling doors, and regulating their size, position, and number.

Figure 18.1 Artist's illustration of a termite mound of *Macrotermes* showing a cross section of a mound (right), a detail of workers building a pillar (left), and a view of the external surface of a large mound (center). (© Bill Ristine 1998)

18

Termite Mound Building

> The great problem of the hive confronts us again in the
> termitary, where it becomes even more insoluble for the
> reason that the organization is more complex. What is it that
> governs here? What is it that issues orders, foresees the
> future, elaborates plans and preserves equilibrium,
> administers, and condemns to death?
> —M. Maeterlinck, *The Life of the White Ant*

Introduction

Some of the largest and most sophisticated of all animal structures are the
mounds built by African termites in the subfamily *Macrotermitinae*, the fungus
growers. These castles of clay, relative to the individual termites that helped
build them, are air-conditioned skyscrapers immensely larger and arguably
more sophisticated than the vast majority of human buildings.

The largest termite mounds are probably those of *Macrotermes bellicosus*.
There has been, considerable taxonomic confusion over the species within the
genus *Macrotermes* (Bagine et al. 1989), and so we will employ species names
as used by the original authors. Certain *M. bellicosus* mounds may reach a
diameter of 30 m and others a height of 6 m (Grassé 1984). Rough calculations
suggest that if termites were the size of human beings, the biggest nests would
be about a mile high—four times the height of the Empire State Building—
and five miles in diameter (Howse 1970). It is not just the size of these mounds
but the complexity and sophistication of their internal structure that is truly
awe-inspiring. We will now consider in detail the main structural elements in
mature nests of members of the genus *Macrotermes*.

The Structure of a *Macrotermes* Nest

A mature nest of a *Macrotermes* species, such as *M. bellicosus* or *M. subhyalinus*, has a number of relatively discrete architectural elements which, with
the exception of the fungus gardens, are built exclusively from tiny pellets of
soil. There are generally six structural components to a mound (Figures 18.1
and 18.2).

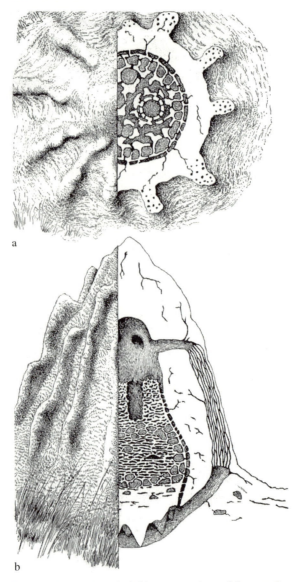

a

b

Figure 18.2 Horizontal (a) and vertical (b) cross-sections of the complex structure of a *Macrotermes* termite mound (From Lüscher 1961). Rising from the bottom of the nest are supporting pillars. The queen's chamber can be seen just to the right of center in the horizontal section. Surrounding it are brood chambers and fungus combs. In the vertical section the hole in the back of the attic, above the nest, leads to an exterior ridge of the mound wall. To the right of this hole are air channels which permeate the ridge and provide air circulation that helps regulate temperature and carbon dioxide levels in the nest. These features of the termite nest can also be seen in Figure 18.1.

1. The roughly cone-shaped outer walls can be up to 60 cm thick. These protective outer walls often have conspicuous ribs containing ventilation ducts that run from the base of the mound toward the summit.
2. The brood chambers within the central nest area have a laminar structure and contain nurseries where young termites are raised. The nest comprises thin horizontal lamellae supported by pillars.
3. The base plate (in some cases) has spiral cooling vents, presumably to promote cooling (Bristow and Holt 1987; Collins 1979). Collins (1979, p. 243) provides the following description of these structures:

   In plan view the plate is circular, up to 3.5 m across and supported by a solid pillar approximately a quarter of the width of the plate. Very small cones protruding from the underside of the plate, fit into cavities on the pillar surface, but plate and pillar are not physically bonded, the plate merely resting on the pillar surface. The underside of the base-plate bears a remarkable series of clay vanes... encircling the plate in a series of spirals. Three or four complete turns of the spiral are common before a break occurs and a new spiral begins. The vane is stalactitic in cross-section, up to 2.5 cm thick at its attachment, 1 mm thick and very fragile at the irregularly wavy fine edge. The vanes are generally coated with a white layer of mineral salts, increasing with age of mound.

4. The royal chamber is a thick-walled protective bunker only slightly bigger than its largest occupant, the queen, and has a few minute holes in its walls through which workers can pass. It tends to be located in the best protected central part of the nest, often just below ground level, beneath the hive. The royal chamber is the aptly named copularium that houses the queen and king of the termite society whose role is to supply the colony with vast numbers of fertilized eggs that will develop into new generations of termites. The queen of *Macrotermes natalensis*, for example, can produce 36,000 eggs in a day (Hegh 1922). Such a queen has been known to increase its length from 35 mm to 140 mm as a result of hypertrophy of the abdomen, increasing its weight 125 times in the process (Bouillon 1958). It is the massive ovarian development of the queen that bloats the abdomen. Like the rest of the termite mound the royal chamber has to grow to accommodate its occupants. For the termite mound as a whole the pressure for new building comes from the growth of the worker population. In the royal cell the pressure for expansion comes from the phenomenal growth of the queen.
5. The fungus gardens consist of special galleries or combs that lie between the inner nest and the outer walls.
6. The peripheral galleries are constructed both above and below ground and connect the mound to its foraging sites.

Figure 18.3 shows the stages in the development of a *Bellicositermes rex* mound (Grassé and Noirot 1949). The colony grows from a single founding pair, the king and queen termite, which excavate the first crude royal chamber themselves, to a colony of perhaps a million or more workers.

Functional Design

Macrotermes rely on their fungus gardens for much of their nutrition. They harvest plants by traveling under cover through their immense, ramifying system of peripheral corridors and arcades to harvesting sites where they cut grass and other vegetable matter after encasing it with their covered runways (Darlington 1982). In this way the termites are able to forage without being exposed to predators and environmental extremes. The plant matter is carried to the mound and kept in special storage chambers adjacent to the fungus gardens before it is used as compost for the termites' symbiotic fungus (*Termitomyces*). The compost consists of the decomposing vegetable matter and the termites' feces. The fungi convert the lignin in plant material to simpler substances digestible by the termites with the additional help of their internal symbionts (Rouland et al. 1991). The fungal gardens can be thought of as the power stations of the termites' economy.

Defense against Biological Enemies

The termite mound grows from its first most vulnerable founding stage into "a factory constructed inside a fortress" (Wilson 1971, p. 342). The fortifications, principally the immensely thick external walls, help protect the termit-producing factory against the dangerous outside world, a hostile climate, and powerful biological enemies—principally the ant-eating aardvark, and Doryline army ants (see Collins 1981, Darlington 1985, and Chapter 14). The larger termite mounds may contain millions of termites—equivalent to several kilograms of insects (Lüscher 1961; Howse 1970; Collins 1981; Darlington 1990; Darlington et al. 1992). This is an immense concentration of animal protein and it is hardly surprising that a number of mammal and ant species have evolved to attack termite mounds. However, termite colonies have, in a sense, outgrown some of their natural enemies. Termite colonies are rarely completely destroyed by mammalian ant eaters. Large termite colonies are, however, sometimes completely destroyed by *Dorylus* army ants (Collins 1981). It is not surprising, therefore, that each society is protected not only by the fortifications of the mound but by physical and chemical weaponry. *Macrotermes* soldiers not only have a powerful bite but many species also discharge a variety of effective repellants, toxins, and glues that thwart invading ants (Prestwich 1983, 1984). The mean life time of a mature *Macrotermes* colony is probably between fifteen and twenty years (Collins 1981).

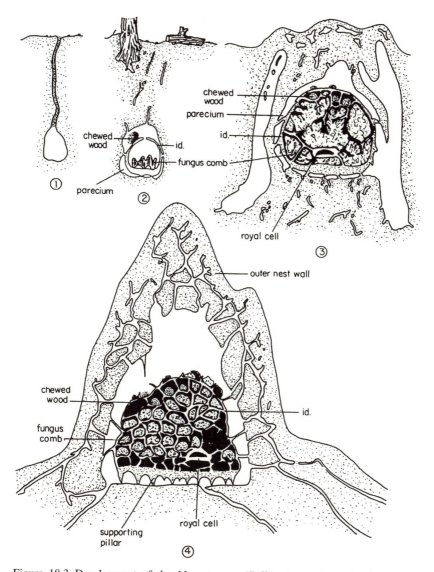

Figure 18.3 Development of the *Macrotermes* (*Bellicositermes*) *natalensis* mound: (1) "copularium," or first chamber made in the soil by the royal couple; (2), (3) intermediate stages of development; (4) fully developed nest. The wall (id.) of the fungus garden (the idiothèque of Grassé and Noirot) surrounds numerous chambers containing masses of finely chewed wood that are used as the substrate for the fungus. The parecium is the airspace surrounding the fungus garden. (From Wilson 1971, redrawn from Grassé and Noirot 1958)

Protection against Climate

The thick outer walls protect the termites and their fungus gardens from desiccation. With the exception of virgin males and females that participate in nuptial flights, all other termites survive in the cloistered, climatically stable environment provided by the nest and the tunnels connecting it to food and water supplies (Figure 18.3). Lüscher (1961) took measurements in the nests of five species in the Ivory Coast that build different types of nest and found that the relative humidity was usually between 98 percent and 99 percent and never dropped below 96.2 percent. This is impressive, given that such termite nests occur not in deep forest but in open savannah where the relative humidity of the atmosphere may be very low indeed, especially in the dry season.

The thickness of the outside walls of the mound do pose a problem: gas exchange. The termites and fungi within the nest consume oxygen and produce high carbon dioxide levels. It is known that termites can survive in unusually high concentrations of carbon dioxide in the range of a few percent (Howse 1970). Lüscher (1955) made the following calculations, which, according to Grassé (1984), are slightly exaggerated but illustrate the general problem for termites. A very large mound of *M. natalensis* may contain about 2 million termites, amounting to about 20 kg of insects, consuming about 500 mm^3 of oxygen per gram per hour, equivalent to 1,200 liters of air per day for the total termite population of the mound (Lüscher 1961). Yet, the nest contains only about 500 liters of air, not enough for the termites to survive even twelve hours. However, Lüscher (1961) found the carbon dioxide content in the center of the nest to be a mere 2.7 percent. Clearly, the colony has a means of renewing the air in the mound.

The air-conditioning system of an *M. natalensis* mound is based on the heavily fluted, substantial "ribs" in the outer walls that run in fairly straight lines from the base of the wall toward their summit (Figure 18.2). The ribs house part of a system of air ducts. Near the top of the nest, six to twelve radial canals, the thickness of a person's arm, each pass into the top of the ribs and, as they descend, divide into smaller branches 2–3 cm in diameter. More or less at ground level these branches reunite to once again form larger ducts 10–15 cm in diameter that open below the nursery area in the cellar region. The microclimate measurements that Lüscher took at various points in this system demonstrated air circulation driven by convection currents. The activity of the termites along with fermentation in the fungus gardens in the central area warms the air and raises its carbon dioxide content. The air rises to the upper air space (the attic) and is then forced out by the slight excess pressure generated through the radial ducts that are very close to the surface of the ribs. The fluted ribs act as lungs; carbon dioxide diffuses out and oxygen diffuses in through their thin walls. The ducted air is cooled in the same process and sinks to the cellar, which has

a

Figure 18.4 Queen cell building. (a) Photographs of the building activities of *Macrotermes subhyalinus* workers as they construct an arched roof (a replacement royal chamber) over their exposed queen. The photographs show the early (upper) and late (lower) stages of the process after workers were placed in a petri dish with soli and their queen. (From Bruinsma 1979) (b) Drawing of the building of the royal chamber. R is the queen; M is the male (king); the black dots correspond to pillars and vaults; T (the stippled area) is a portion of the roof; the white circles represent material deposited at the beginning of the building activity which has subsequently lost its attractiveness as a building stimulus. (From Grassé 1984)

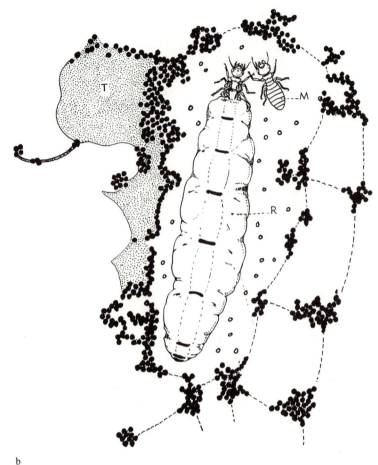

b

Figure 18.4 (continued).

the lowest temperature and lowest concentration of carbon dioxide. Notice that the termite mound is able to exchange gasses through parts of its walls without opening up holes through which enemies could enter. The termite mound is indeed a sophisticated structure. Now we will begin to consider the behavior of the individual construction workers that put it together.

Construction Activity

In this chapter, we will focus in detail on the construction not of the whole mound but of a simpler substructure, the royal chamber. The reason for choosing this structure is that its construction is amenable to detailed experimental investigation.

Building the royal chamber appears to involve the queen as a template for the developing structure. Our discussion will show how building appears to proceed through an interplay between a template and a self-organization mechanism. The queen, as producer of an important building pheromone (see pages 389–391), has a key role but does not appear to act as a leader collecting information and distributing orders. Rather, the queen's pheromone production lays down the chemical template that determines the dimensions of the royal chamber.

The very size and complexity of a *Macrotermes* mound makes quantitative observations of what individual workers do very difficult. A mature *M. subhyalinus* queen is a bizarre-looking creature with a tiny head and thorax and a massively long and bloated white abdomen. The queen is incarcerated in the royal chamber and the average distance between her body and the wall of the chamber is just a few centimeters. As her abdomen grows the royal chamber grows. The fact that the royal chamber must be regularly and quickly enlarged to accommodate and protect the growing monarch probably explains why workers will rapidly build a new chamber around an isolated queen that has been excavated from a mound and laid out in an experimental observation dish (Figure 18.4).

The process of reconstruction of this royal chamber was first observed by Grassé (1939, 1959, 1960, 1967, summarized in Grassé 1984), and discussed in great detail by Bruinsma (1979), who studied *M. subhyalinus* termites taken from mounds in Kenya.

The experiments involved observations of construction carried out by different numbers of workers around a queen and recording the dynamics of the building process. The behavior of individual *M. subhyalinus* workers during building and associated activities appears to be made up of a few, relatively simple activities. Individual workers pick up a soil granule near the queen, transport it to the site of deposition, a zone around the queen located approximately 2–5 cm from her, and deposit and cement the granule in that zone. A worker carrying a pellet of soil mixes saliva into it producing a mortarlike paste that can be molded and more or less glued into place. With time, the workers start to concentrate their deposits in one or more specific areas in the deposition zone. This leads to the construction of incipient pillars or columns a few centimeters from the queen. The pillars are lengthened until they reach a height of between 0.5–0.8 cm. Workers then change the direction of building, expanding the structure laterally to form lamellae. The growing lamellae are extended and connected to one another, to form a roof over the queen. Then the pillars are connected to form a wall.

Dynamics of the Building Process

Bruinsma (1979) described a relationship between the time taken to begin the building process and the size of the termite group. He termed this the *building latency time*, defined as the elapsed time between the introduction of the workers in the arena with the queen and the first observed grasping of a soil pellet within 0.5 cm of the queen. This is appropriate because building cannot begin without material being picked up, and most of the pick-ups occur in the neighborhood of the queen. Bruinsma tested two types of workers, "varnished" and "unvarnished." The varnished workers had been lacquered on the undersides and tips of their abdomens and were unable to lay a trail. Termites probably lay trails continuously during building activities. Such trails almost certainly recruit other termites to the sites of building activity. Figure 18.5 shows the decrease of the latency time as a function of group size. The latency times for the varnished groups were similar to those of correspondingly large groups of unvarnished workers.

One of the most noticeable dynamic effects during the construction process is that the rate of building increases very rapidly, especially in large groups (Figure 18.6). This snowball effect is typical of positive feedback system. Similarly, Figure 18.7 shows how the per capita rate of building increases disproportionately with the number of builders around the exposed queen, but then

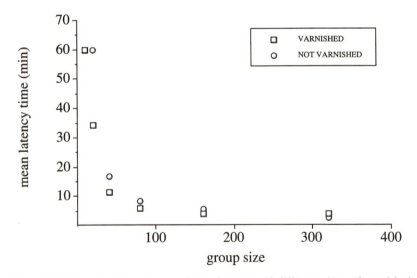

Figure 18.5 Mean building-latency time of groups of different sizes of varnished (squares) and unvarnished workers (circles). Bruinsma (1979) observed the experimental design for 60 min. The value for the 10-worker group and the group of 20 unvarnished workers is at least 60 min; at this stage no building had been observed.

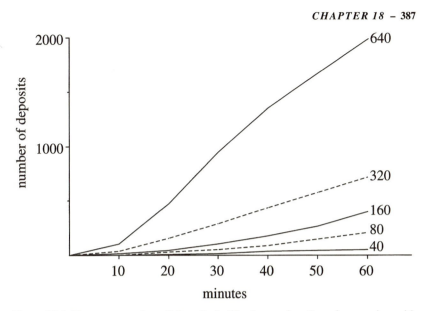

Figure 18.6 The mean number of deposits in 60 min as a function of group size, with groups of 40, 80, 160, 320, 640 (Bruinsma 1979). The activity of groups with only 20 termites is very weak and coincides with the horizontal axis.

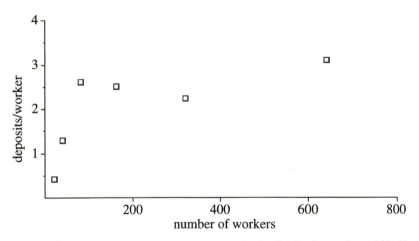

Figure 18.7 The mean number of deposits per worker in 60 min. Group sizes of 20, 40, 80, 160, 320, 640 (Bruinsma 1979).

plateaus. After 60 min, in a group of 40 workers there were about 1.5 depositions/worker on average, whereas after a similar interval a group of 640 workers make about 3 depositions/worker, on average. These experimental results correspond to the observations of Grassé (1984), where he showed that a critical number of workers is required to obtain a coherent structure.

What forms of positive feedback are involved in this building behavior? Bruinsma (1979) has shown clearly three positive feedback mechanisms: The first two involve the cement pheromone and spatial heterogeneities that can work as sources of short-range positive feedback. The third is a long-range positive feedback—the trail pheromone.

Mechanisms of Positive Feedback

After picking up a soil pellet each worker turns it into a paste by kneading it with its mandibles and adding an oral secretion which Bruinsma (1979) showed to contain an attractive cement pheromone that helps to coordinate building activity. The cement pheromone seems to lose its biological activity within a few minutes of deposition. This substance orients workers from a distance of 1–2 cm to a deposition site; it induces workers to pick up pellets in this area and to deposit pellets on other recently deposited pellets.

Bruinsma (1979) also studied the role of tactile stimuli in releasing building activity. His experimental method was to introduce tiny steel ball bearings (diameter 2 mm) into the experimental arena. Within 15 min after the start of an experiment involving 160 varnished workers, the spheres became foci of deposition activity. After an average 33 min (range 28–40 min; $n = 5$) the spheres were used as the foundations for the first pillars, 0.4 cm high, around the queen. These observations accord with those of Stuart (1967), who demonstrated that surface irregularities release building behavior in *Nasutitermes*. This clearly shows that tactile stimuli can play a significant role in coordinating building activity in termites, in the sense of focusing their efforts in space.

Another positive feedback source, the trail pheromone, plays three roles in the building activity. The pheromone trail is itself a source of long-range positive feedback, as such trails recruit workers to building sites (see Stuart 1967 for an example of the use of trails in other termites). Trail pheromones also help orient the movements of workers as they walk around the queen. During the first excursions, workers carrying soil pellets perform a more or less serpentine path around the queen (Figure 18.8). There excursions typically end after 10–15 s when the soil pellets are deposited in a zone a few centimeters from the queen. The serpentine walking patterns are gradually replaced by straighter paths from the queen to the deposition site (the transportation time being reduced to 3–5 s). This transition occurred after about 25 min from the start of the experiment, when 640 workers were used. It suggests that well-marked trails have emerged and can lead workers more quickly to the deposition zone. However, this increase in the rate of carrying and the attraction toward the building zone does not seem to play an important quantitative role in the global rate of building behavior in Bruinsma's experiments. Indeed a comparison between

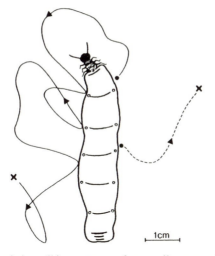

Figure 18.8 Characteristic walking patterns of two soil transporting workers around their queen. The solid line is the path of the first observed transporting worker. The dotted path is that of a transporting worker 30 min after the start of the experiment. Dots show the pick-up site and *X*s show deposition sites. (From Bruinsma 1979)

unvarnished and varnished workers (the latter cannot lay trails) revealed no difference between the rate of building whether trails were present or not. The single difference between varnished and unvarnished workers was the level of their crowding around the queen. The number of workers crowding around the queen is slightly higher for varnished than for unvarnished workers (see Figure 18.9). Varnished workers cannot lay the trails that would otherwise probably lead them to the zone of building activity.

The trail pheromone also plays a role in shaping the royal chamber. Varnished workers did not construct a replacement royal cell nor even the prerequisite pillars which are a key stage in construction work. Instead, in four separate experiments with 320 varnished workers, after 24 hours only two oblong flattened ridges of deposited soil granules had been constructed parallel to, and on either side of, the queen. These results show that the trail pheromone plays a role in the coordination of building activity leading to pillar construction.

However, even the cement pheromone, the growing sites of deposition (i.e., local physical heterogeneities) and the trail pheromones are not alone sufficient to explain the building of the royal chamber. Pheromone emitted by the queen also plays a crucial role.

Strong evidence that queens produce a building pheromone comes from observations of the reconstruction of a royal chamber when the queen is placed in a slight draft. Bruinsma (1979) placed a living queen in a slowly moving air stream and observed that the workers built an asymmetrical shelter that

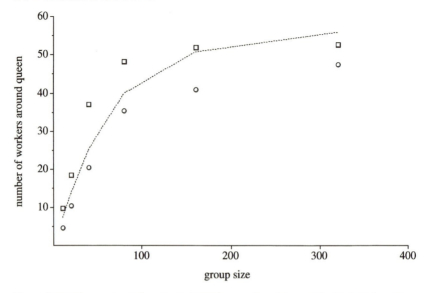

Figure 18.9 The mean number of varnished (squares) and unvarnished (circles) workers crowding around the queen. Five replications were conducted for each experimental group size. (From the data of Bruinsma 1979)

blocked the air stream. The soil pellets were placed nearer the queen in the direction of the wind source rather than on the queen's other side. This suggests that building is inhibited where the concentration of the queen's odor is too high. Bruinsma also showed that workers will attempt to construct a royal chamber around a caged and suspended queen (Figure 18.10). This observation confirms that the queen's odor, rather than physical contact, is important in instigating building activity since the workers could smell but not touch the caged queen. Workers will also begin to construct a royal chamber around a recently killed queen, but not around a wax dummy of the queen. However, the average distance of the reconstructed chamber wall to a live queen was 2 cm compared to 1.25 cm for a dead queen.

Other experiments by Bruinsma involved bioassays suggesting that the source of the queen's building pheromone is the so-called "royal fat bodies" in the queen's abdomen. These fat bodies are located around the tracheal system that supplies oxygen and carries carbon dioxide away from the respiring tissues in the abdomen. Presumably, volatile building pheromones are also carried through the tracheal system to the air around the queen.

All these observations are consistent with the living queen continuously producing a volatile pheromone from the abdomen. This pheromone diffuses into the air and its concentration declines with increasing distance from the queen.

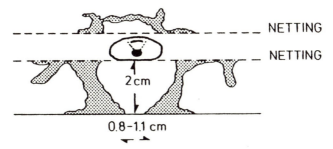

NETTING

NETTING

2 cm

0.8–1.1 cm

Figure 18.10 In this schematic representation a live queen is held between two net screens and the shaded areas represent a cross-section through the structures built around her. (From Bruinsma 1979, figure 12)

The concentration of queen pheromone determines the location of the wall of the royal chamber, evidently through its influence on the workers' tendencies to pick up or deposit soil pellets. Simple positional information is also available to the building termites as some stand on the queen's abdomen while they construct the roof of the royal chamber.

In sum, three pheromones—the cement, trail, and queen pheromone—along with tactile stimuli, play key roles in the organization of building activities during reconstruction of the royal chamber.

Self-Organization and Templates

Our goal in this chapter is not to attempt a complete description of chamber-building behavior but to stress the interaction between templates and self-organization in the building process and the benefits of using models to explore these phenomena. Using the experimental findings described in the previous section, we will explore the importance of group size on the time to initiate building activity, the queen pheromone as a template, and the role of amplification and the interplay between templates and self-organization.

We can summarize the building behavior around the queen, based on the observations described above. In general, workers without soil pellets in their mandibles move toward the queen. Typically, they pick up soil in the neighborhood of the queen and move away from her carrying a pellet. Generally, they deposit the pellets a few centimeters from the queen. We will assume that the termites lay pheromone trails constantly as they walk.

Group Size and the Time to Initiate Building Activity

We have seen that the building latency time is unaffected by the workers' ability to lay trails. However, the latency time declines markedly with increasing group size (Figure 18.5). Is this decrease due only to an increase in the number of workers—a simple statistical or numerical effect—or does it also express a group effect? A group effect occurs when membership in a group modulates the physiology or behavior of individuals in the group by means of signals or cues from other group members. This means that the signals or cues are not triggered by specific events in space or time. This definition is a modification of those provided by Grassé (1946) and Wilson (1971).

To determine if a statistical or group effect is involved in latency time we need to develop a model. In the case of a statistical effect, the probability that a worker picks up a pellet of soil in a unit of time (P) is constant and independent of the group size. So the workers exhibit the same basic behavior regardless of the group size. A decrease in latency time, therefore, arises as a statistical consequence of the greater number of workers.

In the case of a group effect, the individual probability, P, is a function (in this case an increasing function) of the group size, N. The following model clarifies this situation. The probability that no worker picks up a soil pellet during a unit of time is $Q = q^N$, where $q = 1 - p$ is the probability that a worker does not work during a unit of time. Hence, the probability that one or more workers does pick up a soil pellet during a unit of time is $1 - Q$. The probability $P(i)$ that the latency time is i time-units long is the probability that one or more individuals work during this i^{th} time unit, $(1 - Q)$, multiplied by the probability that no individual has worked previously (Q^{i-1}). The general equation for $P(i)$ is:

$$P(i) = Q^{i-1}(1 - Q). \tag{18.1}$$

The mean latency time $\langle T \rangle$ is:

$$\langle T \rangle = \sum i P(i) = \left(1 - q^N\right)^{-1}, \tag{18.2}$$

and it is easy to reorganize (18.2) to obtain:

$$q^N = 1 - \langle T \rangle^{-1}, \tag{18.3}$$

which in logarithmic form is:

$$N \ln q = \ln\left(1 - \langle T \rangle^{-1}\right). \tag{18.4}$$

A linear relation between N and $\ln(1 - \langle T \rangle^{-1})$ is obtained when $\ln q$ is constant, i.e., independent of N. In this case, there is no group effect and only a numerical effect. The acceleration of nucleation results only from the statistical summation of individual contributions, without mutual stimulation. A nonlinear relation would indicate that q depends on the population size and that a group effect exists. The agreement between experimental results and the linear

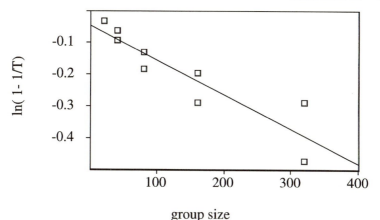

group size

Figure 18.11 The plot is a linear fit of equation (18.4) to available data. The horizontal axis is the total population, N, the vertical axis is $\ln(1 - 1/T)$, $r = 0.91$, $p < 0.001$. The data are combined for varnished and unvarnished termites. (From Bruinsma 1979)

fitting of equation (18.4), (Figure 18.11), suggests that if a group effect exists it is very weak and that the individual probability ($p = 1 - q$) to pick up a pellet is effectively constant ($P = 0.001$/min) and independent of the group size.

A calculation similar to the previous one can be made with the total population, X, of termites in a 0.5 cm wide zone around the queen. Indeed the latency time that Bruinsma measured is related to the activity in this narrow zone. In this case, we obtain $p = 0.01$ min^{-1}. It is not surprising that this value is bigger, because we are now doing the calculation with a much smaller number of termites ($X < N$).

This leads us to conclude that group effects, such as mutual stimulation either are not involved in the probability to begin to work or are at most very weak. We can conclude that in the initiation of building activity individuals appear to act independently of one another.

The Queen Pheromone as a Template

Initially, we neglect amplifying mechanisms, such as the cement pheromone, and only consider pick-up and deposition behavior under the influence of the queen pheromone. We assume that pheromone concentration decreases as we move further from the queen, thus providing a termite with some explicit measure of its distance from the queen. The simplest possible rule to explain the termites' behavior is that workers pick up soil pellets where the concentration of the queen pheromone is high and deposit material where the concentration

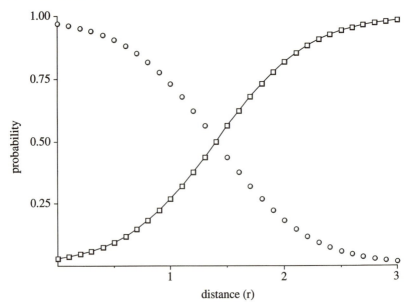

Figure 18.12 $P(r)/P_M$ (circles) and $D(r)/D_M$ (squares) (the probability of picking up and depositing pellets, respectively) are shown as a function of distance from the queen: $r_0 = 1.4$ cm, $\eta = 2.5$ cm^{-1}.

is low. Based on this rule, we can use two functions, $P(r)$ and $D(r)$, to de-scribe the response of an individual worker as a function of distance from the queen. $P(r)$ is the probability that an individual will load at r, the distance to the queen; it increases with increasing proximity to the queen. $D(r)$, the rate at which a loaded termite deposits (unloads), is also a function of r, and increases as the worker moves further from the queen.

The functions $P(r)$ and $D(r)$ can be expressed graphically as monotonic, nonlinear increasing or decreasing functions of r, as shown in Figure 18.12. Mathematically,

$$P(r) = \frac{P_M}{1 + e^{\eta(r-r_0)}}, \tag{18.5}$$

$$D(r) = \frac{D_M}{1 + e^{-\eta(r-r_0)}}, \tag{18.6}$$

where D_M and P_M are the maximum values of $D(r)$ and $P(r)$, respectively. The queen's body is at $r = 0$; r_0 is the distance corresponding to a probabil-ity of 0.5 D_M of depositing a pellet and to a probability of 0.5 P_M of picking up a pellet; and η is a parameter expressing the modulation of the response. The greater η, the steeper the response; when η is very large the response becomes an all-or-nothing step function. These mathematical functions correspond to the functions $D(r)$ and $P(r)$ used for *Leptothorax* (Chapter 17). For the wall of *Leptothorax*, however, the function $D(r)$ $P(r)$ exhibits a maximum (minimum) at r_0. In the case of the royal cell, these functions are monotonic nonlinear increasing or decreasing function of r.

We have seen that the workers' movements are more or less random at the beginning of the experiment and become more oriented with time. Termites carrying soil pellets move increasingly away from the queen whereas those without soil pellets tend increasingly to move toward the queen, rather than following a more circuitous path. This shift from random walks to more systematic walks is influenced by the trail pheromone and leads to a decrease in the carrying time (Figure 18.8). We neglect this effect in the model and assume that the workers perform a random walk throughout the experiment. This simplification is reasonable because the shift occurs only after some time and, as discussed above, the rate of building is unaffected by the presence or absence of the trail.

Formulation of the Model

A simulation can be based on the behavior of termites picking up and dropping material. As in the *Leptothorax* model, the space on which the termites build is represented by a square lattice (with a distance of 1 mm between two nodes in the lattice). The queen is at the center of the lattice and occupies one node. In this model we assume that termites move slowly; at each time step, each termite may only move randomly to one of the first four neighboring sites of the node it currently occupies.

Unladen termites move randomly and reach zones close to the queen, where they have a high probability of picking up material. The probability at which these events appear in a particular zone is $P(r)$. We assume that materials needed for building cover the ground and are abundant, so it does not affect the frequency with which an individual loads. Laden termites come from the neighborhood around the queen and enter zones where their probability of depositing material, $D(r)$, increases as r increases. In this model (but not as in *Leptothorax*), we assume there is no limit on the amount of material that can occupy a node. The coupling between these mechanisms leads to a spatial peak in the deposition of building material.

The model predicts an automatic regulation of the size of the royal chamber. When the queen's size increases, the pheromone concentration increases and r_0 moves towards the periphery. The colony reacts automatically and reorganizes the distribution of material around the queen, as in the case of wall-building in

Leptothorax (see Chapter 17), where the brood acted as a template to regulate the diameter of the nest.

In this model, as presented so far, the template provided by the queen's building pheromone influences the probabilities of picking up and depositing material. It does not include any positive feedback. Feedback is discussed in the next section.

Amplification, Templates, and Self-Organization

It is impossible to reproduce the experimentally observed exponential dynamics of building and the link between worker activity and group size if the functions P and D are kept constant in time. When P and D are constant, the cumulative deposition of soil pellets is linear in time and the activity per worker is independent of group size. To reproduce the relationships seen in Figures 18.5 and 18.6 we need to consider the sources of positive feedback in this system that were described above. These include the trail and cement pheromones, and increased physical heterogeneity of the system due to structures formed during building activities. We will not consider further the role of the trail pheromone because varnished workers are similar in their building rate to unvarnished ones.

As we have seen previously, the cement pheromone stimulates picking up and depositing behavior. So, we assume that the probability of picking up or depositing a soil particle is influenced not only by the queen pheromone but also by the previously dropped material that is still active (M). Experiments indicate that soil pellets eventually lose their stimulating activity, probably due to the evaporation of the cement pheromone.

As in the hypothesis used in the model of *Leptothorax*, we assume that the overall rate of deposition of pellets is the product $D(r)\,G(M)$ where $D(r)$ is under the influence of the queen's pheromone and $G(M)$ accounts for material previously deposited and still active.

We shall use an approach similar to the one we used to model the effect of the queen's pheromone on the workers to link $G(M)$ to the quantity of material deposited previously:

$$G(M) = \frac{G_M}{1 + e^{-\eta'(M - M_c)}}, \tag{18.7}$$

where G_M is the maximum value of $G(M)$.

In this simulation, we assume that the cement pheromone stimulates deposition at short range, and so M in equation (18.7) is the quantity of material both at the node where the termite is located and at the first four neighboring sites. η' is a parameter expressing the modulation of the response and is a measure of the steepness of the termites' response to the material deposited previously. If η' is large, equation 18.7 is an all-or-nothing function similar to the function (17.5) used for *Leptothorax*. M_c corresponds to the value of M for which $G(M)$ reaches 50 percent of its maximum. If $M \ll M_c$, the deposition probability is small. Examining the description of the stimulating effect of the cement pheromone, we estimate M_c to be reached at a density of a few pellets per node. In this model, we assume that the building material does not stimulate pick-up behavior, so the probability of picking up material is independent of M and remains $P(r)$. We also assume that termites do not pick up active (pheromone-laden) materials. Figure 18.13 is a result of a simulation showing the movement of soil as a function of the distance from the queen. Figure 18.14 shows the dynamics of deposition and is in good agreement with experimental results (compare with Figure 18.6). In this case, the sensitivity to worker numbers is clear. The building rate is low for a small number of workers but becomes high for a large population.

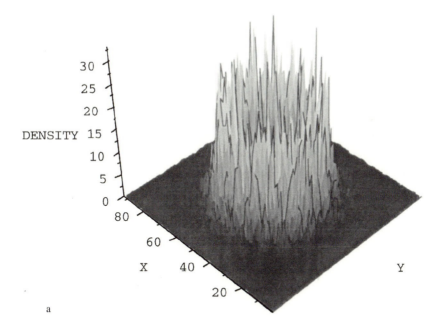

a

Figure 18.13 Distribution of building material around the queen after 8 h of simulated building. Total population, 640 workers. $D_M = 0.1$, $P_M = 0.1$, $r_0 = 2$ cm, $\eta = 0.5$, $\eta' = 0.5$, $M_c = 4$, $G_M = 0.5$. Time step, 2 s. A 3-D plot of the density of material around a central region where the queen is located is shown in (a) while (b) is the same as (a), but expressed as a series of isodensity curves with the queen in the center.

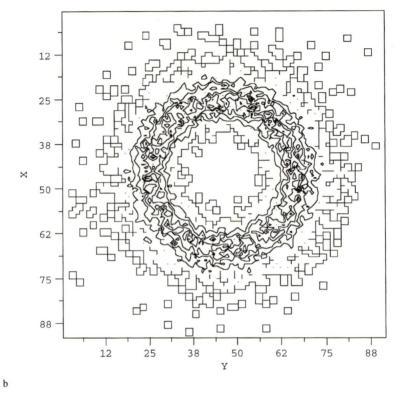

b

Figure 18.13 (continued).

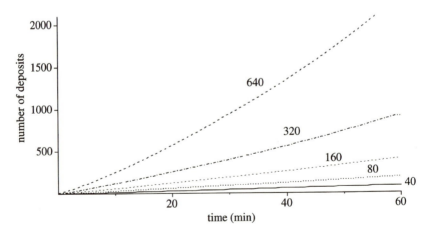

Figure 18.14 Number of deposits for 640, 320, 160, 80, 40, and 20 workers with positive feedback. Parameters are the same as in Figure 18.13. Each curve corresponds to the means from 20 simulations.

Amplification and Competition

We mentioned previously that the cement pheromone can attract workers in a zone of radius of less than 1.5 cm. This chemotaxis is sufficient to induce a regular distribution of pellet deposition. Bruinsma (1979) showed that workers with varnished abdomens can concentrate their activity at certain sites but cannot build vertical pillars. The cement pheromone diffuses freely in space and its odor attracts termites toward the region of high concentration (i.e., a pillar). They arrive in the neighborhood and deposit material. With this accumulation of material, there is a stronger and stronger emission of pheromone that attracts more termites into the neighborhood. Many neighborhoods of attraction can occur simultaneously if they are sufficiently far apart. This positive feedback at different sites in the building area gives rise to competition between different pillars that are close to one another. The by-product of this amplification and competition is a regular distribution of pillars that is not explicitly coded. The regularity results purely from a self-organized process. To illustrate this process, we use a differential equation model rather than a simulation.

A model very similar to that for clustering in social amoebae and bark beetles (Chapters 8 and 9) shows how the different parameters characterizing the random walk of the termites, the attractiveness of cement pheromone, and the diffusion of pheromone determine the regular distance between pillars (Deneubourg 1977). We will describe this mechanism and its mathematical form and then introduce it into the system of equations (18.8). We let H be the amount of cement pheromone in the air where it diffuses freely. Equation 18.8a describes the dynamics of H:

$$\partial_t H = \varepsilon M - k_2 H + D\nabla^2 H, \qquad\qquad (18.8a)$$

where ε is the amount of pheromone emitted per unit of deposited material per unit of time, so that total production is εM (M is the deposited material, still active). The term $-k_2 H$ represents pheromone decay. The third term, $D\nabla^2 H$, is the diffusion of the pheromone, where D is the coefficient of diffusion. To account for the attractiveness of this pheromone, we must describe the dynamics of laden termites, L. As we did in Chapters 8 and 9, we assume that at the level of walking the termites' response is the sum of two processes: the random walk ($\mu\nabla^2 L$) and the response to the gradient of the pheromone. The simplest form of this term assumes that the response is proportional to the gradient; the greater the gradient, the more the termites are attracted toward a peak of concentration corresponding to an important zone of deposited material. So, the term describing this attractiveness of the gradient is $\gamma\nabla(L\nabla H)$. The parameter, γ, expresses the intrinsic attractiveness of the gradient; for the same value of the gradient, the greater γ the greater the attractiveness. The equation describing the spatio-temporal dynamics of the laden termites is:

$$\frac{\partial L}{\partial t} = \Phi - k_1 L + \mu \nabla^2 L - \gamma \nabla (L \nabla H). \tag{18.8b}$$

It is further assumed that there is spatially and temporally constant flow Φ of laden termites into the system and that the rate of unloading per termite per unit of time is a constant k_1. Finally, equation (18.8c) describes the dynamics of the active material M:

$$\frac{\partial M}{\partial t} = k_1 L - \varepsilon M. \tag{18.8c}$$

The amount of material, M, deposited per unit time is $k_1 L$, and the rate of disappearance of M is εM.

In the previous simulation, the spatial distribution of material was essentially under the control of the distance-to-the-queen parameter and positive feedback increased only the rate of dropping. In this model, we do not consider positive feedback that affects the rate of dropping, and no heterogeneities are imposed on the system. We account only for the attractiveness of the cement pheromone.

When material is dropped, the cement pheromone is emitted and diffuses, attracting more termites towards this area—and so the new termites drop more material. This leads to inhibition of additional pillar formation in the immediate neighborhood of the pillar and also facilitates the emergence of another pillar farther away. This automatically produces a regularity without explicit coding of the distance between pillars (Figure 18.15). However, this mechanism of attractiveness does not automatically produce regularity under all conditions. When the termite density is too low the rate of deposition of material may become too low and the development of pillars will be difficult or even impossible. (See Deneubourg 1977 or Bonabeau et al. 1998b for a discussion of this model which considers only the attractiveness of the cement pheromone.) This sensitivity to density should not be surprising, as we have already seen numerous cases of such critical behavior related to positive feedback.

In natural situations the spatial structure results from the interplay between different building mechanisms, including the response to the queen pheromone and the different influences of the cement pheromone. It is easy to understand that the model is highly sensitive both to termite density and to density of the material, and that pillar formation is also influenced by the distance from the queen in her role as a template. Although pillars have the same attractiveness to builders wherever they occur, they can only develop in a favored region such as a zone in which a high rate of deposition occurs.

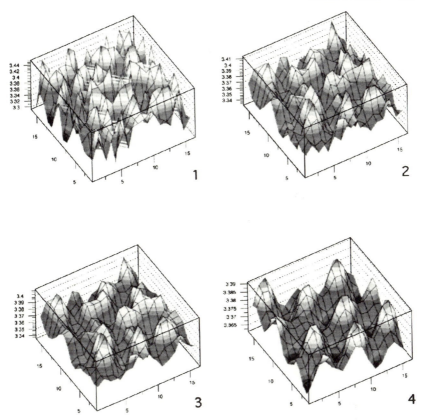

Figure 18.15 Spatial distribution of the material, M, for a 2-D system. $\varepsilon = 0.8888$, $k_2 = 0.8888$, $D = 0.000625$, $\Phi = 3$, $k_1 = 0.8888$, $\mu = 0.01$, $\gamma = -0.004629$. Distributions over time (t) are shown for $t = 1$; $t = 2$; $t = 3$; $t = 4$.

Discussion

In the models discussed in this chapter and the previous one, individual behavior may be interpreted as a result of simple stimulus-response reactions involving templates and self-organization. Although most of this chapter was devoted to a discussion of the building of the royal chamber, the same mechanisms may also apply to the construction of galleries, arches, and passageways. Although Grassé (1984) suggested that a succession of qualitatively different steps is needed to produce an arch overlying a passageway within a termite mound, Bruinsma's experiments suggest a much simpler mechanism may be involved. As termites lay trails, the trail pheromone diffuses in space; the maximum concentration is at the center of the trail and decreases from this center

toward the periphery. This concentration regulates pick-up and deposition behavior in much the same way as the termite queen's pheromone did. A high
concentration inhibits deposition. We speculate that a high concentration also
stimulates the picking up of material. This provides a very simple mechanism
for building a gallery of the right size. If traffic increases, the trail pheromone
concentration increases and construction work moves outward. The result is
the regulation of gallery size to traffic intensity. In this case, the profile of the
trail pheromone concentration constitutes a self-organizing template that regulates the distance between two pillars or two walls just as the queen pheromone
determines the size of the royal chamber. This behavior not only regulates
the width of the gallery but also explains arch formation. The isoconcentra-
tion lines are semicircles that radiate from the trail on the ground and provide
a template for a complete arch above the trail. We might, therefore, obtain
both regulation of the distance between pillars and arch formation using a self-
organizing template that arises from diffusion of the trail pheromone. Taking
such processes into account it is easy to develop a mathematical model to simulate this behavior. We do not do this here, because it would be very similar,
in principle, to the model for the queen's pheromonal template. The same fundamental logic that we have elucidated for pillar formation around the queen
during the formation of the royal chamber probably has a role in pillar formation along a trail.

Taking pheromone diffusion and its effect on building one step further, consider how air currents and ventilation patterns within the developing structure might be able to modify the distribution of pheromone. Howse (1966)
has shown that termite movement patterns are influenced by the insects' detection of minute air movements, which suggests that the ventilation channels
in *Macrotermes* mounds may be structured by the termites' tendency to build
around air-flow patterns. Recall Bernouilli's principle that changes in the speed
of air-flow over obstacles of various shapes cause changes in air-flow pressure;
as a result, taller naturally ventilated buildings typically have speedier internal
air streams than shorter naturally ventilated buildings. (Croome and Roberts
1981; see also Vogel 1981.) This may explain some aspects of the building
behavior of termites. When their nest is small, the air stream is below the
termites' threshold of response. Their response appears only when the nest
reaches a critical height that initiates an air stream that the workers are able to
detect. The influence of such scale factors could explain why large nests are
generally more complex than small ones (Figure 18.3).

The purpose in this chapter was to show that using available data along
with hypotheses about the building behaviors of individual termites we can
simulate some of the termites' global-level building behavior. At this point,
however, we cannot go further and attempt to explain the formation of more
complicated structures within the termite mound. The reason is the lack of data
at both the individual and colony levels. The kinds of data one would need to

develop more detailed models include a quantitative description of individual termite behavior as it relates to the deposition and gathering of building material, movements within the mound, and the effects of various pheromones on individual building activity. In addition, almost nothing is known of the physical properties of the pheromones (for example, their diffusion and evaporation constants). This lack of data would provide too many degrees of freedom in a model and too many alternative hypotheses to consider for how building occurs.

At this point, the main benefit of modeling termite building behavior is to identify the kinds of experiments that need to be conducted and the type of data that need to be collected.

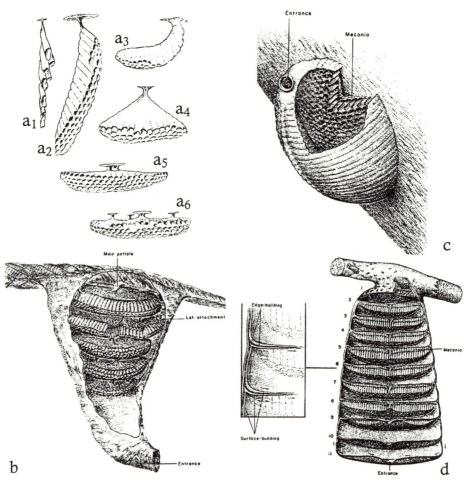

Figure 19.1 Various stelocyttarus gymnodomous *Polistes* nest architectures are shown in (a): a₁ *P. goeldii* builds nest with cylindrical cells attached to the stalk and only a few walls are shared by cells; a₂ *P. canadensis* builds hexagonal cells that are extended together and share common walls, but the pedicel is asymmetric to the comb; a₃ *P. annularis*; a₄ *P. major*; a₅ *P. flavus*; a₆ *P. fuscatus* as well as *P. dominulus* (not represented) builds combs that are symetrical to the pedicel. (Adapted from Brian 1983) *Angiopolybia pallens* nest in shown in (b). Typical stelocyttarus, calyptodomous mature nest is depicted with a portion of the envelope and second comb cut away. The first comb is suspended from an initial, more or less central petiole (main petiole). As the comb grows, this petiole is thickened and secondary petioles are added. Where the combs come close to the envelope, lateral attachments are added. The envelope is constructed by means of the edge-building technique only, and is delicate and paperlike

(*Figure 19.1 caption continued on page 405*)

19

Construction Algorithms in Wasps

Scatter on the ground hundreds of sheets of paper and ask
workmen to glue them together randomly, that is to say in
any which way, and nonetheless build a hot-air balloon.
Without a working plan, in the absence of precise
coordination of individual actions, building a hot-air
balloon is as unlikely as Emile Borel's imaginary typist
monkeys writing the Holy Bible
—P.P. Grassé, *Termitologia*

Introduction

Wasp building has been well studied only in species such as *Polistes* that
build the simplest architectures. We will examine hypothetical mechanisms
that may be used by these wasps to build their nests and we will see to what
extent the same building principles might be applied to more complex architectures. We focus here on stigmergic building algorithms, meaning collective-
building algorithms in which individuals communicate indirectly only through
the local perceived environment. The building algorithms that we present are
still hypothetical mechanisms that wasps may or may not actually use to coordinate their building activity. Nonetheless, the overall logic of this type of

---

Figure 19.1 (Adapted from Jeanne 1975)

Synoeca surinama nest is shown in (c). Typical astelocyttarus nest has sessile cells built
directly on the undersurface of a large branch, then covered with a domed envelope.
The ribbed appearance of the envelope reflects the ribbed outer margin of the comb, of
which the envelope is an extension. The oldest cells, in the center of the comb, are seen
in this cutaway view to contain meconia and thus to have produced adults. (Adapted
from Jeanne 1975)

Chartergus chartarius nest is shown in (d). Typical phragmocyttarus nest. The swarm
constructs the first (uppermost) comb on the underside of a branch, then covers it with
an envelope. An entrance is left in the center of the envelope. The cells of the second
comb are built directly on the lower surface of this envelope and, in turn, are covered.
As the nest is enlarged, the workers strengthen the upper portions by means of surface-
building. This is shown in cross-section in the inset and in surface view by the blotchy
appearance of the attachment of the nest around the branch. Direction of growth of the
nest are shown by arrows. (Adapted from Jeanne 1975)

building behavior is simple, plausible, and serves as a novel alternative to mechanisms based on self-organization.

The Structure of Wasp Nests

Social wasps have the ability to build nests with architectures ranging from extremely simple to highly complex. Figure 19.1 shows some examples of nest architectures built by different species of paper wasps. Depending on the way in which different parts of the nest are organized, several main architectural types among different wasp species, have been distinguished by the French naturalist Saussure (1853–1858) and modified later by Richards and Richards (1951):

Stelocyttarus nests have combs connected to the substrate and, when there are several combs, to each other by stalklike structures called pedicels. Two types of *stelocyttarus* nests are found. In one group the combs are enclosed with an external envelope sometimes named *involucrum*. These nests are termed *calyptodomous* (covered with an envelope). Such architectures are built by several families and genera of wasps, such as *Vespa, Vespula, Provespa, Angiopolybia, Chaterginus, Leipomeles*, and *Parachartergus*. The second group does not possess an envelope and the nests are termed *gymnodomous*. This type of architecture is seen in *Polistes, Mischocyttarus, Belonogaster*, and some species of *Ropalidia* and *Stelopolybia* (Jeanne 1975).

Astelocyttarus nests are calyptodomous, but the whole comb is built directly on the substrate. The position of the entrance hole varies from one species to another and the structure of the envelope is simpler than the envelope structure in *stelocyttarus* nests.

Phragmocyttarus nests have modular structures. The first module is a horizontal comb covered with an envelope that has an entrance. The second module is built directly on the outer face of the first module and cells are added on the undersurface of the envelope. Subsequent modules are added in the same way and the resulting structure appears to be a highly organized succession of combs regularly spaced inside a cylinder. Depending on the position of the entrance hole in the module, a central or peripheral communication opening goes through successive combs, allowing the wasps to move from one floor to another. This is the kind of architecture built by *Polybia, Chartergus, Epipona, Brachygastra*, and *Protonectarina*.

Wenzel (1991) has classified wasp-nest architectures and found more than sixty different subtypes, with many intermediates between the extreme forms described above. A mature nest can have from a few cells up to a million packed in stacked combs, the latter being generally built by highly social species.

The Evolution of Nest Design

In Vespidae, the functional unit is the cell. Cells are generally hexagonal and sometimes circular, and provide a container for the development of only one offspring, from the egg to the imago (Michener 1964; Jeanne 1975; Starr 1991). In contrast to honey bees, cells are not used by wasps to store food such as honey or pollen. An examination of all kind of nests built by Vespidae reveals the existence of three major architectural changes that lead to highly structured nests in the most advanced eusocial species (where a substantial size dimorphism exists between queen and workers as well as polyethism between workers). These changes include the clustering of isolated cells in organized comb, detachment of the comb from the substrate by means of the pedicel, and the protection of combs with an envelope.

Among these changes, the organization of cells into a comb has played a crucial role in the subsequent evolution of nest design. Clustering of cells to make a comb can already be observed in certain solitary wasp species that build aerial nests. Even if the walls of cells are not yet joined together, this appears to be a first attempt to aggregate cells in the same place and thereby favor efficient provisioning. It has been argued that a selective factor that could have favored the evolution of the comb arrangement was a general principle of energy conservation (Jeanne 1975). From a foraging and manufacturing point of view, it is less costly to produce adjacent cells that are in contact, sharing a common wall, and organized into a comb than to place isolated cells scattered on the substrate. The observation of the frequent use of a pedicel in Vespidae has also brought some authors to assume the existence of an ancestral pedicellate nest architecture common to all Polistinae and Vespinae (Wenzel 1989, 1991; Jeanne 1975). According to this hypothesis, all nest shapes that are now known in Polistinae and Vespinae would have been derived from this primitive form. Finally, the envelope contributes to this clustering and isolation process that makes a clear distinction between the inner nest and the outside world. Envelopes appeared several times independently in the Vespidae, but in the Vespinae it is a constant architectural feature. In nine species of Polistinae its presence has induced the loss of the pedicel and reversion to sessile cells (Wenzel 1991). This is undeniably the structure reaching the highest degree of evolution in wasp nest design.

The Adaptive Functions of Wasp Nests

The primary biological functions of the wasp nest are reproductive, to protect the brood against predators and assure the proper conditions for the development of the brood. The nest also possesses a social function facilitating the cohesion of the society, a colonial identity, and communication among nestmates.

Protection against Climatic Changes

In species that build calyptodomous nests, the envelope plays an important role in thermoregulation (Montagner 1964; Berland and Grassé 1951; Spradbery 1973; Makino and Yamane 1980). In *Vespa crabro*, for example, temperature variations inside the nest do not exceed 1.9°C, while variations of the external temperature reach more than 10°C during the same time. The structure and the quality of the material comprising the envelope may also affect temperature regulation of the nest. Important changes in the envelope can be seen depending on where wasps build their nest. For example, in nests built in subterranean cavities the envelope is generally thin and made with a single layer of building materials; the same kinds of structures are found in aerial nests built by tropical species (Sakagami et al. 1990). On the other hand, in temperate zones the envelope in aerial nests is thicker and has a laminar structure with several independent layers overlapping one another, such as in *Dolichovespula media* (Figure 19.2). In other cases such as nests built by hornets (*Vespa crabro*) the envelope is made of numerous scales that overlap one another (Figure 19.3). Overlapping layers contribute to creating a protective air layer between the internal part of the nest and the external environment that minimizes temperature variations. In addition to this insulating role, the envelope protects the brood against the rain. The secretions used as cement to mix with the building materials contain hydrophobic compounds and the resulting structures are highly water-resistant (Schremmer et al. 1985, cited by Downing 1991). Some *Vespa* species even smear the upper part of the envelope with a hydrophobic varnish (Sakagami et al. 1990; Wenzel 1991).

Defense against Biological Enemies

In his remarkable work on the adaptive functions of wasps nests, Jeanne (1975) considered that predation by ants was a driving force in the evolution of nest structures in tropical social wasps. The devastating attacks of predatory ants and, among them, voracious army ants (see Chapter 14), would have exerted a strong selective pressure on the nest design, favoring the appearance of new protection devices. Improvement of nest defense against ants has been achieved in the course of evolution with the aid of two new structural elements, the envelope and the pedicel. The pedicel provides protection against predation by walking insects; the contact surface between the nest and the substrate is reduced to the minimum, giving the wasps the opportunity to easily watch this small potential invasion zone (West-Eberhard 1969). In addition, salivary secretions are mixed with building material allowing a strong fixation of nest to the substrate. In this way, nest development can be more important and nest size can be much greater. A second line of defense against ants is the envelope. Wenzel (1991) believes that the envelope has evolved independently at least seven times. It has been adopted by all tropical swarm-founding Polistinae

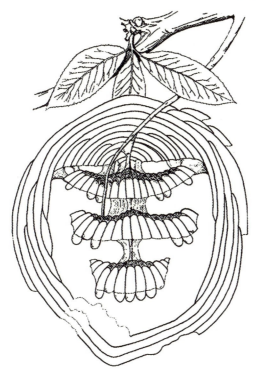

Figure 19.2 Longitudinal section of a *Dolichovespula media* nest, showing the successive envelopes, the entrance hole, and the pedicels connecting successive combs. (Adaped from Berland and Grassé 1951)

(Jeanne 1991) and by all Vespinae. An argument in favor of the development of both structures as a response to ant predation is that no wasp uses an ideal nest that minimizes the amount of building material needed for its construction. Jeanne showed that such a nest would be a round flat comb, whose upper surface is built directly on the substrate (Jeanne 1975, 1977). This would be a perfect astelocyttarus gymnodomous nest. All existing nest structures possess either an envelope or a pedicel.

Along with the incorporation of new substructures into nest design, wasps often use ant-repellent substances that are laid down in some places in the nest. In *Mischocyttarus drewseni* (Figure 19.4), Jeanne (1970b) observed that after wasps have rubbed their abdomen against the pedicel, ants were unable to cross it. It has been established in a large number of wasps species that ant-repellent substances are applied to the pedicel during this rubbing behavior (Turillazzi and Ugolini 1979; Jeanne 1996). The use of such substances is

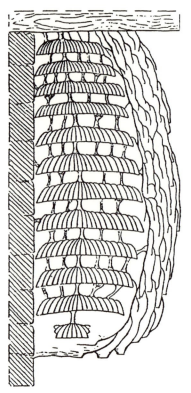

Figure 19.3 Longitudinal section of a *Vespa crabro* nest with twelve combs. (Adapted from Berland and Grassé 1951)

common to all independent founding Polistinae (Kojima 1993) and it appears that all species building stelocyttarus gymnodomous nests use ant-repellent substances (Jeanne 1975). In these species, before the emergence of the first adults, this use can greatly reduce the risk of predation when the foundress is foraging away from the nest. As shown by Kojima (1993), in *Polistes* rubbing is done before the wasp leaves the nest and after the pedicel is enlarged. The use of repulsive secretions has also been described in two tropical swarm-founding Polistinae, *Nectarinella championi* and *Leipomeles dorsata* (Jeanne 1991; Wenzel 1991). It seems that these two species have maintained this behavior because of their small size (Jeanne 1991). In fact, defense in tropical swarm-founding Polistinae relies primarily on a great number of adults at the beginning of the foundation, and secondarily on the presence of the envelope.

Building Materials

In the vast majority of wasp species, paper is the main building material, hence their common name of paper wasps. Depending on what kind of raw ma-

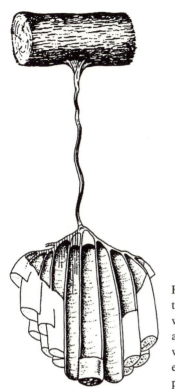

Figure 19.4 The *Mischocyttarus drewseni* nest, a typical stelocyttarus gymnodomous nest, is shown with the front half cut away. As the nest grows, adults extend the walls of peripheral cells downwards. The wasps strengthen the petiole by repeatedly adding layers of oral secretion over the original pulp core.

terial is used, as well as the way it is manufactured by wasps, the quality of the paper changes a lot (Wenzel 1991). For example, *Polistes* use long wood fibers and plant hairs mixed with salivary secretions. The resulting carton is easily shaped and is robust, though it is extremely fine and light (Hansell 1985). In *Mischocyttarus*, the carton is thicker and made with long fibers joined with coarse wood shavings and occasionally mixed with mud (Richards 1978). In Vespinae, *Dolichovespula* uses long plant fibers while the three other genera (*Vespa*, *Vespula*, and *Provespa*) make their carton with little vegetable chips mixed with wood bark and rotten wood (MacDonald 1977; Wenzel 1991), adding secretions sparingly. This leads to a more compact and rigid paper. The particular nest colors reflect the wasps' preferences for a particular kind of building material.

Paper is not the only building material used by wasps. Oral secretions, mud, and other inorganic compounds are also used, but less frequently. The transition from mud to paper is considered a major transition in the evolution of nest shapes and of eusociality in wasps. Currently, most mud nests are encountered in the three genera of the subfamily Stenogastrinae. In this case the main ad-

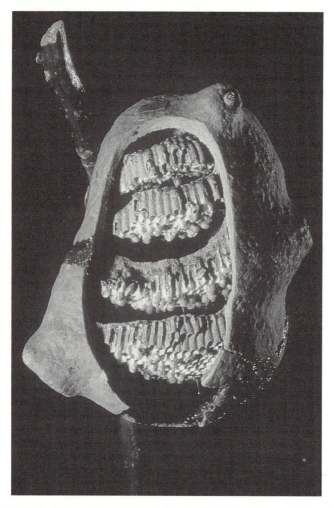

Figure 19.5 Part of the envelope of the mud nest of *Polybia* cut away to show the inner organization of the nest.

vantage of a mud nest, as mentioned by Hansell (1984) and Turillazzi (1989), is its strength when colonies are violently attacked by other wasps, especially *Vespa*. But these authors disagree when it comes to evolutionary consequences. While Hansell (1985) argues that the heaviness of building material is responsible for the reduced nest size in Stenogastrinae, Turillazzi (1989) emphasizes that numerous solutions could have occurred to overcome this constraint and to favor the evolution of large structures. Indeed, there exists at least two species of *Polybia* that use mud to build a nest about 20 cm in diameter that weighs up to 5 kg (Figure 19.5) (Wenzel 1991; Berland and Grassé 1951). In contrast to mud, the use of glandular secretions can be considered a sign of a higher level of adaptation even if their production appears to be more costly than foraging for materials.

Two uses of secretions are distinguished (Kojima and Jeanne 1986). The first is common to all Polistinae except the Polistini, and involves the replacement of the bottoms of cells after the wasps have removed the meconiums left by the newly emerged individuals. Most tropical independent founding species do this cell cleaning because it diminishes parasitism, especially that caused by certain fly species. Meconiums are extracted after the cells have been opened from their back surface. The new back is then fashioned as a translucent window made of secretions. The second use of secretions is limited to a small number of species, such as *Ropalidia orifex* (Kojima and Jeanne 1986) and *Pseudochartergus fuscatus* (Jeanne 1970a). These species build an envelope of secretions directly upon leaves. This envelope is strong and has the high flexibility required to cope with substrate mobility. Finally, other inorganic materials are occasionally used to strengthen or to conceal nests.

Alternative Hypotheses for Nest Construction

The first naturalists who observed wasps in the field considered their behaviors were fixed and highly stereotyped (Fabre 1879–1907). It was difficult to understand how simple chains of behavior performed by a group of insects could lead to the production of amazing architectures. How was the coordination of all these multiple, independent actions achieved? This was even more incomprehensible if the insects had no overall concept of the nest they were building. This explains why the first hypotheses to account for these problems assumed that individual wasps had an internal (that is, innate) representation of the global structure to be produced. In this way, all the decisions taken by wasps could be made on the basis of that representation.

In his blueprint theory, Thorpe (1963) used the results of Hingston's (1926, 1927) experiments to develop the idea that wasps build nests by continually comparing the emerging structure with an innate image of how the completed nest should look. Each insect would adjust its building activity as needed to produce the proper structure. In this scenario, nest complexity arises from the complexity in the individual insect's behavior. However, the experimental tests that followed failed to confirm this intriguing hypothesis. In particular, when nest repairs are made, the appearance of the nest is often altered in such a way that it is no longer typical for that species, indicating that construction is not directed toward some innate nest image (Olberg 1959). The results of further experiments performed by Smith (1978) on the mud wasp *Paralastor*, where nest structures were modified by experimental perturbations, revealed the inadequacy of the approach proposed by Thorpe and outlined, on the contrary, the fact that construction behavior was based on an innate building program. What

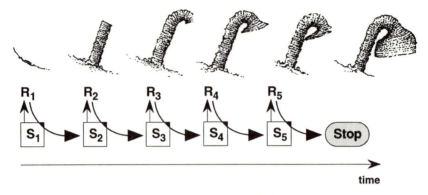

Figure 19.6 Stimulus-response sequence leading to construction of the mud funnel in the nest of the eumenid wasp *Paralastor*. Each new building stage, n, is associated with a stimulus, S_n, that triggers a set of building actions, R_n. The completion of each building stage, results in a new stimulus, S_{n+1}, that triggers new building actions, R_{n+1} leading to the construction of the next building stage, $n + 1$. When the fifth stage has been completed, no more stimuli appear associated with the funnel to trigger new building actions, and construction stops.

are the main features of this building program and how does coordination arise when several insects together build their nest?

From Sequential to Stigmergic Activity

To understand better how multiple independent building acts are coordinated, we must come back to the mechanisms underlying nest construction in solitary wasps. The experiments performed by Smith (1978) shed some light on how building activities are coordinated and deal more generally with pre-adaptation of this behavior. These experiments were first summarized in Chapter 4. Let us describe the main results obtained by Smith.

Nest construction in the eumenid wasp *Paralastor* occurs as a stimulus-response sequence in which the completion of one stage provides the stimulus for commencement of the next (see Figure 19.6). A wasp begins with the excavation of a narrow hole approximately 8 cm long and 8 mm wide. When the nest hole has been completely lined with mud, the wasp begins the construction of a large and elaborate mud funnel above its entrance. The funnel is built in five distinct stages from a series of mud pellets applied in a highly stereotyped sequence.

Stage one involves the building up of the funnel stem by application of a series of mud pellets until it reaches a length of 3 cm. In stage two, the wasp ceases to build uniformly upwards and, by adding more mud to one side, begins construction of a uniform curve in the stem of the funnel. Once the curve has been completed, stage three begins with the formation of a bell with the splaying of the stem to form a uniform flange of approximately 2 cm diameter. In stage four, the flange is widened more on the side nearest to the stem than elsewhere, thus giving the bell a characteristic asymmetry in one direction. In stage five, finally the sides of the bell are formed by building uniformly downwards from the edge of the flange.

At the end of each building stage, the stimuli for responses that lead to the completion of the next stage are encountered (by the wasp) as a consequence of its earlier behavior. What happens when the wasp encounters these stimuli was studied by Smith. He made spherical holes in the necks of the nests' funnels just after completion of stage 3 construction (see Figure 19.7). After examining the damage several times, the wasp began construction of a second funnel over the hole and on the top of the first funnel. This result is extremely important for anyone who wants to understand the coordination of building activities in social wasps, and more generally in social insects. In a solitary species such as *Paralastor*, the indirect coordination of behavior through the previous consequences of its building actions result in a sequencelike behavior. One consequence of this behavior is that the order in which stimuli arise in the course of the construction must follow a precise sequence. If by chance a stimulus triggers building actions that cause a previous sub-element of the architecture to occur at a later stage, the result is the automatic construction of a redundant structure and an abnormal nest architecture. This has important consequences that we will encounter when we examine the coordination of building activity in social wasps.

One could easily imagine that if two distinct wasps do not distinguish the product of their own activity from that of the other, they could complete the same nest structure. One wasp could continue the work done by the other wasp whatever the stage of construction of the nest. Such a mechanism may then in turn be a step towards indirect cooperation between individuals. This is precisely the kind of mechanism Grassé had in mind when he introduced the concept of stigmergy. The term was initially introduced to explain task coordination and regulation in the context of nest reconstruction in termites of the genus *Bellicositermes* (Grassé 1959, 1984). Grassé showed that the coordination and regulation of building activities do not depend on the workers themselves but are achieved mainly by the nest structure: A stimulating configuration triggers the response of a termite worker, transforming the configuration into another configuration that may trigger in turn another (possibly different) action performed by the same termite or any other worker in the colony. Stigmergy offers an elegant and stimulating framework to understand the coordination and regulation of building activities. But the main problem is to determine how stimuli are organized in space and time so as to lead to a robust and coherent construction: colonies of a given species build qualitatively similar architectural patterns.

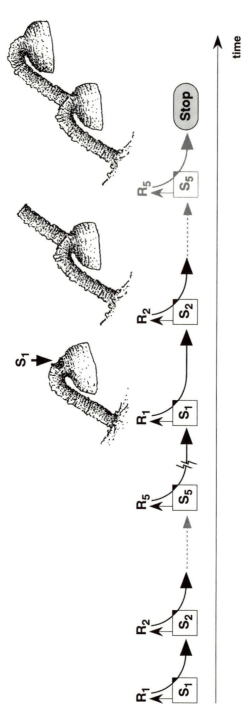

Figure 19.7 The construction of an abnormal mud funnel in the nest of the Eumenid wasp *Paralastor*. After the funnel has almost been completed, a spherical hole (shown by the arrow) has been made. The hole is equivalent to the kind of stimulus, S_1, that triggered the beginning of funnel construction. The result is that the wasp builds a second funnel over the hole and on top of the first funnel that was already built.

Stigmergy Revisited

The term stigmergy merely refers to a mechanism that mediates worker-worker interactions. Thus stigmery must be supplemented with an additional mechanism that uses these interactions to coordinate and regulate collective building in a particular manner. There exist at least two such mechanisms, one quantitative (Deneubourg and Goss 1989; Bonabeau et al. 1997) and the other qualitative (Theraulaz and Bonabeau 1995a, b) (see Figure 19.8). With quantitative stigmergy, successive stimuli do not differ qualitatively but only modify the probability of the termites' response to particular stimuli. Although we did not refer to it as such, in Chapters 16, 17, and 18 we saw how self-organization using quantitative stigmergy could explain comb organization or the construction of walls in *Leptothorax* and termite nests. When termites build pillars, for example, they respond to *quantitative* differences in particular stimuli, such as the magnitude of pheromone concentrations that affect the probability of performing a particular task.

Qualitative stigmergy differs from mechanisms based on self-organization in that individuals interact through and respond to *qualitatively* different stimuli. Qualitative stigmergy is based on a discrete set of stimulus types. For example, an insect responds to a type-1 stimulus with action A and responds to a type-2 stimulus with action B. It is easy to see how such a mechanism is still compatible with stigmergic interactions. For example, a type-1 stimulus triggers action A by individual I_1; action A transforms the type-1 stimulus into a type-2 stimulus that triggers action B by individual I_2. More difficult to see, however, is how coordination and regulation can be achieved in qualitative stigmergy. In this respect building in social wasps provides a good example of qualitative stigmergy. Although, strictly speaking, we do not consider stigmergy to be a building mechanism based on self-organization, we present this example of wasp-building behavior because it provides a useful comparison with other building behaviors of social insects, and because it illustrates the relationship between self-organization and stigmergy that we have emphasized in this book. We examine a particular example of stigmergic building, in the primitively eusocial wasp, *Polistes*, which has been best studied.

Dynamics of *Polistes* Building Activity

The building behavior of *Polistes* wasps has been the most extensively studied because of their small colony size and simple nest architecture. This permits investigators to easily manipulate the building behavior in the laboratory with simple techniques, such as sequentially offering workers different colored papers as construction material. *Polistes* nests consist of a single stelocyttarus gymnodomous comb made of plant fibers (Wenzel 1991). The comb is round and mature nests can contain about 150 cells (Reeve 1991). Nest construction

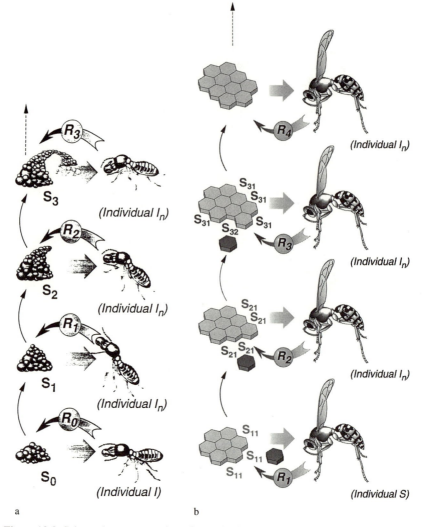

a   b

Figure 19.8 Schematic representation of quantitative and qualitative stigmergic mechanisms implementing indirect interactions between workers during construction (see text). Self-organization or quantitative stigmergy is illustrated (a) by successive stages leading to the emergence of pillars in termites nests. In order to build their nest termites workers use soil pellets impregnated with pheromone. The existence of an initial deposit of soil pellets stimulates workers to accumulate more material through a positive feedback mechanism, since the accumulation of material reinforces the attractivity of deposits through the diffusing pheromone emitted by the pellets. Qualitative stigmergy (b) is illustrated by the successive stages leading to the construction of the comb in wasps. Wasps use wood pulp to build the various elements of their nest. In the present case, each building stage correspond to the addition of a new cell to the pre-existing comb. At the beginning (bottom of the figure) all potential building sites are equivalent.

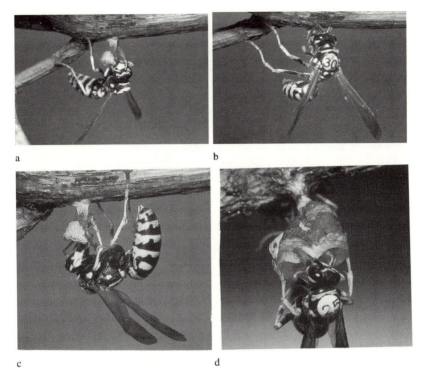

Figure 19.9 The first steps in construction of a *Polistes dominulus* wasp nest: (a) A wasp prepares a pulp load before shaping it (papers of different colors are offered to the wasps at regular intervals, to track the building dynamics); (b) the nest is initated with construction of a pedicel; (c) the first cell is constructed; (d) subsequent cells are constructed. (Photos © Guy Theraulaz)

occurs through a succession of three building phases: the pre-founding phase, in which the wasp selects the nest site; the initial phase ending with construction of the first cell; and a development, or nonlinear, phase in which new cells are added to the growing comb (see Figure 19.9).

The Pre-Founding Phase

In this stage wasps prepare the substrate to receive the first paper pellets. The foundress wasp increases the roughness of the substrate by licking and scraping the surface. Then it progressively concentrates its efforts on a small area approximately its own size (Karsai and Theraulaz 1995). This zone is termed the *foundation spot* (Deleurance 1957). During this phase some species, such as *P. fuscatus*, add a chemical marking by rubbing the tarsi of their hind legs on the substrate (Downing and Jeanne 1988). These physical and chemical landmarks serve as guides for the wasp when it comes back to the nest site.

The Linear Phase

The first building stages occur in a specific sequence. A strip of material is built on the substrate; the pulp is drawn into a small spike and lengthened into a pedicel; a flat sheet is built on the tip of the petiole to become the base for the first two cells; the first cell is initiated. In the subsequent phase the comb is enlarged and new cells are added to the structure, but building is no longer constrained to follow a predefined sequence, as pulp now can be added to several different locations at the same time.

Once the site of the foundation spot has been established, the first stages of the building activity appear somewhat disorganized. The first collections of pulp are hesitant and frequently interrupted by short flights, grooming, and restless walks during which the wasp often loses its pulp load (Deleurance 1957). Likewise, the first pulp deposits lack precision. Some wasps spread pulp over a surface that is much larger than the size of the initial foundation spot. Deposits then become scattered. In contrast, other wasps rapidly succeed in concentrating deposits at a single site and eventually build the strip with only two or three loads. This phase appears to follow the same pattern as that observed in termites during the period preceding the emergence of pillars. This aspect of the construction may therefore be based on a self-organized positive feedback. On average, six to eight pulp loads are used to build the first cell and complete this first phase.

Figure 19.10 shows the succession of building stages that occurs during the linear phase in *P. fuscatus* (Downing and Jeanne 1988). Downing and Jeanne found that in this species the linear sequence of construction does not rely on

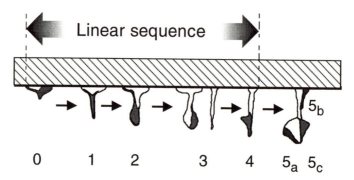

Figure 19.10 The sequence of building stages occuring during the linear phase in *Polistes fuscatus* (adapted from Downing and Jeanne 1988) begins with construction of the pedicel (1), followed by the construction of a flat sheet (2), and ending with initiation of the first cell (4). Once the first cell has been started, the wasp can lenghten it (5a), strenghten the pedicel (5b), or initiate a second cell (5c). The last building material added to the structure is indicated in gray.

a strict stigmergic process. In some cases the wasp not only uses stimulating cues coming from the preceding stage to build the next one but also evaluates multiple cues coming from different previous stages. In particular, the flat sheet is not simply added to the distal end of the pedicel, it is built at a specific distance from the substrate and the wasp repeatedly re-evaluates this distance before starting the next building stage. Regarding the crucial role of the pedicel in nest defense against ants, selection would have favored wasps that could detect changes in this critical distance to the substrate and correct them. Conversely, other building stages do rely on a stigmergic process. For example, when a completed flat sheet is experimentally moved so that it is off-center from the petiole, the wasp does not correct it and does not pay attention to its relative distance to the pedicel. Furthermore, the completion of each building stage is not regulated by the total amount of construction performed, which would be the case if building were guided by an internal representation of the structure.

The Nonlinear Phase

Once the first cell has been built, the wasp starts a second cell on the side of the first one. Now, there will be two building activities at the origin of the enlargement of the comb, including the addition of new cells and cell lengthening. From time to time the wasps engage in a third activity—thickening of the pedicel. These three types of building activity are performed with specific behavioral postures and motor sequences. The particular types of building activities were distinguished by Deleurance (1957). He described three different types of carton construction: cell construction, cell lengthening, and pedicel construction. Cell construction corresponds to the initiation of a new cell (Karsai and Theraulaz 1995):

> During cell initiation, the new cell is added as an arch symmetrically around the groove between two or three adjacent cells. The builder usually begins to apply the pulp on the top of the comb in the upper part of the groove. Meanwhile its body turns strongly, its head and thorax are almost horizontal, and its abdomen is bent. All legs rest upon the side of the comb, which is formed by the outer walls of peripheral cells. The wasp lays down the pulp working backward across the top of the groove and down the side forming a half arch. Meanwhile the horizontal position of the wasp becomes vertical to the petiole. Just as before, the wasp climbs back to the top of the comb quickly and takes the horizontal position again, but in the opposite direction. By repeating this procedure generally three to four times all pulp is used up. The wasp performs similar movements in 3 to 4 additional series while it forms the rim of the arch and chews it together with the outer walls of the old cells.

Karsai and Theraulaz described two different lengthening procedures— alternate and circular—that wasps perform:

> During the alternate lengthening, the wasp is situated on the side of the comb and applies the pulp to the arch that was initiated before. Contrary to initiation, the wasp does not start pulp addition on the middle of the arch, but rather almost at the end of the arch near the point where the new cell is connected to its neighbor. At this moment the wasp hangs on the nest almost upside down: its head faces down and its body and neck is strongly bent, while its abdomen is nearly horizontal. During pulp addition the wasp walks slowly backwards, and its posture changes considerably. In the last moment of one subact (building without lifting up the mandibles from the cell), when the wasp adds the pulp to the lowest concave wall of the new cell in one direction, the head of the wasps faces up, its abdomen down and its middle and hind legs are on the mouth part of the cells. In the next moment the wasp returns to the side of the comb and takes the upside down position again and continues building as the next subject in the opposite direction. . . . During circular lengthening, the wasp remains hanging on the surface of the comb. It straddles the cell walls and attaches pulp around the rim of a single cell with the mandibles while moving backward and antennating the neighboring cells. Then the builder with similar movements works the walls to the required thickness with successive passes of the mandibles along the rim, each time tamping the wall thinner. The body of the wasp is slightly bent, but its head is turned considerably towards the built cell. When the wasp reaches the connection of two walls, it makes a side step and turn around 60° towards the built cell and continues building the cell wall (Karsai and Theraulaz 1995).

Finally, the pedicel is constructed at a fixed point. The wasp does not have to change the position of its body to complete this building action. It stands perpendicular to the substrate, facing upward, and has its four posterior legs resting on the back of the nest with its forelegs near the pedicel base. The wasp puts its pulp load at the connection point between the base and the pedicel, cranes its neck forward strongly, and adds pulp to the substrate with considerable mandibular scissoring. A rootlike strip is usually built with some ramifications. No particular movements of the antennae have been noticed during pedicel construction (Downing and Jeanne 1988; Karsai and Theraulaz 1995).

As we have said, this construction phase is characterized by a nonlinear unfolding of building activities. Many activities can be performed at the same time and building acts are not performed a priori in a well-defined sequence. The ability of a swarm to build at more than one location on a growing nest is certainly an important evolutionary step for the development of complex nest architectures (Downing and Jeanne 1990). But this fact has, in turn, induced new constraints since local cues regulating building behavior must be organized appropriately to ensure a coherent construction. We will now focus on the individual behavior that governs the addition of new cells to the pre-existing comb, since cell lengthening is strongly correlated with growth of the larvae and is not a stigmergic process.

Individual Construction Rules

The number of potential sites where a new cell can be added increases as construction proceeds and thus permits building activities to proceed in parallel. However, this raises the question of where new cells should be added to build a species-specific nest architecture. Because of parallel activities, construction may become disorganized if conflicting actions are performed simultaneously (see, e.g., Darchen 1959, in the case of bees). The issue of where to put new cells may be resolved by the constraints of the architecture itself, which appears to properly canalize building activity. But in what way? Let us consider for example, the comb shown in Figure 19.11 with six cells already built. This comb has twelve potential building sites: seven sites with one adjacent wall (labeled S_1), four with two adjacent walls (S_2), and another with three adjacent walls (S_3). Figure 19.12 shows how the mean number of potential building sites (with up to three adjacent walls) varies with the number of comb cells already built. Since the growth of the comb is regular (it occurs

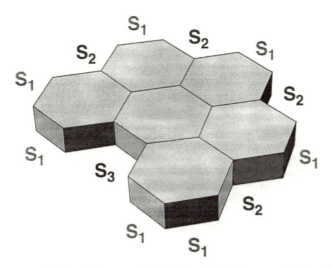

Figure 19.11 Definition of the potential building sites having one (S_1), two (S_2), and three (S_3) walls in common with the new cell added to the comb.

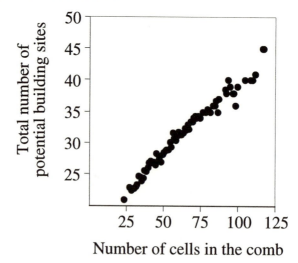

Figure 19.12 Mean number of potential building sites in the comb in *Polistes dominulus* having one, two, and three adjacent walls as a function of the number of cells already built. The curve is based on 155 measurements made on thirteen colonies reared in laboratory and observed at various stages of development.

equally in all directions). The question arises: How are cells added to ensure that the comb grows fairly evenly in all directions from the pedicel?

Previous studies showed that in *Polistes* individual building algorithms consist of a series of "if-then" and "yes-no" decisions (Brockman 1986; Downing and Jeanne 1988; Wenzel 1996). Figure 19.13 shows the typical nest development in *P. dominulus*. With a careful study of the dynamics of local configurations of cells during nest development, it can be seen that there are not equal numbers of sites with one, two, or three adjacent walls. The majority are sites with two adjacent walls. The number of sites with one wall, evidently is strongly correlated with the number of sites with three walls (Figure 19.14); in fact each time a site with three adjacent walls is created two sites with one wall are automatically inserted in the structure.

Studies on *P. dominulus* and *P. fuscatus* revealed similar findings: Cells are not added randomly to the existing structure. Wasps tend to add new cells to a corner area where three adjacent walls are present rather than start a new row by adding a cell on the side of an existing row (Downing and Jeanne 1990; Karsai and Theraulaz 1995). Thus wasps obviously are influenced by previous construction. Building decisions seem to be made locally based on perceived configurations in a way that possibly constrains the building dynamics.

One building rule is that wasps tend to finish a row of cells before initiating a new row. But this rule is not deterministic. Figure 19.15 shows the probability of adding a new cell as a function of the ratio, S_3/S_2, between the number of sites with three walls and the number with two walls. As can be seen, the probability of adding a cell in a three-wall site is higher than the probability for a cell in a two-wall site. The disparity between the two probabilities becomes even higher as the value of S_3/S_2 increases. Figure 19.16 shows the overall probabilities for each case as computed from the preceding data. The probability of adding a cell in a three-wall site is about ten times higher than that for a two-wall site. What are the consequences of this probabilistic discrete stigmergy on the development of the comb structure? In order to study this point we introduce a model of qualitative stigmergy to help us understand in a more general way the kinds of constraints associated with nests built in a distributed, discrete way.

The Lattice-Swarm Model

Inspired by the observations and empirical data described in the previous sections, Theraulaz and Bonabeau (1995a, b) introduced a class of algorithms

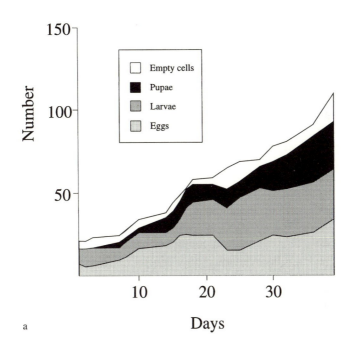

Figure 19.13 Development of a wasp nest structure in *Polistes dominulus*: (a) dynamics of the number of brood items, and (b) dynamics of the number of potential building sites with one, two, and three walls.

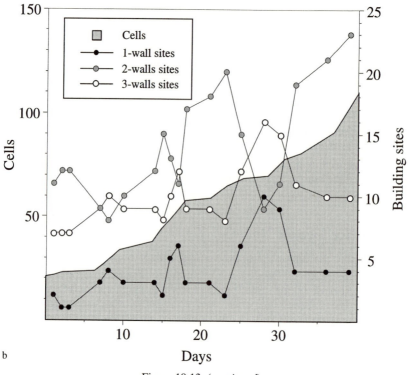

b

Figure 19.13 (*continued*).

to explore qualitative stigmergy as a model of nest construction. Conceptually, their model of nest construction could hardly be simpler. It envisioned a number of automata that moved asynchronously in a three-dimensional discrete space and behaved locally as pure, stimulus-response machines. The automata were builders that responded to specific stimuli presented by the developing comb structure. The three-dimensional space in which the automata moved was a discrete cubic or hexagonal lattice—hence the name lattice swarms. The deposit of an elementary building block (hereafter called a brick) by an automaton depended on the local configuration of bricks in the cells surrounding the cell occupied by the automaton (see Figure 19.17). Several types of bricks could be deposited. No brick could be removed once it was deposited. Increasing the number of types of bricks increased the richness of the local configurations to which the automata could respond.

In the present case, we will use two types of bricks, the minimal number of types needed to build complex architectures. All simulations start with a single initial brick. A micro-rule is defined as the association between a particular,

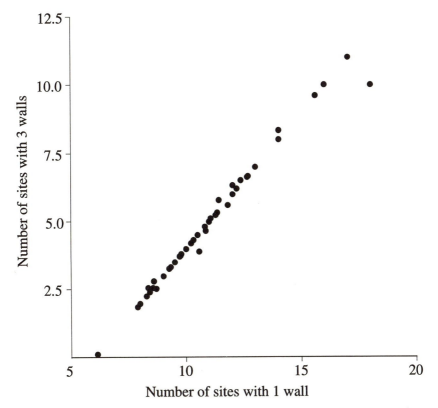

Figure 19.14 Relationship between the number of sites with one wall and the number of sites with three walls. Results are based on 155 measurements made on thirteen colonies reared in laboratory and observed at various stages of development. The linear correlation is best fit with the curve: $y = 0.925x - 5.225$ ($r = 0.989$).

stimulating configuration and a brick that is to be deposited. An algorithm is defined as any collection of compatible micro-rules. Thus, the algorithm consists of a set of if-then rules where the "if" part tests for a particular configuration of bricks in the structure, and the "then" part results in the deposition at that site of one of two types of bricks. Two micro-rules are incompatible within an algorithm if they correspond to the same stimulating configuration, but lead to the deposition of different bricks. An algorithm can be characterized by its micro-rule table: a look-up table comprising all its micro-rules comprising all stimulating configurations and associated actions. Finally, an algorithm can be deterministic or probabilistic; in the first case all micro-rules are applied with the same probability equal to unity; in the second case, each micro-rule is applied with an assigned probability between 0 and 1.

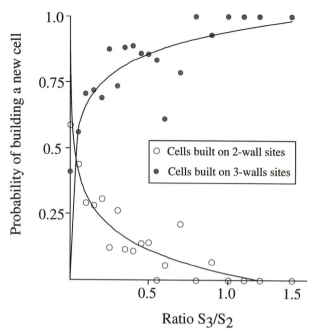

Figure 19.15 The relative probability of building a cell on sites with two or three walls (expressed as the fraction of cells built on these sites) depends on the ratio between the number of sites with three walls and the number with two walls. Results are based on 155 measurements on thirteen colonies reared in laboratory and observed at various stages of development.

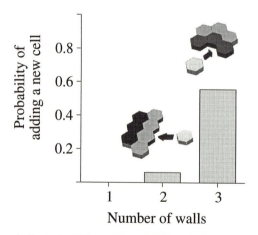

Figure 19.16 Bars indicate the differential probability of adding a new cell to a corner area where three adjacent walls are present, and to the side of an existing row where two adjacent walls are present.

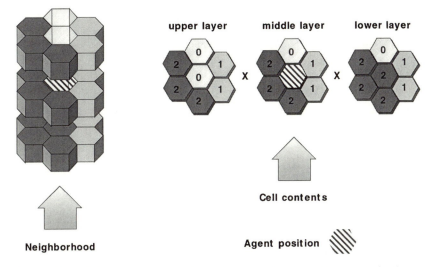

upper layer   **middle layer**   **lower layer**

Cell contents

Neighborhood

Agent position

Figure 19.17 Schematic representation of the agent's perception range in three-dimensional hexagonal lattice swarms. The range of local perception of the agent is limited to the 20 cells in the agent's immediate beighborhood. Each neighboring cell can be empty or contain a brick of type 1 or 2. Each local configuration can be associated with a building action involving deposit of brick. There exist $3^{3^{20}}$ possible elementary micro-rules, but the building behavior of the agent relies only on a small subset of several micro-rules.

A single automaton in this model can complete an architecture. In that respect, building is a matter of purely individual behavior. But individual building behavior is determined by local configurations that trigger building actions and has to be organized so that a group of automata can produce the same architecture as well. *Polistes* wasp species face the same problem since nest construction is generally first achieved by one female, the foundress, and is then taken over by the group of offspring workers. We saw previously that stigmergic behavior, as it appears in solitary species, can be considered a pre-adaptation and the first step toward cooperation. In the case of social species, the group must be able to build a nest architecture without the combined actions of the different agents interfering and, possibly, destroying the whole activity of the swarm. This requirement imposes strong constraints on the way in which the stimulating configurations have to be organized in space and time.

Collectively Induced Constraints

When each of n automata in a swarm joins in the construction of an architecture, it acts on the environment at time t. At time $t + 1$ it can sense feedback resulting from its own action, as well that of any other agent in the near environment. Since the number of potential building sites increases with time, how do the agents of a swarm following these rules coordinate their activities? Empirical conclusions about the conditions for coordination can be drawn on the basis of many computer simulations.

A first condition is that all automata must respond in a uniform manner to the entire set of local configurations they encounter on the architecture. At any given time, there exists an active set of local configurations that triggers the same qualitative type of brick deposit. Let C be the set, $(C_1, C_2, \ldots, C_n)$, of all local stimulating configurations; that is, the configurations that trigger the specific deposition of a brick when they are encountered.

The second condition is that the architecture must be built through a succession of qualitatively distinct building stages, $(S_1, S_2, \ldots, S_n)$. Each of these stages, S_p, is characterized by a subset of local stimulating configurations of C, $C(S_p)$, with the following necessary condition: A particular stimulating configuration must not belong to two distinct building stages. Considering each building stage, S_p, when a brick is put down the result produces one or more configurations belonging to $C(S_p)$.

The construction process may then go on in one of two ways. One way is a succession of single configurations that allow automata to deposit a brick producing the next stimulating configuration. The other is a parallel process where several configurations allowing the deposition of a brick are simultaneously present (see Figure 19.18). Completion of the building state, S_p, corresponds to the appearance of new configurations that belong to the next building stage, S_{p+1}. As we can see, the construction process follows some emergent sketch and can produce recurrent states. Such states are at the root of the modular structures that appear in the architecture. One may have the two following cases (where R_i denotes the set of responses generated in state S_i):

a linear chain of building states,

$$S_1 \xrightarrow{R_1\downarrow} S_2 \xrightarrow{R_2\downarrow} S_3 \xrightarrow{R_3\downarrow} S_4;$$

on a chain of recurrent building states:

$$S_1 \xleftarrow[\quad R_4\uparrow \quad]{} \xrightarrow{R_1\downarrow} S_2 \xrightarrow{R_2\downarrow} S_3 \xrightarrow{R_3\downarrow} S_4.$$

Two conditions must be met in order to build a given architecture using a swarm of automata with qualitative stigmergic behavior: The construction must be decomposed into a finite number of building stages, and local stimulating configurations created in a given stage must differ from those created by a previous or forthcoming building stage. Such a building algorithm, with

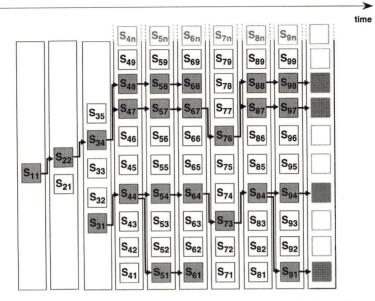

Figure 19.18 A hypothetical nest development in a lattice swarm is shown where the stimulating configurations are indicated in grey. Building an architecture amounts to finding the right sequence of stimulating configurations. The construction begins with an initial stimulating configuration, S_{11}. The second brick added to the structure creates two new configurations, S_{21} and S_{22}. Only one of these configurations is stimulating and triggers a new building action. When the third brick has been deposited, five potential building sites are present on the structure, $S_{31}, \ldots, S_{35}$. At this point two of these configurations are stimulating and simultaneously present on the structure. Thus construction can continue in a parallel way. In the fourth step, the deposition of brick in response to the stimulating configuration S_{34} leads to the creation of two new stimulating configurations, S_{47} and S_{48}. These two configurations in turn induce a sequential-like construction until the sixth step is reached. At this point the deposition of new bricks leads to the creation of only one stimulating configuration, S_{76}, which is the same configuration as S_{34}. A first module has been completed and a new one can be built through a similar sequence. At the same time another module is built starting from the configuration S_{31}.

time evolution that obeys this nonoverlapping condition is called a *coordinated algorithm*; that is, all individuals cooperate in the current building state at any time (Theraulaz and Bonabeau 1995b).

Construction of Polistes Nests

Having presented our tools, we can now return to investigating the algorithms *Polistes* wasps may use to build their nests and study the effect of prob-

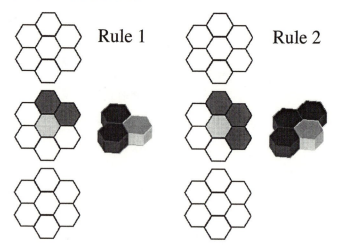

Figure 19.19 Local micro-rules for building the comb of a *Polistes*-like nest (see also Figure 19.17): the three rows of cells show the 20 cells that comprise the immediate neighborhood of the builder (agent). Rule 1 indicates the deposition of a new cell at a site where two walls are the stimulus. Rule 2 refers to the deposition of a new cell at a site where three walls are already in place. The cell being built is indicated in light gray, while adjacent cells (with walls that provide stimuli) are shown in dark gray.

abilistic stigmergic rules. For simplicity we only consider the enlargement of the comb. The kind of comb structure produced with deterministic rules will be compared to the structure produced with probabilistic rules. Only two rules are considered: add a new cell in a two-wall location, and add a new cell in a three-wall location (see Figure 19.19). In the deterministic case, both rules will be applied each time the automaton encounters the correct stimulating configuration. In the probabilistic case, rules will be applied with the probabilities found in the experiments: 0.057 for the first rule and 0.55 for the second.

Figure 19.20a shows the comb that resulted from using deterministic rules. The comb is indented in many places, creating several lobes. On the other hand (Figure 19.20b), the use of probabilistic rules led to a round-shaped comb similar to that found in natural *Polistes* nests. Similar results were obtained by Karsai and Penzes (1993). The different outcome with probabilistic rules can be explained because in this approach, three-wall sites are more likely to trigger cell building and lead to closely packed, parallel rows of cells, and the nest grows fairly evenly in all directions.

Deterministic Rules Complex Architectures

What kind of nest architectures can be produced using simple qualitative stigmergic rules? Using the rules defined above, could the automata build much

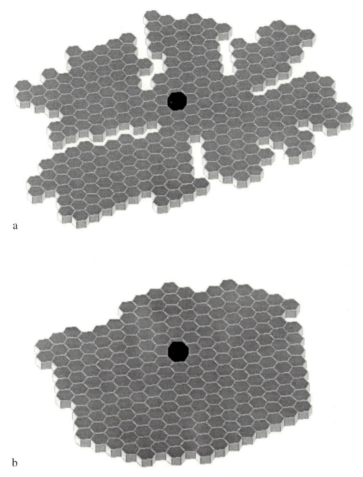

a

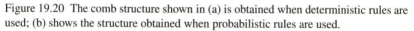

b

Figure 19.20 The comb structure shown in (a) is obtained when deterministic rules are used; (b) shows the structure obtained when probabilistic rules are used.

more complex architectures then those shown in Figures 19.19 and 19.20? We present here a few architectures that were obtained using deterministic rules in cubic and hexagonal environments. According to their neighborhoods and their lookup tables, the automata may deposit type 1 or type 2 bricks. Although the rules are simple, Figure 19.21 shows that complex architectures can be produced, some of which closely match those found in nature. All these architectures were obtained with coordinated algorithms. Figure 19.22 shows the result of three simulations using a similar but noncoordinated algorithm. A striking difference is seen between these architectures and those shown in

Figure 19.21. A particular coordinated algorithm always converges toward architectures that possess similar features. The structures produced by an uncoordinated algorithm, in contrast, seem to diverge. The same algorithm leads to different global architectures in different simulations. The tendency to diverge arises because the simulated configurations are not organized in time and space and many of them overlap. The architecture grows incoherently in space. As an illustration of this, architectures f and g in Figure 19.21 result from two successive simulations using the same coordinated algorithm. Moreover, even in shapes built with coordinated algorithms, some degree of variation may occur, which is greater in cases where a large number of different choices exist within a given building state. But the result may also be deterministic in some cases; that is, all simulations lead to exactly the same architecture despite the random component in individual behavior.

Some of the structured architectures in Figure 19.21 are reminiscent of natural wasp nests. Among these nests, plateaus are present in c, d, e, f, and g and different levels of the nests are linked together with either a straight axis (d, e) or a set of pedicels (f, g). Others possess an external envelope; nest i, shown with a portion of the front envelope cut away to reveal the nest's interior, corresponds to nests found in the genera *Chartergus*. These results do not prove that all or any natural nests are built with qualitative stigmergic rules, but they do show the power of such rules to produce complex stable architectures.

A Closer Look at Coordinated Algorithms

To understand how a coordinated algorithm operates, we focus on the example depicted in Figure 19.23. This figure represents the successive steps in the construction of a nest with a design similar to nests built by *Chartergus*. The transition between two successive building stages is shown to depend on a certain number of local stimulating configurations associated with micro-rules triggering the deposit of a brick. Once all the bricks in the current step have been deposited, the building process goes to the next step. Steps 1 to 5 correspond to the enlargement of the top of the nest, including the first sessile comb of cells (step 3). Step 6 represents the beginning of the construction of the external envelope and step 7 its lengthening, from which parallel combs will be built (steps 8 and 9). These steps determine the distinction between this nest, where the entrance and access holes at the different levels lie in the middle of the comb, and an *Epipona* nest where this hole lies in the periphery of the nest. Steps 8 and 9 in Figure 19.23 represent the final architecture. The important point to notice is that some of the stages are completed in a sequential way whereas others are completed in a completely parallel way.

Figure 19.24 illustrates the concept of modularity. We said previously that recurrent states may appear, inducing a cyclicity in the group's behavior. From the architectural viewpoint, this corresponds to a modular pattern, where

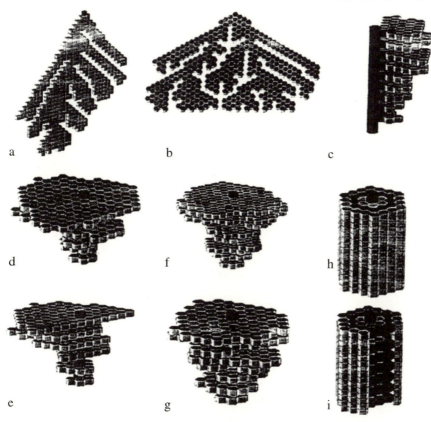

a
b
c
d
f
h
e
g
i

Figure 19.21 These comb structures were obtained by simulations of collective building on a 3-D cubic (a) and hexagonal (b)–(i) lattice. Simulations were based on a
40 × 40 × 40 lattice for architectures (a) and (b); a 20 × 20 × 20 lattice for architecture (c); and on a 16 × 16 × 16 lattice for architectures (d)–(i). These architectures
are similar to those of natural wasp nests. Each architecture is labeled with the name
of the wasp species that builds a similar, corresponding nest. In parenthesis is the total
number of micro-rules used the to build the nest. (a) *Agelaia* (13). (b) *Parapolybia* (12).
(c) *Parachartergus* (21). (d) *Vespa* (13). (e) Same architecture as (d) shown in front
section. (f), (g) *Stelopolybia* (12). (h) *Chartergus* (39). (i) Same architecture as (h), but
part of the external envelope has been removed to show the internal nest structure.

each module is built during a cycle (all modules are qualitatively identical).
This modularity is a simple way for the group to enlarge the architecture.
Figure 19.24a shows the final structure of a *Stelopolybia*-like nest, and Figure 19.24b shows the module that contributes to the structure. Likewise Figure 19.24c shows a complete *Chartergus*-like architecture, and Figure 19.24d
shows how the modules are combined to produce the final structure.

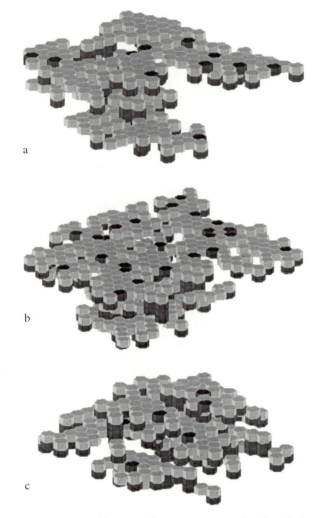

Figure 19.22 These three different architectures were produced with the same algorithm in three different simulations, using the noncoordinated mechanism.

Several authors (Downing and Jeanne 1988; Stuart 1967) have claimed that a problem with qualitative stigmergy is that it fails to explain how construction ends. A careful examination of the structure of coordinated algorithms combined with data provided by observation of natural nest development provides some answers. At least two ways exist to stop nest construction.

One way is to have a rule separate from the building rules themselves. When the structure reaches a critical size in relation to the population of wasps, build-

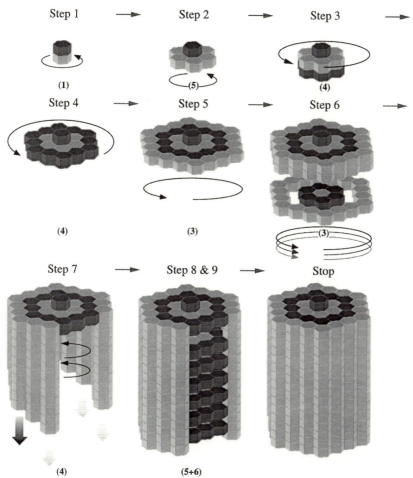

Figure 19.23 Successive building steps in the construction of *Chartergus*-like nest with a hexagonal-lattice swarm. The completion of each step results in stimulating configurations that initiate the next step. All stimulating configurations are organized to allow a regular building process. In any given step the stimulating configurations need not be spatially connected and can occur simultaneously at different locations. In steps 7 to 9, the front and right portions of the external envelope have been cut away. The total number of micro-rules used to carry out each step is given in parentheses.

ing behavior could be inhibited. This situation might obtain in the case of modular constructions, such as those built by *Polybia* (Jeanne, personal communication), where the initial size of the colony could be an important cue regulating building activity. The entire nest structure of *Polybia* wasps is built

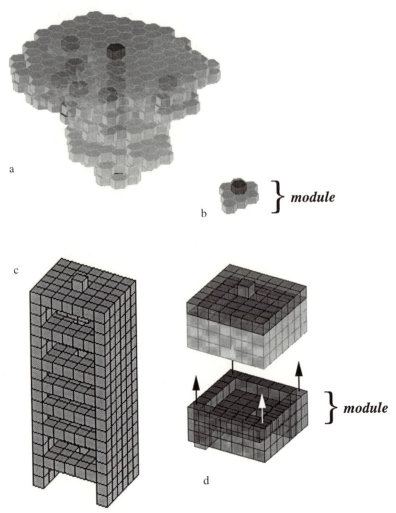

Figure 19.24 Examples of modular structures formed as a by-product of recurrent stimulating configurations in the architecture. Depicted are a *Stelopolybia*-like nest, (a) and (b), and a *Chartergus*-like nest, (c) and (d).

before the cells are filled with eggs, and a strong correlation has been observed between the final number of combs built and the initial size of the swarm. Conceivably, building activity stops when the entire swarm fits inside the nest structure. The size of the colony would then act as a negative feedback controlling the overall construction. In the case of single module constructions as in *Polistes*, the growth of the nest is regulated by an empty-cells rule. New cells are built by wasps only when there are no more empty available cells in which to lay eggs. The size of the nest results from a complex dynamic that takes into account the egg-laying rate of the foundress and the duration of brood development (Karsai et al. 1996). The growing size of the nest may also curtail the colony's building activity by inhibiting particular positive feedback amplification mechanisms (see, e.g., Chapter 21 and Deneubourg and Franks 1995).

A second way to stop construction might be that a sequence of stimulus response activities leads to a final configuration that is not stimulating. This is the case in *Paralastor* wasps, where achievement of the architecture is signalled by the absence of new stimulating configurations when the last stage has been completed.

Validating the Qualitative Stigmergy Model

The algorithmic structure of coordinated algorithms as well as their consequences for the logical organization of local stimulating configurations on the nest were drawn mainly on the basis of a theoretical study of simulation results. We therefore cannot claim that wasps actually use this mechanism in building their nests. What we have developed is a plausible, bottom-up model that provides an attractive starting point for the study of nest building in wasps. The crucial next step will be to test the model. But how is it possible to test the validity of such a model, and what kind of testable predictions can be made on the basis of the theoretical results we have so far obtained?

It is important to point out that the model represents the logical organization of building mechanisms needed to achieve the coordination of distributed activities. Thus it must be adapted to each species-specific nest building activity. Since the vast majority of building studies have been conducted with *Polistes* species, we know almost nothing about the precise individual building mechanisms that other wasp species use to build the most complex architectures. A first step would be to examine the building behavior of other species to determine whether certain stimulating configurations trigger particular building activities. The second step would be to look at the way in which these configurations are organized in space and time and to see if this organization fits the general requirements of the coordinated algorithms drawn from theoretical studies with lattice swarms. Quantitative data could also be used to generate a particular building algorithm that could be compared to the results of a lattice-swarm simulation.

Particular experiments could be designed to test the existence of local stimulating configurations on a nest. A key test would transfer a particular building stimulus to a site where it does not occur normally during construction. A complementary test would remove from the nest structure some key stimulating configuration to see if construction stops. In *Polistes*, it should be possible to use artificial 'chimeric' nests made of pieces of two different young nests arranged one below the other with the pedicel of the second nest stuck on the cell walls of the first one. In that way the two combs would have the same spatial organization as occurs in *Vespa* nests. The model predicts that if the structure is accepted by *Polistes* wasps both combs will develop normally. An important problem is that local carton configurations that wasps perceives on the nest are often combined with other information, such as gravity, that could act as an external cue. Since other mechanisms in addition to qualitative stigmergy may simultaneously be operating, experimental tests will have to be carefully designed to interpret the results of experimental manipulation.

Qualitative Stigmergy as an Alternative to Self-Organization

As discussed in the chapters on ants and termites and this chapter on wasps, the building activity of social insects has been the focus of many studies. During the construction of an architecture, social insects are involved in a collective, cooperative phenomenon usually considered to be of great complexity. Only a few models of collective building behavior have been developed that attempt to explain how these structures are built. The models based on self-organization suggest that the amplification of small fluctuations may explain some aspects of building behaviors, such as the initiation of pillars in termites (Deneubourg 1977), or of parallel combs in bees (Belic et al. 1986; Skarka et al. 1990). In particular, Belic et al. (1986) and Skarka et al. (1990) introduced a mathematical model for the construction of parallel combs in a beehive, where the bee-wax and bee-bee interactions are coupled through a nonlinear feedback mechanism. These authors showed the importance of competition between bees and the bee-wax interactions for the growth of parallel and equidistant combs.

In the present chapter we have shown that qualitative stigmergy provides a viable alternative to self-organization for coordinating the building activities of a group of insects. In this case, individual actions are directed solely by the dynamically evolving components of the construction. When the regulation of the building behavior operates in a strict qualitative stigmergic mode, the only way to build a coherent structure is to use a particular class of algorithms that we call *coordinated algorithms*. In such algorithms, local patterns that result from previous construction and are encountered by individuals moving randomly on the nest structure provide the exclusive cues necessary to direct and coordinate the building activities of the group. Another important issue was to show that strong behavioral constraints exist when insects use qualitative stigmergy. In particular we saw that the sequence of stimulating configurations produced by the distributed construction activities over time has to follow a logical path in order for the colony to build a coherent architecture.

When a group of individuals builds a particular architecture using a coordinated stigmergic algorithm, the construction must be decomposed into a finite number of building stages. A necessary condition is that the local stimulating configurations that are created at a given stage must differ from those created by a previous or a forthcoming building stage.

The logic behind this form of qualitative stigmergy is similar to that of other models developed in this book such as the model that describes the collective sorting of honey, pollen, and brood on the comb of honey bee colonies (Chapter 16).

Figure 20.1 Two paper wasps, *Polistes dominulus*, engaged in a hierarchical interaction. (© Guy Theraulaz)

20

Dominance Hierarchies
in Paper Wasps

My reasons for working with social hierarchies came in part
from my immediate amusement about certain superficial
similarities between the peck-order in flocks of hens as
described by Schjelderup-Ebbe and some human social
hierarchies in which I was enmeshed, certain college
faculties in the U. S. A., for example.
> —W.C. Allee, "Dominance and Hierarchy
> in Societies of Vertebrates."

Introduction

This chapter deals with the social organization of insect colonies, specifically the hierarchical organization of the primitively eusocial paper wasp,
Polistes dominulus (formerly *Polistes gallicus*), a common wasp in temperate regions of Europe. *P. dominulus* exhibits an annual cycle. Overwintered
females often cooperate in the founding of colonies in the spring, and these
females—the foundresses—form associations characterized by a linear dominance hierarchy (Pardi 1942, 1946, 1948; Gervet 1962, 1964; Röseler 1991). In
a linear dominance hierarchy, the dominant female is called the α-female. The
second-in-rank female, the β-female, is dominant over all females except the
α-female. The rank-order continues in a similar manner to the bottom of the
hierarchy ($\alpha > \beta > \gamma > \ldots$), where the lowest-ranking female is dominated
by all other females.

Although any overwintered female is able to found its own nest, the α-
female becomes the principal egg-layer of the colony. The aggressive behavior
of the α-female toward the other foundresses causes their ovaries to regress
(Pardi 1946; Deleurance 1946) and they take on the role of workers for the
colony. Other females may lay fewer eggs, all or most of which are eaten by the
dominant. This phenomenon, first observed by Heldmann (1936), was called
differential oophagy by Gervet (1964). The α-female recognizes the eggs laid
by her subordinates and eats them within a day after they are laid (Gervet
1964).

The dominance hierarchy is established through aggressive interactions
among foundresses within the first days of colony foundation. The intensity

of these interactions decreases with time, from severe fights when the females meet for the first time to the simple recognition of a dominant by a subordinate after several days. The resulting hierarchical organization determines not only the partitioning of reproduction but also the division of labor among nestmates. The dominant female remains in the nest where she lays eggs and contributes to nest building, while other females perform other tasks, especially outside the nest. Although division of labor will not be discussed in this chapter, it is clear that the reproductive success of a social insect colony depends on the efficiency of its mechanism for task allocation. In *P. dominulus*, task allocation is coupled with the hierarchical organization (Theraulaz et al. 1990, 1992).

In this chapter we consider whether the hierarchical organization of a *P. dominulus* colony is a self-organized structure or reflects a preexisting order. Does the hierarchical structure emerge as a result of interactions among females, or is it predetermined by the characteristics of the wasps before they actually interact? The answer to this question is not obvious. After reviewing experimental evidence we will construct a self-organization model based on assumptions derived from experiments. Although this model makes predictions consistent with experimental observations, a different model based on initial differences between females explains the same observations equally well. Because no experimental evidence is available to determine which of these models is the better one, we suggest experiments that *would* discriminate between these models. The possibility that other models may apply cannot be ruled out, but the two competing assumptions presented in this chapter are the most consistent with what is known about dominance hierarchies in these wasps.

Dominance hierarchies, also called dominance orders, are common throughout the animal kingdom whenever animals live in group. A short section is therefore dedicated to discussing the adaptive significance of dominance orders in *P. dominulus*, as well as in other wasp species and animals. If the self-organization model does not appropriately describe the emergence of dominance orders in *P. dominulus*, preliminary experimental evidence suggests that it may apply to dominance orders in bumblebees (van Honk and Hogeweg 1981; Hogeweg and Hesper 1983, 1985), primates (Hemelrijk 1996), or crayfish (Gössmann and Huber 1998). But in these cases also, more experiments are needed.

One situation where the self-organization model applies without ambiguity is jamming avoidance in weakly electric fish (Heiligenberg 1977, 1991). This is described later in this chapter. Imagine radio stations operating in the same city using different carrier frequencies assigned by a federal agency; but in weakly electric fish, the assignment of the electrolocation frequency is a self-organizing process in which fish shift their frequencies after pairwise interactions. The process results in different fish having distinct and well-separated frequencies; that is, the interacting population of weakly electric fish becomes differentiated. The self-organization model is important by itself because it is a general model of differentiation with potential relevance to the many situations in which there is differentiation in biology. The present chapter is instructive in showing how this model produces differentiation and also offers an alternative hypothesis and experiments to discriminate between various explanations for *P. dominulus* dominance hierarchies.

Social and Reproductive Dominance

Among social insects paper wasps have played an important role in the study of dominance hierarchies and the development of methods for characterizing dominance hierarchies (Heldmann 1936; Pardi 1942, 1946, 1948). Foundress associations of *P. dominulus* usually contain a small number of individuals, typically between one and ten. Hierarchical interactions are more or less ritualized pairwise contests during which a female physically dominates another female (Figure 20.1). Let us at this point distinguish *social* or *behavioral* dominance from *reproductive* dominance. While behavioral dominance has to do with physical dominance behavior, reproductive dominance is related to the partitioning of reproduction. These two dominance forms are almost always associated—the physically dominant female also monopolizes reproduction in normal conditions—but they can be distinguished experimentally. Ovariectomized foundresses of *P. dominulus* can still become and remain behaviorally dominant, but cannot lay eggs. When an α-female is ovariectomized, the β-female becomes the principal egg-layer and her eggs are tolerated by the α-female (Röseler et al. 1985; Röseler and Röseler 1989). The α-female remains socially or behaviorally dominant, but the β-female achieves reproductive dominance (Röseler 1991). As will be discussed later, the physiological correlates of both types of dominance are tightly coupled, but it is not clear exactly how they interact.

Formation and Characterization of the Hierarchy

Dominance Index

Rank is not enough to characterize hierarchical activity because it does not consider the number of interactions, dominances, and subordinations in which an individual has been involved. In order to characterize the dominance order of a group of *P. dominulus* wasps, Pardi (1946, 1948) introduced a variable that can be measured in experiments: the dominance index, X, of an individual, defined by $X = D/(D + S)$, where D is the number of times the individual has been dominant in pairwise contests since the formation of the group, and S is the number of times the individual has been defeated since the formation of the group. X is therefore the proportion of successful contests with respect to the total number of interactions, $D + S$. $X = 1$ when all contests have been won, and $X = 0$ when all contests have been lost. The relationship that describes X as a function of rank is called the *hierarchical profile*.

X is a biased measure of an individual's hierarchical activity. It does not give enough weight to the individual's total number of interactions (an individual that has been involved in only one successful contest has an index of $X = 1$); it does not include the identity of the individuals with which interactions have taken place (some pairs of individuals may have frequent interactions while other pairs only rarely interact); and it gives as much weight to recent contests as to contests that took place a long time ago. However, these biases do not seem to affect the value of X as an indicator of hierarchical activity in *P. dominulus*. The biases mentioned above are limited because the hierarchical profile quickly stabilizes after all possible pairwise interactions have taken place at least once.

Experimental Studies

A number of laboratory studies have contributed to our understanding of the formation of the dominance order in *P. dominulus* (Theraulaz 1991; Theraulaz et al. 1989, 1990, 1991a, 1992). Queens were collected in the Marseille area (south of France) during hibernation in December 1987 and January 1988 and placed in a room at a temperature 10°C. Starting on the first of March, 1988, they were placed in groups of five in transparent plastic cages ($16 \times 19 \times 24$ cm) at a mean temperature of 27°C, and provided with prey (caterpillars), blotting paper, water, and sugar. The cages were continuously refilled with a sufficiently large amount of food, but not directly on the nest so that foraging was necessary to get the food. Light was provided for 12 h every day (from 6:00 AM to 6:00 PM). As soon as the first nest was founded, surplus individuals were removed so that only monogynous colonies could develop. At emergence, each wasp was individually marked. The total number of females was kept at thirteen by removing additional females. All males were removed to eliminate the effects of group size and the presence of males on the establishment and maintenance of social interactions. At the beginning of the experiment, each colony therefore consisted of one foundress and the first twelve newly emerged females. In two control nests, no intervention took place apart from replacing dead females. In two experimental nests, the first queen was removed after five days, and the subsequent α-females were systematically removed every eight days, five times in a row. Eight days were assumed to be sufficient for a new hierarchy to become established and settled. The two experimental colonies provided ten experimental conditions to study the formation of the hierarchy.

The females were observed for 4 h/day (2×2 h), in 10 series of one-week observations. Thirty-one behavioral items were recorded. The dominance index, X, was computed after observation of the outcomes of individual encounters (dominance or subordination), using standard cues to identify the win-

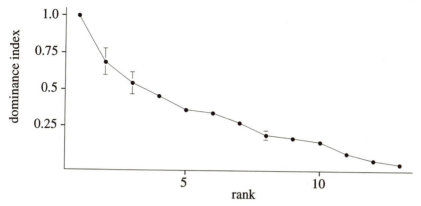

Figure 20.2 Dominance index is shown as a function of rank in *P. dominulus*. Sample size: ten experimental colonies of thirteen individuals each. (After Theraulaz 1991)

ner and the loser of a hierarchical interaction (Pardi 1942, 1946, 1948; Gervet 1964; Reeve 1991; Röseler 1991; Theraulaz et al. 1992). At first contact, two females intensively antennate one another and begin to fight, straightening up, grappling with their forelegs, and attempting to bite one another. After such a fight, one of the females escapes or adopts a subordinate posture in which she remains motionless, antennae and head lowered, while the dominant climbs on her body and intensively antennates and mouths her. In subsequent encounters, the subordinate female has a crouching posture, with antennae lowered, and sometimes spontaneously regurgitates fluid. When the hierarchy is settled, the interactions between a dominant and a subordinate are often limited to avoidance behavior by the subordinate. Abdominal wagging, a high frequency vibration of the gaster from side to side (Reeve 1991), is sometimes associated with dominance behavior during antennation of a subordinate by a dominant, but is also performed during cell inspection. The function of abdominal wagging is not clear.

Figure 20.2 shows the dominance index as a function of rank—the colony's hierarchical profile—averaged over ten experiments. The profile is remarkably stable under fixed experimental conditions. There is little variance among the ten profiles obtained experimentally. In all experiments, older wasps that had spent more time on the nest became dominant. Generally, the rank of each female reflected the order in which it was introduced into the nest.

Probability of Interaction

The frequency and intensity of the aggressive interactions vary with hierarchical rank. As Pardi (1946, 1948) and Theraulaz et al. (1989, 1992) established, the stronger individuals of a hierarchy tend to interact more frequently

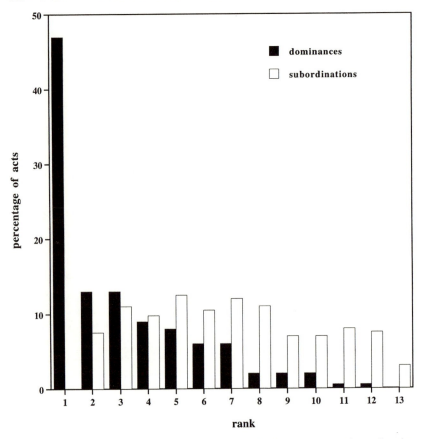

Figure 20.3 Mean proportion of dominances and subordinations plotted as a function of hierarchical rank in colonies of *P. dominulus*. Sample size: ten experimental colonies of thirteen individuals each. (After Theraulaz 1991)

than others. Figure 20.3 shows the percentage of all dominances and subordinations accounted for by each individual as a function of the individual's rank. The percentage of dominances is characterized by a rapid decay as a function of rank, from 47 percent for the α-female to 0 percent for the lowest-ranking female. The percentage of subordinations first increases and then decreases as a function of rank. This is because lower-ranking individuals, although they are almost always defeated, are involved in few interactions (see Figure 20.4).

Several factors may explain the differential probability of interaction observed in the experiments:

1. Foundresses have different motivations as a result of neurophysiological differences associated with dominance. Indeed, when two individuals

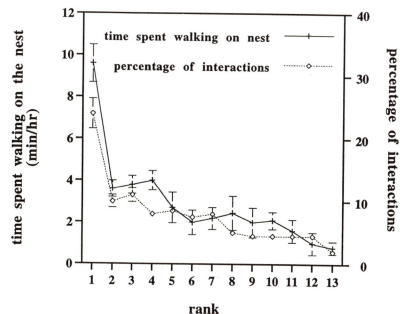

Figure 20.4 The plot shows the fraction of interactions, $PI(R)$, involving an individual and the time, $T(R)$, spent walking on the nest as a function of rank R. The best fit to $PI(R)$ is given by $23.03 \cdot R^{-0.721}$, and the best fit to $T(R)$ is given by $8.9 \cdot R^{-0.758}$. Sample size: ten experimental colonies of thirteen individuals each. (After Theraulaz 1991)

meet the initiative to interact is usually taken by the higher ranking wasp. Conversely, the motivation of a subordinate to engage in a contest with a dominant may be weak.

2. The dominance order is tightly coupled to the division of labor in the colony. The α-female spends most of the time on the nest where dominance interactions take place. Other females act as workers for the colony and frequently leave the nest to retrieve food and water, and thus have fewer opportunities to interact.

3. Dominant females exhibit increased mobility on the nest. The α-female moves more per unit time then the other females and has more opportunities to interact.

The probability of interaction and the time that a female spends walking may be estimated as functions of rank. Let $T(R)$ be the time spent walking on the nest by an individual of rank, R, and $T(X) = T[X(R)]$ the time spent walking on the nest by an individual of dominance index X. Figure 20.4 shows $T(R)$ and the percentage of all interactions, $PI(R)$, in which each individual

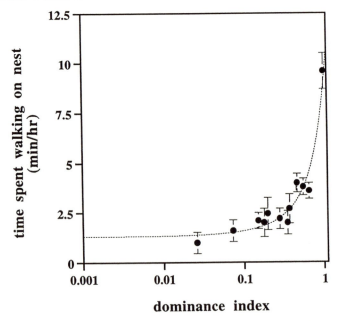

Figure 20.5 Time spent per hour walking on the nest is shown as a function of the dominance index, X. The dashed curve represents the best exponential fit, $T(X) = 1.32 \cdot 10^{0.836X}$. Sample size: ten experimental colonies of thirteen individuals each. (After Theraulaz 1991)

was involved as a function of rank. $T(R)$ and $PI(R)$, although represented on different scales, clearly have similar shapes. One may therefore assume that the probability for an individual to interact with any other individual is proportional to the time, T, it spends walking on the nest. This assumption is simple, but T results from complex factors such as those mentioned above: motivation, spatial location, and mobility. The best fit to $PI(R)$ is given by $23.03 \cdot R^{-0.721}$ (df $= 128, r = 0.918$), whereas the best fit to $T(R)$ is given by $8.9 \cdot R^{-0.758}$ (df $= 128, r = 0.918$).

A well-defined relationship also appears to exist between X and T. Figure 20.5 shows that $T(X)$ increases in an exponential-like manner as a function of the dominance index. The best fit to the data is $T(X) = 1.32 \cdot 10^{0.836X}$ (df $= 128, r = 0.944$). Each point in Figure 20.5 represents the average $T[X(R)]$, and error bars are given for both T and X. Error bars in the y-direction correspond to the standard deviation of T over 10 experiments, whereas those in the x-direction correspond to the standard deviation of X over 10 experiments for individuals of a given rank. The fit obtained for $T(X)$ is slightly better than for $PI(R)$ or $T(R)$, suggesting that T may be determined by the dominance

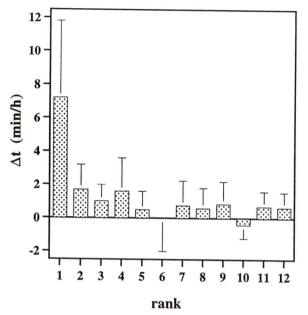

Figure 20.6 Change in the time, T, spent walking on the nest after and before reaching a higher rank, as a function of the rank reached. Sample size: nine experimental colonies of twelve individuals each. (After Theraulaz 1991)

index rather than by rank alone. Again, the dominance index contains more information than rank about the hierarchical status of an individual.

Removal of the α-Female

When the α-female is removed from a group and replaced with a newly emerged individual, a burst of hierarchical activity is observed. The hierarchical profile involving the remaining individuals obtained after this burst of interactions is exactly the same as the one observed before the perturbation (Theraulaz et al. 1989, 1992). The mean number of hierarchical interactions per unit time in the perturbed colonies is significantly larger than in the control colonies, especially for the top individuals of the new hierarchy. The new α-female accounts for 45 percent of all the dominance scenes recorded in these bursts of hierarchical activity, while the immediate subordinate individuals— newly promoted to ranks two to four—account for approximately 35 percent of these scenes. In addition, the time, T, spent walking on the nest increases when individuals reach a higher rank. Figure 20.6 shows the difference between values of T after and before reaching a higher rank as a function of the rank reached.

Table 20.1 Mean number of interactions per 5 min between α-females and subordinates in four-foundress nests after return of the α-female

Duration of Absence of α-Female	Number of Colonies	Mean Number of Interactions/5 min
1 hour	1	0.0
2 hours	2	0.5
4 hours	4	2.8
24 hours	7	3.4
7 days	1	5.0

Source: Röseler et al. 1986, p. 288, Table 4. Reprinted with permission. (©Monitore Zoologico Italiano)

Other Experiments

Röseler et al. (1986) also removed the α-female and made the same observations as above. They also observed the colony's response to reintroduction of the withdrawn α-female after variable delays and noticed that the longer the α-female was absent, the greater the number of interactions between the reintroduced α-female and the other females (Table 20.1).

This suggests either that other females in the group are increasingly willing to interact as removal time increases, or that the α-female interacts more and more after returning. The reintroduced α-female always regained her rank, according to Röseler and his colleagnes. A longer time before reintroduction might have prevented the α-female from regaining her rank, but new experiments need to be done to check this hypothesis. It is possible, however, that rank reversals are more likely to occur at an early stage of development after emergence. Early rank reversals were observed in another experiment conducted by Röseler et al. (1986). Among twenty-four pairs of individuals, immediately after hibernation, pairwise interactions occurred only once a day during a meeting lasting less than five minutes, for seven days. Dominance did not change in nineteen pairs, but the β-female became dominant over the α-female in five pairs. Comparable results were obtained when the two females of a pair were separated for three days before reuniting them.

Determinants of Dominance

The determinants of dominance in social wasps in general, and *P. dominulus* in particular, is extensively surveyed by Röseler (1991). Pardi (1946) was the first to seek a physiological basis for dominance behavior. He observed that the α-female has well-developed ovaries and that egg-formation is inhibited in subordinates: the lower in the hierarchy, the greater the inhibition. Oogenesis is correlated with high titers of two hormones in the hemolymph (the equivalent of blood), the juvenile hormone (JH) produced by the corpora allata (CA) and ecdysteroids produced by the ovaries (Röseler et al. 1980, 1984; Turillazi et al. 1982). The CA are a pair of small compact glands of tightly packed cells located in the neck region and connected to the brain by a nerve (Nijhout 1994). The synthetic activity of the CA is lower in subordinates than in dominant females (Röseler et al. 1980), resulting in lower JH titers in the hemolymph. The size of the CA in subordinates is also smaller (Röseler et al. 1984, 1985). Once the hierarchy is settled, oogenesis and endocrine activity become high in the dominant female and progressively decrease in subordinates. A good survey of the nature and role of insect hormones can be found in Nijhout (1994).

The role of JH and ecdysteroids was further demonstrated by hormone treatment (Röseler et al. 1984). More foundresses with relatively small CA and oocytes became dominant after they had been injected with juvenile hormone (JH I) or ecdysteroids (20-hydroxyecdysone), or both simultaneously, than did controls (Röseler et al. 1984; Röseler 1991). The two hormones together did not have additive effects, but when the titer of one hormone was low—not too low—it could be compensated by the other hormone to achieve dominance. When the CA activity was very low, as in parasitized wasps, injection of ecdysteroids was not sufficient to achieve dominance.

The interaction between the two types of hormones is not clear. In ovariectomized foundresses, the ecdysteroid titers are low but still result in dominance if their CA activity and JH titers are high (Röseler et al. 1985). Dominance behavior is more strongly associated with the activity of CA than in unmutilated females. These findings suggest that both JH and ecdysteroids influence dominance behavior directly. JH is sufficient to induce dominance behavior in ovariectomized foundresses and is believed to stimulate the production of ecdysteroids in unmutilated foundresses. On the other hand, ecdysteroids alone are also sufficient to induce dominant behavior. Ecdysteroids may also be responsible for aggressive behavior, although it has not been shown in *P. dominulus*. Aggressive honey bee workers have more highly developed ovaries than less aggressive workers, but they are not as developed as egg-laying workers (Velthuis 1976).

The endocrine activity of the α-, β-, and γ-females is greater in the presence of more subordinates females and the synthetic activity of the CA is greater in the α-female of a multiple-foundress association than in a lone foundress. This suggests that interactions with subordinates promote endocrine activity and, possibly, reinforce the hierarchical status of dominant individuals (Röseler et al. 1984). Similar results were obtained by Turillazi et al. (1982). The basis for this observation may be that β-females in two-foundress associations cannot dominate subordinates and exploit them by forcing them to forage and obtain food from them. Moreover, a β-female in a larger group of foundresses experiences less subordinations, because all subordinates are dominated by the α-female, responsible for much of the interactions.

Hierarchical interactions not only inhibit oogenesis and endocrine activity in subordinates, they also amplify the dominant females' reproductive physiology. Another factor related to division of labor may further amplify differences in reproductive abilities between dominant and subordinate females. Subordinates often leave the nest to forage; foraging flights are energy-expensive and may slow oogenesis. Moreover, the dominant female frequently receives food from subordinates.

Body size has been found to play a role in determining the outcome of an encounter. Although large females are favored in aggressive interactions, a dominant foundress is not always the largest female in the association. Turillazi and Pardi (1977) showed that 30 percent of *P. dominulus* associations are dominated by a female that is not the largest one. It seems that the influence of body size is mainly due to a relation between large size and high endocrine activity (Turillazi and Pardi 1977; Noonan 1981; Dropkin and Gamboa 1981).

Most measurements of endocrine activity, CA size, and ovarian development have been performed after the establishment of the hierarchy. But are all females identical at first contact with respect to these factors, or are they already differentiated? Experiments by Röseler et al. (1985) showed that after hibernation foundresses were not physiologically equal right after the first pairwise relationships between females had been formed after hibernation, the investigators measured morphological and physiological factors. They found that size did not play a significant role in determining the rank of a female, but that 91 percent and 83 percent of the dominant females had larger CA and oocytes, respectively, than their subordinates. When foundresses in a group were tested in pairs, the resulting hierarchy reflected the volume of their CA. Figure 20.7 shows the CA volume of differently ranked foundresses at the time the dominance order was established in four-foundress associations (Röseler 1991).

Differences in endocrine activity and ovarian development, and therefore in the ability to exhibit dominant behavior, exist at the end of hibernation even between individuals from the same hibernation site (Röseler 1985). This initial differentiation results both from asynchrony in emergence from hibernation and differential exposure to environmental signals. The first foundress of a group often achieves dominance; however, this may result from higher endocrine activity of females that leave hibernation early; from progressive increase of endocrine activity after hibernation in the absence of any interaction with other females; or from a "prior residence" effect whereby females familiar with a nest or a location more easily achieve dominance. It is not clear, moreover, whether females leaving hibernation early have a higher endocrine activity or if females with higher endocrine activity tend to leave hibernation early.

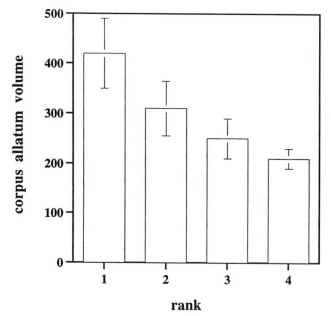

Figure 20.7 CA volume of differently ranked foundresses at the time of the establishment of the dominance order in four-foudress associations. Sample size: eight for ranks 1, 2, 3 and six for rank 4. (After Röseler 1991, figure 9.3)

Some foundresses may have an inherently higher endocrine activity resulting from a combination of factors, including genotype, feeding during development, or other environmental variations. The disadvantage of emerging late, with lower endocrine activity and smaller CA than potential cofoundresses, decreases within a few days if the female has not been dominated. This suggests that being dominated prevents the CA from increasing and causes decreases in dominance-related endocrine activity. This also indicates that the endocrine activity increases naturally after hibernation in the absence of interactions.

The Adaptive Significance of Dominance Orders

Animals that live in group may benefit from reduced predation risk, easier access to food, or increased per capita productivity, and availability of mates. But conflict may result from group living because of increased competition for critical resources. "Because self-interests will never entirely coincide, social relationships are expected to reflect a certain degree of coercion or compromise" (Pusey and Packer 1997). In many animal species, when several unacquainted individuals are placed in a group they often engage in (usually pairwise) contests for dominance. Some contests are violent fights, some do not lead to serious injury, and some are limited to the passive recognition of a dominant and a subordinate. For an initial period ranging from hours to weeks, depending on such factors as group size and animal species, contests are extremely frequent before becoming less and less frequent and ultimately replaced by stable dominance-subordination relations among all group members (Chase 1974; Wilson 1975). Guhl (1968) coined the term *social inertia* to label the decrease of aggressive hierarchical interactions over time. Threat displays quickly replace fights. Real fights can be energetically expensive, damaging, or even lethal. Ritualized fights in which individuals assess their opponents' abilities in order to decide whether to escalate or retreat, may be evolutionarily stable strategies (Maynard-Smith and Price 1973; Maynard-Smith 1974). The larger the group, the longer it takes for relations to settle. Once they are settled, they usually last for a long time, with few, generally unsuccessful, attempts by subordinates to take over. When such a network of dominance-submission relationships—a hierarchy—arises in a stable group, the group becomes organized in such a way that conflicts do not completely offset the advantages of group living.

Dominance behavior has been described in hens (e.g., Schjelderup-Ebbe 1913, 1922; Allee 1942, 1951, 1952; Guhl 1968), cows (Schein and Fohrman 1955; Barton et al. 1974), ponies (Tyler 1972), fish (Wilson 1975), crabs, lobsters, and crayfish (Bovbjerg 1956; Bovbjerg and Stephen 1971; Jackowski 1974; Glass and Huntingford 1988; Huber and Kravitz 1995; Lowe 1956), lizards (Evans 1951, 1953), frogs when they are crowded together (Haubrich 1961; Boice and Witter 1969), rats (van de Poll et al. 1982), primates (Kummer 1968; Baldwin 1971; Candland and Leshner 1971; Mendoza and Barchas 1983; Thierry 1985), and social insects (Wilson 1971), especially wasps (Gervet 1962, 1964; Pardi 1942, 1946, 1948; West-Eberhard 1969; Evans and Eberhard 1970; Röseler 1991; Theraulaz et al. 1992), ants (Cole 1981; Franks and Scovell 1983; Heinze 1990; Heinze et al. 1994; Bourke 1988; Oliveira and Hölldobler 1990; Medeiros et al. 1992); and bumblebees (van Honk and Hogeweg 1981).

Dominance hierarchies in social wasps are more widespread than originally thought. Such behavior is not limited to the conspicuous, overt dominance observed in *P. dominulus*, but includes more subtle forms through chemical interactions (Jeanne 1991). This list is far from complete. The literature on dominance orders contains hundreds of references dating back to the first systematic investigations of Schjelderup-Ebbe (1913, 1922) on the domestic fowl *Gallus domesticus*. Linear dominance orders, in which there are no loops (such as A dominates B, B dominates C, C dominates A), are by far the most common in group-living animals. Linear dominance orders seem to result in higher group efficiency (Wilson 1975). In a group of hens for instance, food is eaten more quickly when the hens form a linear dominance order than when the order includes intransitive triadic relationships where individuals frequently displace one another.

Dominant behavior is usually positively correlated with reproduction, although this rule has exceptions (Alcock 1993). The dominant individual is often the one that reproduces most. But why do some animals in a group accept a subordinates rank and sacrifice part or all of their reproductive potential? Their first choice is to reproduce, but they may be better off waiting for the more dominant individual to leave the group or die rather than fight as fighting may prove costly (Vehrencamp 1983). In addition, if group members are related it may be advantageous for an individual with a low fighting ability to forego reproduction in favor of another individual with higher fighting ability if that increases the per capita output of the group (with the help of the subordinate).

Hamilton's theory of inclusive fitness (Hamilton 1964) provides a framework for computing the costs and benefits of being a subordinate helper rather than a direct reproducer. Even in a kin-structured group fighting abilities may influence the propensity of an individual to challenge the dominant. The evolutionary theory of peace incentives (Reeve and Ratnieks 1993; Keller and Reeve 1994) is an extension of Hamilton's (1964) theory; it predicts how reproduction should be partitioned among cofoundresses according to their relative fighting abilities, their degree of relatedness, ecological conditions, and the output of the group, assuming that the dominant controls reproduction. An individual may leave the group if the expected benefit of foregoing reproduction or escalating a fight is lower than the expected benefit of reproducing outside the group.

Dominance hierarchies in social insects in general and social wasps in particular are especially interesting because they relate to both individual and group selection (Wilson and Sober 1989). One manifestation of individual selection is intragroup competition, including dominance behavior. Because of competition between groups, the group as a whole is also a unit of selection. These two levels are not incompatible but the cost of intragroup competition must be sufficiently low for the group to be able to compete with other groups.

West-Eberhard (1981) argued that intragroup competition was a major force in the evolution of worker or helper behavior. The three points of her argument have been summarized by Jeanne (1991) as follows: an individual's reproductive success depends increasingly on the ability to win in social competition; intragroup competition may lead to significant disparities in reproductive success among group members; although the preferred option of individuals that do not reproduce would be direct reproduction, these individuals may still get some indirect reproductive success by becoming helpers. It follows from this argument that intragroup competition, that results from individual-level selection, is the organizing force that makes the group an adaptive unit.

In primitively eusocial insect species with no predetermined caste, the dominance hierarchy also organizes division of labor, making the group as a whole more efficient. First, there is reproductive division of labor, which is, the partitioning of reproduction that was just mentioned; one individual is the principal egg-layer while others assume worker roles. Another division of labor among workers (Jeanne 1991) occurs in the groups of *P. dominulus* described above. High-ranking females other than the α-female spend most of their time on the comb in close contact with the brood and engage in building and brood care, while intermediate-ranking females engage in foraging, and lower-ranking (mostly newly emerged) females remain idle most of the time (Theraulaz et al. 1990, 1992). The dominance order is therefore tightly coupled with the organization of labor, which is believed to be an important factor in the evolution of social behavior (Oster and Wilson 1978).

The Goal of a Dominance Model

Before developing a model, we have to determine what there is to explain in the formation of a dominance order in *P. dominulus*. The key question is whether the dominance order emerges as a result of simpler interactions among females. To resolve this question we need to construct a self-organization model based on interactions among individuals that is consistent with empirical data and that reproduces the shape of the hierarchical profile (Figure 20.2) and the pattern of dominance and subordination (Figure 20.3). The model should also explain the stability of the hierarchical profile; that is the final profile produced by the model should remain the same despite changes in initial conditions, such as CA volume. The same remark holds true for the pattern of dominances and subordinations. If the dominant individual is removed, the profile should reorganize so as to become nearly identical to the profile before the perturbation.

Knowing *how* the hierarchy forms in *P. dominulus* is important in view of understanding *why* social behavior evolved in the genus *Polistes*. This is because hierarchy is connected to the partitioning of reproduction among cofoundresses, the characteristics of which are predicted by evolutionary theories (Reeve and Ratnieks 1993; Keller and Reeve 1994), and to the division of labor, the ergonomics of which is expected to have been optimized by natural selection (Oster and Wilson 1978). If the self-organization model, described in the next section, is an appropriate description of how the dominance order organizes in *P. dominulus* foundress associations, it may provide a simple proximate mechanism for strategies predicted by evolutionary theories of how labor is organized and reproduction is partitioned. Understanding the proximate mechanisms underlying the formation of the dominance order in *P. dominulus* thus has far-reaching implications.

Model 1: Self-Organization

The model presented in this section (Hogeweg and Hesper 1983, 1985; Theraulaz et al. 1991b, 1995; Jäger and Segel 1992; Bonabeau et al. 1996) makes several assumptions about the dynamics of hierarchy formation. These assumptions are described in detail in Box 20.1. The model is based on a fundamental hypothesis initially introduced by Chase (1982a, b): An individual that wins or loses a contest is more likely to win or lose subsequent contests. Assume that all individuals are initially almost equally likely to win contests. The outcomes of the first contests are relatively unpredictable because either individual in a pairwise contest can win. But as the number of interactions increases, individuals progressively differentiate; those that won the first contests are more likely to win future contests, and those that lost the first fights are unlikely to make it to the top. This simple model clearly uses the self-organizing, principle of positive feedback, as winners tend to win more and more, and losers tend to lose more and more. This reinforcement mechanism amplifies small initial differences between individuals, and generates a group of differentiated individuals—a social structure—out of an initially homogeneous group.

Such loser and winner effects (Chase 1982a, b, 1985, 1986; Chase and Rohwer 1987), have been reported in chickens (McBride 1958; Chase 1980, 1982a, b, 1985), crickets (Alexander 1961; Burk 1979), fish (Francis 1983; Beaugrand and Zayan 1984), mice (Ginsburg and Allee 1975), rats (Van de Poll et al. 1982), rhesus monkeys (Mendoza and Barchas 1983; Barchas and Mendoza 1984), bumblebees (Van Honk and Hogeweg 1981), wasps (Theraulaz et al. 1989, 1992), and more recently crayfish (Gössmann and Huber 1998).

Apparent winner-loser effects of course can result from initial differences between individuals (Slater 1986): if the strength of an individual is an intrinsic property of that individual that does not change over time, and if unacquainted individuals with different strengths oppose each other, the stronger individual wins the first contest and all subsequent contests, whereas the weaker one loses the first contest and all subsequent contests. Winner-loser effects are indeed observed, but they do not result from a reinforcement. Later on, we will determine whether a model based on intrinsic initial differences among individuals can explain the *P. dominulus* data.

We will now describe in more detail a model that generates a hierarchical structure which, for particular parameter values, exhibits the hierarchical structure observed in *P. dominulus*, such as the shape of the profile, the stability of the profile in response to removal of the dominant individual, and the robustness of the profile despite different initial conditions. We will use Monte Carlo simulations (see Chapter 6 for an introduction to this method of studying self-organization), in which wasps are represented by agents characterized by a variable, F, called force.

Reinforcement

If two individuals, i and j, engage in a contest, the outcome is assumed to be governed by probability. Individual i is dominant over j with a probability given by

$$Q_{ij}^+ = \frac{1}{1 + e^{-\eta(F_i - F_j)}}, \tag{20.1}$$

where F_i is individual i's force, F_j is individual j's force, and η is a positive parameter, the meaning of which is discussed below. The probability for i to lose is equal to the probability for j to win:

$$Q_{ij}^- = Q_{ji}^+ = \frac{1}{1 + e^{-\eta(F_j - F_i)}} = 1 - Q_{ij}^+. \tag{20.2}$$

F_i is increased by a constant value, δ^+, in case of victory, and decreased by a constant value (δ^-) in case of defeat. This force can be seen as an indicator of the physiological state of the animal, such as the synthetic activity of the CA, JH, and ecdysteroid titers in the hemolymph, or the size of the ovaries. It is part of the model's assumptions to start from the existence of an aggregate variable, denoted by F, which directly reflects the ability of an individual to dominate in a hierarchical interaction. It is assumed here that F_i does not depend on which nestmate individual i is currently facing; the same force is used to compute the outcome of any contest with any other member of the group. Recognition on an individual basis has no influence and the outcome of a contest depends only on the respective forces of both individuals, as given by equation (20.1). See Box 20.2 for a self-organization model with individual recognition, which is an important extension for analyzing intransitive loops.

F is subject to a feedback mechanism, the sign of which depends on how well an individual with force, F, is performing, given the current state of the colony's dominance order. When $\delta^+ = \delta^-$, F_i is simply proportional to the number of times individual i has been successful minus the number of times it has been dominated. F_i can in principle take any negative value (for subordinate individuals) or positive value (for dominant individuals), but it is bounded by the maximum number of interactions, I_{max}, allowed in the simulations. In reality, F_i may be bounded by physiological limits; for example, hormonal titers cannot become infinitely large. One may also assume that hierarchical interactions stop whenever an individual's F exceeds some threshold, reflecting the unquestioned, dominant status of that individual, or that the individual starts laying eggs.

The choice of the sigmoid function in equation (20.1) is arbitrary; however, it is a classic example of a function with deterministic and stochastic features that are easily modulated by a "tunable" parameter and exhibits saturation at large values. When $F_i - F_j \gg \eta$, $Q_{ij}^+ \approx 1$. When $F_i - F_j \ll \eta$, $Q_{ij}^+ \approx 0$. When $F_i = F_j$, $Q_{ij}^+ = 1/2$. In the simulations described in the next section, $\delta^+ = \delta^- = 1$ and $\eta = 1$, so that the outcome of a contest is 'almost' deterministic. As soon as F_i is greater than F_j by little more than the minimum amount, δ^+, individual i is almost sure to win.

Probability of Interaction

Let P_{ij} denote the probability that two individuals, i and j, interact per unit time. We assume that P_{ij} is given by

$$P_{ij} = Y_i Y_j, \tag{20.3}$$

where

$$Y_i = \frac{1}{1 + e^{-F_i/\theta}} \tag{20.4}$$

denotes individual i's 'likelihood' to interact with any other individual, and θ is an interaction threshold. $Y_i \approx 1$ when $F_i \gg \theta$, $Y_i \ll 1$ when $F_i \ll \theta$, and $Y_i = 1/2$ when $F_i = 0$. In the simulation, we set θ to 100. In other words, the probability of interaction remains relatively close to $1/2$ during the first 100 interactions per individual, the time needed for some individuals to reach a force equal to θ. P_{ij} is a symmetric function of i and j, with a value that is at a maximum when both individuals in the pair are strong, and at a minimum when both individuals are weak. The middle value taken when a strong individual meets a much weaker one reflects the fact that the strong individual will try to engage in a contest while the weak one will tend to escape, or that the more dominant individual spends more time walking on the nest than its subordinate.

Results of Model 1

In the simulations identical initial forces were assigned to all individuals. All individuals started in the same state; $D = 1$, $S = 1$, $F = 0$, $D/D + S = 0.5$, so that a given individual had an equal probability of winning or being defeated. The total number, I, of hierarchical interactions among cofoundresses was initially equal to 0.

An algorithmic description of the Monte Carlo simulations is given in Box 20.3. At each time step, a pair of individuals (i, j) is randomly selected and tested to determine if they will interact. This is done according to the probability P_{ij} [equations (20.3), (20.4)]. If the answer is positive, the total number of interactions, I, is incremented by one and individual i wins with probability

Q_{ij}^+. D or S is updated according to the fight's outcome. The simulation stops when I reaches a predefined maximum number of hierarchical interactions, I_{max}.

Simulations have also been run to study the removal of the dominant individual. When the profile has stabilized, the α-individual is removed.

Model 1 requires tuning the parameters η, θ, and δ^+ (assuming that $\delta^+ = \delta^-$). Since θ and η can be scaled in units of δ^+, only two parameter values must be determined.

Convergence Properties

Figure 20.8 shows the hierarchical profile obtained from the simulation, together with the experimental profile. As can be seen, the simulated profile is difficult to distinguish from the experimental profile. In addition, the obtained profile is robust over experiments, as shown by the small error bars in Figure 20.8. As I increases, the fluctuations shown in the error bars decrease and the profile converges to a higly stable rank-order. Figure 20.9 shows that error bars decrease with increasing I. For $I = 10,000$, error bars are within the

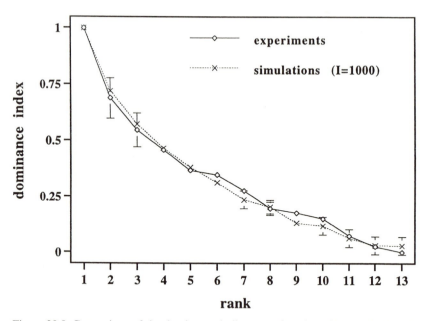

Figure 20.8 Comparison of the dominance indices as a function of hierarchical rank as observed in experiments (same as in Figure 20.2) and in Monte Carlo simulations of model 1. Error bars for the simulation data were obtained by performing twenty simulations of $I = 1000$ interactions for each group of thirteen individuals. Parameter values: $\eta = 5$, $\theta = 100$, $\delta^+ = \delta^- = 1$.

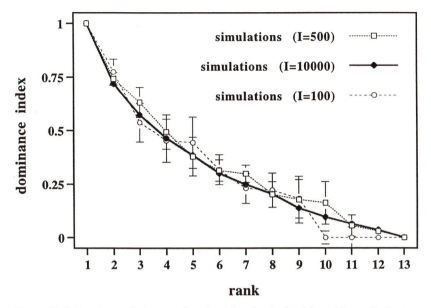

Figure 20.9 Dominance index as a function of rank, obtained from Monte Carlo simulations for different numbers, I, of simulated interactions: $I = 100$, 500, and 10,000. Error bars correspond to averages of 20 simulations. Parameter values are the same as in Figure 20.8.

thickness of the line. After 500 interactions, the average profile obtained is already very close to both the experimental and stable profiles. Note that the experimental profile is always stable after a few hundred interactions.

The final profile is independent of initial conditions. If instead of starting from initially undifferentiated individuals, forces are assigned to individuals according to a Gaussian distribution, the profile to which the system converges is unchanged. Initially stronger individuals, however, become higher-ranking individuals, and initially weaker individuals become lower-ranking. Assuming that F grows regularly over time after hibernation in the absence of social interactions, this result is consistent with the observation that emerging early is an advantage to achieve dominant status in a foundress association.

Figure 20.10 shows the proportion of all dominances and subordinations accounted for by each individual as a function of the individual's rank. This figure is to be compared with Figure 20.3. Obviously, the curves are very similar in Figures 20.10 and 20.3. The percentage of dominances is characterized by a rapid decay as a function of rank, from 43 percent (47 percent in Figure 20.3) for the α-female to 0.02 percent (0 percent in Figure 20.3) for the female at the bottom of the hierarchy. The percentage of subordinations first increases and then decreases as a function of rank. This is because lower-ranking indi-

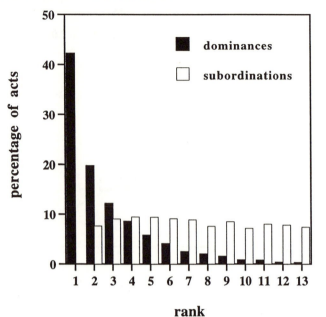

Figure 20.10 Proportion of dominances and subordinations as a function of hierarchical rank were obtained from Monte Carlo simulations of model 1. Parameter values are the same as in Figure 20.8.

viduals, although they are almost always defeated, are not involved in many interactions.

Model 1 generated differentiation between individuals. Figure 20.11 shows how the forces of the 13 individuals varied with the total number of hierarchical interactions, I. As can be seen, the force of the dominant individual increased quickly with I, whereas the forces of its subordinates either decreased, remained in the vicinity of zero, or increased more slowly with I. The process was stopped at I = 500 interactions. In real foundress associations it is possible that when F (for example, hormonal titers) reaches a threshold, the differentiation process stops—for instance when the α-female starts laying eggs.

Model 1 is "endogeneous" in that the social organization emerged from interactions among the females. But the parameters of the model are certainly influenced by external factors. The probability of interaction between two females depends on the time each spends on the nest, which, in turn, depends on how much foraging time is necessary to satisfy colony needs. When food is abundant and easy to find, foragers spend more time on the nest. When food is scarce and hard to find, foragers spend more time outside the nest. There will be fewer interactions between the dominant and its subordinates, but the

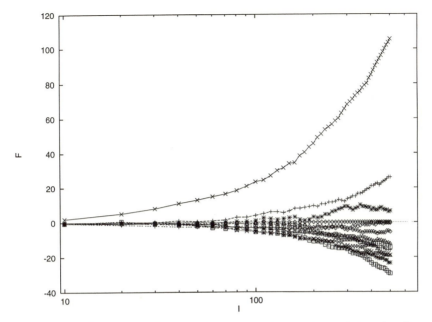

Figure 20.11 "Forces" of the thirteen individuals simulated with model 1 as a function of the number, I, of simulated interactions.

amplification effect may be stronger; a subordinate will have to spend more energy foraging, and that may lead to diminished ovarian development and lower endocrine activity.

Removal of the α-Individual

The removal of the α-individual resulted, in all situations, in recurrence of the initial hierarchical structure before removal (Figure 20.12). A careful study of how ranks changed in the simulations after the removal of the α-individual showed that individuals did not swap ranks in the restructuring process provided initial force differences were sufficiently high: a global leftward shift of the hierarchy is observed. This result is consistent with experiments, in which individuals swapping ranks may be observed in the bottom of the hierarchy, where there is little hierarchical differentiation.

Critique of Model 1

The self-organization model described above reproduces the experimental data quite satisfactorily. It goes beyond mere curve fitting and reproduces fea-

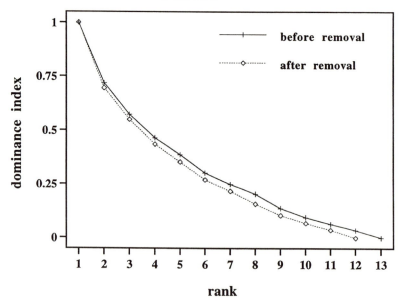

Figure 20.12 Dominance index is plotted as a function of rank before and after the simulated removal of the α-individual in Monte Carlo simulations of model 1. The α individual was removed after 10,000 simulation steps and the simulation was then run for another 10,000 steps to obtain the new profile. Parameter values are the same as in Figure 20.5.

tures that do not directly follow from the assumptions. However, the assumptions behind the model do raise some questions.

Although we assumed that self-organization is responsible for the differentiation, we know that emerging wasps already exhibit initial differences. Can we reproduce the data equally well by assuming that the wasps are already differentiated, without invoking self-organization?

We also know that some differentiation takes place after the group has been formed—because the ovaries of the α-female develop while those of the other females tend to stagnate or regress—and that interactions with subordinates amplify this differentiation. Moreover, differention affects the production of JH and ecdysteroids. However, this has to do with *reproductive dominance*. Social dominance and reproductive dominance can be separated in experiments with ovariectomized females. In any case, the respective roles of JH and ecdysteroids are not at all clear. Is the endocrine activity a cause or a consequence of *social dominance*? We do not know whether a reinforcement process also takes place in the context of social dominance. Again, can we reproduce the data by assuming the wasps are already differentiated, without invoking a further reinforcement process?

Finally, we assumed that the probability of interaction between two individuals, i and j, was determined by a function of their abstract forces, F_i and F_j [equations (20.3) and (20.4)]. But we do not know whether the factors influencing the outcome of a fight directly influence the probability of interaction. Indirect effects, such as the spatial distribution of individuals resulting from the division of labor, may explain most of the probability of interaction. One of the problems is that F_i cannot be measured. Can we construct a model in which the probability of interaction is based on measurable quantities?

The answer to these above three questions is "yes." An alternative model (model 2) detailed in Box 20.4 can be constructed along lines suggested by the criticisms.

Model 2

Model 2 is based on an alternative to self-organization that is not easily classified as one of the mechanisms described in Chapter 4 (leader, blueprint, recipe, and template). In this model, individuals are assumed to be predifferentiated; that is, initial differences exist between individuals before their first interactions, and these differences determine both the outcome of pairwise contests and the global dominance order. Because the ordering of the individuals in the dominance hierarchy directly reflects their initial differences, this alternative to self-organization is most closely related to a template mechanism. We will therefore describe model 2 as a template-based model, for lack of a more accurate term.

For simplicity it is assumed that individual i has rank i. When two individuals i and j interact, i wins if and only if $i > j$. The probability that i and j interact is assumed to be proportional to the product of the time each previously spent walking on the nest. The time spent by i walking on the nest is assumed to depend solely on its dominance, X_i according to the empirical function $T(X_i) = 1.32 \cdot 10^{0.836 X_i}$. An algorithmic description of the Monte Carlo simulations of model 2 is given in Box 20.5. Note that model 2 is a parameter-free model: all assumptions are based on quantitative experimental observations.

Results

The results obtained with model 2 are very similar to those obtained with model 1. Figure 20.13 shows the hierarchical profile obtained from Monte Carlo simulations of model 2 after $I_{\max} = 10,000$ interactions, and compares it with the experimental profile (Figure 20.2). The two profiles are very similar with not much difference between that obtained with model 2 after $I = 10,000$ interactions and the model 1 profile obtained after $I = 10,000$ interactions (Figure 20.9). Model 1 required tuning the parameters, η, θ, and δ^+, whereas

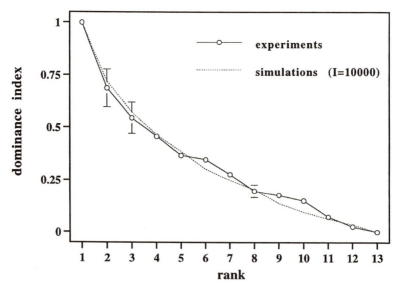

Figure 20.13 Comparison of the dominance indices as a function of hierarchical rank as obtained in experiments (same as in Figure 20.2) and in Monte Carlo simulations of model 2. Data from the simulations are averages of twenty simulations of $I = 10,000$ interactions for each group of thirteen individuals. Error bars cannot be distinguished from line thickness.

model 2 has no parameters. Figure 20.14 shows the fraction of dominances and subordinations of every individual as a function of rank. Figure 20.14 is almost identical to Figure 20.10. More generally, all other properties of model 1 are also exhibited by model 2 (including stability of profile over different runs, decreased fluctuations with increasing I, and robustness with respect to removal of α-individual). The major difference is that model 2 does not generate differentiation but relies on predifferentiation. Robustness of the profile with respect to initial conditions is not relevant here, since individuals are preordered at the beginning of each simulation, and the outcome of a pairwise contest depends only on the rank of both individuals, not the individuals' previous contests.

Although model 2 appears more parsimonious than model 1, there remains the need to explain why the probability of interaction is given by the function $T[X(R)]$. It seems that the probability of interaction is the most important factor in determining the shape of the hierarchical profile and the overall pattern of dominances and subordinations. Finally, if model 2 accounts for the data on social dominance, it cannot describe the kind of differentiation observed in the context of reproductive dominance.

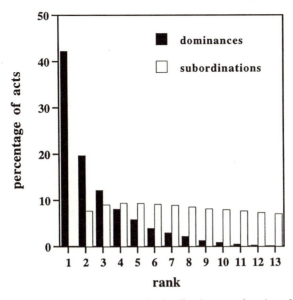

Figure 20.14 Proportion of dominances and subordinations as a function of hierarchical rank obtained from Monte Carlo simulations of model 2.

Discussion of the Models

Both the self-organization model, model 1, and the template-based model 2 yield predictions consistent with available empirical data. Model 2 requires fewer assumptions, but only model 1 produces differentiation. Which model is right (assuming one of them is)? Obviously, more experiments are needed to answer this question. In the absence of further data, model 1 is speculative and model 2 should be preferred. No doubt some form of differentiation occurs in the formation of the *reproductive* dominance order. The fundamental issue is whether a parallel differentiation underlies formation of the *social* dominance order. Although both types of dominance order are most often associated, and therefore probably correlated in natural conditions in *P. dominulus*, it is uncertain that the observation of reproductive differentiation implies the existence of an equivalent social differentiation. Whether there is such a social differentiation, we cannot tell from empirical data.

The existence of initial differences among females makes it difficult to look for the contribution of a possible reinforcement process. The physiological correlates of dominance behavior have been shown to be the causes of dominance behavior but they may as well be the effects of dominance behavior through a feedback loop. For example, JH titers determine whether an individual is likely to achieve dominance, but JH titers may, in turn, be determined by the dominance status of the individual.

As argued in Chapter 6, modeling a phenomenon should consists not only of fitting existing data with a biologically plausible model but also of testing alternative models and carefully comparing among the models. In the present case, the difficulty in discriminating between the models results from the fact that, although the ordering of individuals in model 2 does not rely on interactions among individuals, the ordering can be observed only if individuals interact! The only way to find out which model is better is to monitor the physiological correlates of dominance *during* the establishment of a hierarchy. Since measuring hormonal titers and the size of glands requires killing the wasps, this is problematical.

An experiment to help discriminate between the models would use groups of individuals of the same age raised in the same laboratory conditions (to reduce as much as possible the differences between wasps), and measure relevant variables, including hormone titers and volume of CA at various stages in the formation of a dominance order. Working on ovariectomized females would eliminate the confounding effects of reproductive dominance but may create more problems than it solves, since social and reproductive dominance are tightly coupled. Another study would investigate the distribution of CA-volume changes over time.

If the assumptions of model 1 are valid, one expects the CA distribution to be relatively peaked at emergence and to become wider as the number of hierarchical interactions increases (because small initial differences get amplified). The distribution could be compared to the distribution obtained with control wasps of the same age raised in isolation. In the latter wasps, the distributions of CA volume and hormone titers are expected to remain peaked, though at a higher level. If such were the case, it would strongly suggest that social interactions are a cause of differentiation by amplification of small differences, a major assumption of model 1. A variation of this experiment would consist of systematically plotting the hormone titers and CA volume of an individual as a function of the number of dominances, D, and subordinations, S, of that individual. If the assumptions underlying model 1 are correct, one expects to see a positive correlation (even possibly a linear relationship over some range of values) between D, S, and the measured variables. Regarding the probability of interaction, one has to determine whether and how the physiological correlates of dominance influence the probability of interaction. For example, does mobility or the probability of engaging a pairwise interaction increase with JH titers?

Application of Model 1 to Other Animals

Model 1 has been invoked to explain the emergence of the hierarchical structure in the bumblebee, *Bombus terrestris* (Van Honk and Hogeweg 1981; Hogeweg and Hesper 1983, 1985) and primates (Hemelrijk 1996). In both cases, however, empirical evidence is too weak to justify the assumptions of model 1. Recent experiments on the crayfish, *Astacus astacus*, are promising (Huber and Kravitz 1997; Huber and Delago 1998; Gössmann and Huber, 1998).

When juvenile crayfish are put together, they engage in pairwise contests that may only rarely involve the unrestrained use of the claws. Gössmann and Huber (1998) studied five groups of five crayfish in which a linear dominance order always emerged, and observed a number of behaviors suggesting that model 1 may apply. Reversals in position are common early after the formation of the group but their number decreases significantly over time. The propensity to engage in fights is a function of rank, irrespective of the identity of the individual that has that rank, with the more dominant crayfish being more likely to start fights. The first fights determine the final rank and dominance relationships become polarized, with higher-ranking individuals growing more dominant and lower-ranking individuals becoming more subordinate. Ranks may change from day to day, influenced by the outcomes of contests early in the day. Although these observations are suggestive, they are not sufficient to support model 1.

Studies showing that the neuromodulator serotonin plays an important role in regulating aggression in crayfish indicate that it might be possible to elucidate the neurophysiological basis of that behavior (Huber et al. 1997; Huber and Delago 1998).

Electrolocation in Electric Fish

A phenomenon that has been well studied in the last twenty years is electrolocation in the weakly electric South American fish, *Eigenmannia virescens* (Black-Cleworth 1970; Heiligenberg 1977, 1991). This fish emits weak (typically millivolt) electric pulses at frequencies between 250 and 600/s for electrolocation and communication. In the absence of visual cues, the fish can sense objects and navigate in its surroundings by detecting perturbations that nearby objects make in the self-generated electric field. When two fish operating on similar frequencies meet, they should jam each other's sensory apparatus and blind each other. When the frequencies of the two fish overlap enough to interfere, the fish can sense this collision and respond by shifting their pulse frequencies out of each other's range.

In the jamming-avoidance response the fish with the higher frequency raises its frequency still higher, and the fish with the lower frequency lowers its frequency. For example, let us assume that two fish, A and B, meet, with respective frequencies $f_A = 260$ Hz and $f_B = 450$ Hz. In this case no avoidance response occurs because the two frequencies are sufficiently far apart. But if, $f_A = 280$ Hz and $f_B = 284$ Hz, fish A will shift its frequency downward and fish B will shift upward because the frequency difference is small. After the jamming avoidance response, fish A will have (for instance) $f_A = 272$ Hz while $f_B = 292$ Hz. The 20 Hz frequency difference does not affect electrolocation.

This resolution of one frequency conflict may cause frequency conflicts with other fish in the neighborhood. Such conflicts are resolved using the same algorithm. To implement their jamming avoidance response, the fish need to detect time disparities between the two sets of signals that are less than one microsecond long. Since their individual electroreceptors cannot handle such small time differences, the response must originate in a highly sophisticated signal-processing system in the fish's central nervous system (Heiligenberg 1991; Kawasaki 1993).

What is interesting for us in this example is that the jamming avoidance response is clearly a double reinforcement mechanism similar to the one described in this chapter, where the force of an individual is replaced by the electrolocation frequency. Imagine a group of electric fish in a tank. Pairwise interactions lead to readjustment of individual pulse frequencies; individuals at slightly above average frequencies are more likely to end up at the top end of the spectrum, and those with slightly below average frequencies are likely to be found at the lower end of the spectrum. Even the probability that two individuals engage in a hierarchical contest has a counterpart here: two fish interact only if their characteristic frequencies differ by less than a certain amount (e.g., 20 Hz); otherwise, their frequencies do not overlap sufficiently to justify a jamming avoidance response. In the context of dominance orders, this would be equivalent to individuals interacting only when they have similar forces.

Jamming avoidance is a self-organizing mechanism that relies on the amplification of small frequency differences between individuals to generate clear differentiation. Several aspects of this phenomenon are relevant to the topic of this chapter, and to this book in general.

Although the neurobiological basis of the phenomenon is extremely complex and its explanation at this level still incomplete, the self-organizing mechanism is very simple and corresponds to a higher level of description. In other words, the high-level description of the logical mechanism of differentiation is simple, whereas its biological implementation is still not fully understood. That realization is important for this book (see also the introduction of Chapter 4 for a brief discussion of levels of description and misconception #4 in Chapter 7 for a discussion of the "simplicity" of self-organization models). When we say that a certain structure emerges from simple interactions between individuals that exhibit simple behavior, it does not mean that the underlying biological implementation of the high-level description is simple. It means that the logical mechanisms that describe the emergence of the structure are simple to formulate.

The same is true in the context of hierarchies in *P. dominulus*. The assumptions that were made about the reinforcement process or the probability of interaction were simple at the level of description that we chose; however, the underlying physiological and behavioral correlates of these assumptions clearly are very complex. For example, how does a wasp know that it should withdraw in a first fight or act submissively later on? We have no idea. Such behavior may involve sophisticated perception tools. But we made the assumption that a wasp is capable of determining whether it is a subordinate or a dominant in a contest—which seems to be true—without worrying about how it does so. This is reasonable as long as one understands that one is then limited to the selected level of description.

The example of the weakly electric fish is also relevant in that it shows that differentiation—here, frequency differentiation—probably relies on a combination of predetermined features and interactions among fish. Initial frequency differences exist between members of a population of weakly electric fish because of genetic, developmental, or environmental factors. Interactions among the fish amplify these differences through a self-organizing mechanism. The self-organizing mechanism ensures that individuals become sufficiently differentiated. The parallel with the self-organizing model of hierarchy formation in *P. dominulus* is striking: A possible scenario is that both predetermination and self-organization coexist so that initial differences between wasps, which are known to exist, become amplified by self-organization to generate a clear differentiation.

Finally, it suggests that the same principles may be able to explain differentiation in extremely different contexts and taxa. Of course, it is not at all certain that self-organization plays a role in forming the hierarchical structure in wasps. It is intriguing, however, that we are able to discuss the possibility that the same logical mechanisms underlie certain behavior in both weakly electric fish and primitively social wasps.

Another type of logical mechanism that has general relevance to morphogenesis and pattern formation is the combination of local activation and long-range inhibition. Such a mechanism can be implemented in many different substrates and could explain the production of the same family of patterns in these widely varying substrates, from patterns of ocular dominance in a primate's cortex to stripe patterns on the coat of a zebra (Figure 1.2d and e).

Conclusion

We have presented two alternative models for the formation of a hierarchical dominance structure in *P. dominulus*: a self-organization model, in which small differences between individuals are amplified by a double-reinforcement process and produces differentiation, and a predifferentiated model, where the initial differences between individuals explains the data of social dominance. Experiments have been suggested to distinguish these alternative explanations. Probably, however, both explanations coexist. Small differences between individuals are sufficient to explain the social dominance data, but the amplification of these small differences may be the only way to explain the organization of the reproductive dominance order. As Röseler et al. (1984, p. 141) point out:

> At the end of hibernation differences in the endocrine activity even exist between the foundresses from the same hibernation site (Röseler 1985). The small differences in the beginning become more and more pronounced during colony development by inhibition of subordinate females, as well as by the trophic advantage and lessened external activities of the α-female. This system, originally postulated by Pardi (1946) for ovary development, ensures the reproductive exclusiveness of the α-female.

From an evolutionary perspective, the combination of both mechanisms—predetermination and self-organization—is satisfactory because the reinforcement process makes differentiation robust and ensures a proper partitioning of reproduction. If post-hibernation differences between two females are too small to determine a winner and a loser consistently over time, none of the females will become dominant or subordinate in the absence of a reinforcement process. Successive encounters will produce a sequence of wins and losses for each individual.

Box 20.1 Model 1 Assumptions

1. Each individual is characterized by a force, F, that influences its ability to win contests. F reflects endocrine activity, ovarian development, and other factors.
2. Individuals have initially identical or almost identical forces. This is rarely true in nature but may be approached in the laboratory.
3. Individual recognition, if it exists, plays no role in the outcome of a contest. In other words, the force of an individual does not depend on the individual it is encountering.
4. A pairwise contest between i and j is won by i with a probability given by a rapidly increasing function of the difference between the force F_i of i and the force F_j of j. If i and j have equal forces, the probability that i wins is 1/2. If F_i is larger than F_j, i wins almost certainly, otherwise i loses. The outcome of a contest is probabilistic because reversals of dominance can occur. But such reversals are highly unlikely after some time, so that the outcome should be almost deterministic (rapid increase as a function of F_i and F_j).

Box 20.1 continued

5. When i wins, F_i increases. When i loses, F_i decreases. This assumption relies on three observations: The ovaries of the dominant females develop whereas those of the subordinates regress; ovarian development is induced by JH and generates an increase of ecdysteroids in the hemolymph. Both hormones increase the probability of dominating. More interactions with subordinates increases endocrine activity. The opposite is true at least to some extent; lone foundresses reproduce, but may not reproduce if they are subordinates in a group. Also, the disadvantage of emerging late, with lower endocrine activity and smaller CA than potential cofoundresses, decreases within a few days if the female has not been dominated, suggesting that in the dominated wasp CA is prevented from growing and dominance-related endocrine activity decreases.

6. The probability of interaction of an individual i is an increasing function $Y_i(F_i)$. Y_i indicates that the probability of interaction increases with rank, and that ecdysteroids titers reflected in F may be responsible for aggressive behavior.

7. Two individuals, i and j, interact at a frequency proportional to the product $Y_i Y_j$. This indicates that the frequency of interaction between i and j may be related to the time they spend walking on the nest, where interactions take place.

Box 20.2 Model 1 with Individual Recognition

No assumption is made in model 1 regarding the ability of an individual to recognize its nestmates as individuals. In model 1 encounters are anonymous in that the past experience of an individual, reflected in F, contains no information about the precise identities of individuals it has interacted with. How does individual recognition alter the profile of the hierarchical structure? A model based on anonymous encounters, such as model 1, cannot explain the presence of intransitive loops (A dominates B, B dominates C, C dominates A). To overcome this limitation, we introduce the *effective force* F_{ij}^{eff} of individual i, which is used in computing the probability of individual i winning when interacting with individual j:

Box 20.2 continued

$$F_{ij}^{\text{eff}} = \epsilon F_{ij} + \frac{1-\epsilon}{N} \sum_{\substack{k=1 \\ k \neq i}}^{N} F_{ik},$$

(20.5)

where F_{ij} is the relative force of individual i with respect to individual j, and ϵ is a parameter that characterizes the weight of individual recognition in the outcome of the interaction. Depending on the value of ϵ, loops may be observed. The most natural form for F_{ij} is to take $F_{ij} = Dom_{ij} - Sub_{ij}$ where Dom_{ij} is the number of interactions in which i has defeated j. The probability that i is dominant over j is

$$Q_{ij}^{+} = \frac{1}{1 + e^{-\eta(F_{ij}^{\text{eff}} - F_{ji}^{\text{eff}})}}.$$

(20.6)

When $\epsilon = 0$, the model is similar to model 1. When $\epsilon = 1$, the outcome of a pairwise contest between i and j is determined on the sole basis of the history of the interactions between i and j.

The Landau number is aimed at characterizing the degree of linearity of the hierarchy, or, conversely, the number of loops (Landau 1951; Chase 1974). As a first approximation, i dominates j if $F_{ij} > F_{ji}$, and j dominates i if $F_{ij} < F_{ji}$. Let us denote by n_i^{+} the number of individuals dominated by individual i according to that criterion. The Landau number h of the hierarchy is defined by

$$h = \frac{12}{N^3 - N} \sum_{i=1}^{N} \left(n_i^{+} - \frac{N-1}{2} \right)^2.$$

(20.7)

For a colony h ranges from 0, where all individuals dominate the same number of other individuals, to 1 for a perfectly linear hierarchy (containing no loops). But the outcome of a contest is not deterministic: i dominates j with probability Q_{ij}^{+}. Landau has extended his index (Landau 1951; Chase 1974) to probabilistic cases. The expectation of h is given by

Box 20.2 continued

$$\langle h \rangle = \frac{12}{N^3 - N} \sum_{i=1}^{N} \left\langle \left(n_i^+ - \frac{N-1}{2} \right)^2 \right\rangle$$

$$= \frac{12}{N^3 - N} \sum_{\substack{i=1}} \sum_{\substack{j=1 \\ j \neq i}} \sum_{\substack{k=1 \\ k \neq i \\ k \neq j}} Q_{ij}^+ Q_{ik}^+ - \frac{3(N-3)}{N+1}, \tag{20.8}$$

so that $\langle h \rangle = 3/N + 1$ if all individuals have a 50 percent probability of winning irrespective of the other fighting individual, and $\langle h \rangle = 1$, once again, if there is a perfectly linear hierarchy. Figure 20.15 shows the variation of $\langle h \rangle$ as a function of the individual recognition parameter, ϵ. A departure from pure linearity is observed at $\epsilon \approx 0.4$. At $\langle h \rangle \approx 0.21$ $\langle h \rangle$ saturates as ϵ approaches 1; this corresponds exactly to $\langle h \rangle = 3/(N+1)$, with $N = 13$, the number used in the simulations, so that for ϵ close to 1 all individuals have a 50 percent probability of winning or losing an arbitrary fight. The hierarchy is then composed only of loops.

Experiments by Gervet et al. (1993) suggested that individual recognition takes place at least in very small groups of *P. dominulus* and plays a role in the genesis of loops. In natural foundress associations, however, linearity is the rule. In the context of the reinforcement model, linearity results from a low level of individual recognition. In small groups, it is possible that all wasps recognize one another, but not in large groups—which would explain the observation of Gervet et al. (1993).

Box 20.2 continued

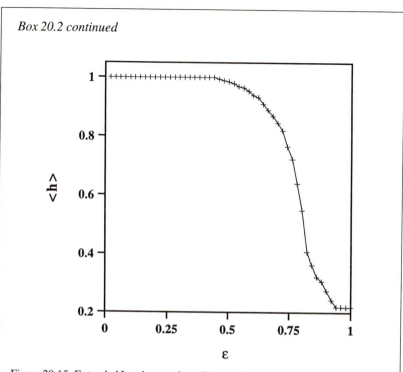

Figure 20.15 Extended Landau number, $\langle h \rangle$, as a function of the individual recognition parameter ϵ. Parameter values: $\eta = 5$, $\theta = 1$, $\delta^+ = \delta^- = 1$. Each point is the average of $\langle h \rangle$ over 20 simulations of 10,000 steps each run with thirteen individuals. At each time step, a pair of individuals was selected. Errors bars are not shown.

Box 20.3 Monte Carlo Simulations of Model 1

```
/*  Initialization  */
I=0     /* number of hierarchical interactions initially equal to
0 */
For every individual i do
D_i =0 /* number of dominances initially equal to 0 */
S_i =0 /* number of subordinations initially equal to 0 */
F_i =0 /* force initially equal to 0 */
X_i =0 /* by convention dominance index initially equal to 0 */
End-for

/*  Main loop  */
While I<I_max do
Randomly select two individuals i and j
P_ij=Y_iY_j /* probability of interaction between i and j */
Draw random real number R between 0 and 1
If (R<P_ij) do /* i and j interact */
  I<-I+1
  Draw random real number R' between 0 and 1
  If (R'<Q_ij^+) do /* i wins */
    D_i <-D_i+1 /* number of dominances of i increased by 1 */
    F_i <-F_i+d^+ /* force of i increased by d^+ */
    X_i <-D_i/(D_i+S_i) /* dominance index of i updated */
    S_j <-S_j+1 /* number of subordinations of j increased by 1 */
    F_j <-F_j-d^- /* force of j decreased by d^- */
    X_j <-D_j/(D_j+S_j) /* dominance index of j updated */
  Else /* j wins */
    D_j <-D_j+1
    F_j <-F_j+d^+
    X_j <-D_j/(D_j+S_j)
    S_i <-S_i+1
    F_i <-F_i-d^-
    X_i <-D_i/(D_i+S_i)
  End-if-else
  End-if
End-while
Sort individuals according to their dominance index
Print dominance index as a function of rank

/*  Parameter values  */
δ^+ = δ^- =1, η =1, θ =100
```

Box 20.4 Model 2 Assumptions

1. Each individual is characterized by a force, F, that influences its ability to win contests. F reflects endocrine activity, ovarian development, and other factors.
2. Individuals initially have different forces, as found in both nature and experiments. This initial differentiation results from asynchrony in emergence from hibernation, genotypic factors, and differential exposure to environmental signals.
3. Individual recognition, if it exists, plays no role in the outcome of a contest. The force of an individual does not depend on the adversary.
4. A pairwise contest between i and j is won by i if and only if the force F_i of i is greater than the force F_j of j.
   If i and j have equal forces, the probability that i wins is 1/2.
5. The probability of interaction of an individual, i is proportional to the time spent that i walking on the nest, as given by the empirical function, $T(X_i)$ of its dominance index X_i: $T(X_i) = 1.32 \cdot 10^{0.836 X_i}$.

Box 20.5 Monte Carlo Simulations of Model 2

```
/* Initialization */
I=0    /* number of hierarchical interactions initially equal to
0 */
For every individual i do
Dᵢ=0 /* number of dominances initially equal to 0 */
Sᵢ=0 /* number of subordinations initially equal to 0 */
Fᵢ=i /* intrinsic force equal to i */
Xᵢ=0 /* by convention dominance index initially equal to 0 */
End-for
```

```
/* Main loop */
While I<Iₘₐₓ do
Randomly select two individuals i and j
Pᵢⱼ=f(Xᵢ)f(Xⱼ) /* probability of interaction between i and j */
Draw random real number R between 0 and 1
If (R<Pᵢⱼ) do /*i and j interact */
  I<-I+1
  Draw random real number R' between 0 and 1
  If (Fᵢ>Fⱼ) do /* i wins */
    Dᵢ<-Dᵢ+1 /* number of dominances of i increased by 1 */
    Xᵢ<-Dᵢ/(Dᵢ+Sᵢ) /* dominance index of i updated */
    Sⱼ<-Sⱼ+1 /* number of subordinations of j increased by 1 */
    Xⱼ<-Dⱼ/(Dⱼ+Sⱼ) /* dominance index of j updated */
  Else /* j wins */
    Dⱼ<-Dⱼ+1
    Xⱼ<-Dⱼ/(Dⱼ+Sⱼ)
    Sᵢ<-Sᵢ+1
    Xᵢ<-Dᵢ/(Dᵢ+Sᵢ)
  End-if-else
  End-if
End-while
Sort individuals according to their dominance index
Print dominance index as a function of rank
```

```
/* Parameter values */
No parameters
```

Part I I I

Conclusions

21

Lessons, Speculations, and the Future of Self-Organization

> By itself, emergence can be no explanation at all if you
> don't have any insight into the mechanisms of the system,
> and it may seem to be an appeal to mysticism.
> —E.O. Wilson in an interview by Roger Lewin in
>   *Complexity: Life at the Edge of Chaos.*

The examples we have reviewed in this book suggest that self-organization may be an important mechanism underlying the emergence of structures comprised of, or resulting from, the activities of groups of organisms. In Chapter 1 we introduced self-organization as a property of certain dynamical mechanisms whereby structures, patterns, and decisions appear at the global level of a system based on interactions among its lower-level components. The rules specifying interactions among the system's constituent units are executed on the basis of purely local information, without reference to the global pattern. The overall pattern, then, is an emergent property of the system rather than a property imposed on the system by an external ordering influence.

We listed the ingredients of self-organization in Chapter 2: the main engine that drives self-organization is positive feedback and the amplification of fluctuations, implemented through direct or indirect interactions among the organisms and between the organisms and their environment. Positive feedback has important consequences—the signatures of self-organization—which we described in Chapter 3. Such signatures include the creation of structure and multistability (the coexistence of several possible states of the system), and the existence of bifurcations, which are discrete qualitative changes in the system's behavior as a particular parameter of the system is gradually varied. The significance of self-organization is in its ability to explain such phenomena as the formation of complex group-level patterns of behavior, including structures such as nests and pheromone trail networks, without invoking the same degree of complexity at the individual level; the observation of different group-level patterns in identical experimental conditions; and qualitative or quantitative changes in group-level behavior without requiring any such change at the individual level.

In addition to emphasizing the role of individual interactions within a self-organizing system, we have also tried to demonstrate the importance of external factors in the form of environmental constraints. Especially in the examples involving social insects, we emphasized that global order can arise through interactions among individuals, but we also discussed how initial conditions, external factors, and modifications of the environment by the individuals act as strong constraints that shape the ultimate form of the self-organizing process. Throughout the book, we have tried to be objective, and not over-zealous in our claims of what self-organization can accomplish. Even in those situations in which self-organization is believed to be an important mechanism for creating structure, self-organization alone usually does not provide sufficient explanation.

Critics of the self-organization approach may complain about the lack of empirical testing of alternative hypotheses for the processes and structures that were presented. Of course they are right. Good science is time consuming. This is the price to be paid for being able to say something valuable about how nature works. One of the goals of this book was to define a set of alternative hypotheses (presented in Chapter 4) that could be tested when studying a system claimed to be based on self-organization mechanisms. A more systematic approach to self-organization and careful, logical hypothesis-testing are certainly needed for self-organization to gain better acceptance.

For this reason, we devoted Chapter 6 to describing a general approach to the study of self-organization, the testing of hypotheses, and the development of models. Chapter 4 introduced what we termed the precursors of self-organization, concepts such as stigmergy and heterarchies. These concepts are precursors of self-organization in that they rely on ideas that form the core of the self-organization approach.

Chapter 5 explained why we think self-organization is an important and widespread mechanism used by groups of organisms. That self-organization requires relatively little to make it work and generates so much in return makes most of our case. Finally, some of the most widespread misconceptions are treated in Chapter 7, the concluding chapter of the book's first part. Perhaps the most unfortunate misconception is the belief of certain biologists that self-organization is an alternative to natural selection. In fact, we view self-organization as a set of phenotype-generating mechanisms that are as much subject to natural selection as are any other biological mechanisms. Of course we contend that self-organization is a critical principle for understanding evolutionary thinking. Some arguments will be developed later in this chapter, but much remains to be done to answer the question of why self-organization is an important conceptual contribution to the evolutionary framework.

Chapters 8 through 20 provided examples of self-organization used by a great diversity of organisms, including slime molds and bacteria (Chapter 8), bark beetles (Chapter 9), fireflies (Chapter 10), fish (Chapter 11), honey bees (Chapters 12, 15, 16), ants (Chapters 13, 14, 17) and termites (Chapter 18). Alternative theories and empirical tests were presented in these chapters, but in most cases such empirical tests remain to be undertaken.

Chapter 19 described a speculative model of nest building in wasps that relies on an alternative to self-organization, qualitative stigmergy. The paucity of experimental data makes it difficult to argue for or against that model, but the model does show that it is possible to generate complex patterns without self-organization and without invoking a large degree of individual-level complexity. The model should be taken not as a definitive explanation of how wasps actually build their nests but as a thought-provoking exercise to question our view of the building behavior of insects.

Finally Chapter 20 showed that self-organization may explain the formation of dominance hierarchies in primitively eusocial wasps, but an alternative mechanism—a predifferentiated process—can account equally well for the empirical data and makes many similar predictions. Chapter 20 clearly showed that we should be cautious before invoking self-organization as the explanatory mechanism—or, for that matter, any kind of explanatory mechanism. Other hypotheses consistent with experimental data can be equally predictive. We must then look for empirical ways of distinguishing among several possible hypotheses.

In concluding, it would be useful to address the following questions: What have we shown? What have we learned? Why is self-organization important? What are the limits of self-organization? Where do we go from here?

What Have We Shown?

We have shown that complex structures appear, at certain levels of description, to be based on self-organizing mechanisms. It is another thing, however, to demonstrate that self-organization is the actual underlying mechanism. In most cases, self-organization and alternative hypotheses remain to be tested. Until then we can only say that we strongly suspect that self-organization plays a role in the phenomena described in this book and in many other biological processes. We hope that this book will stimulate present and future researchers to investigate further the phenomena that we have described.

What Have We Learned?

If nothing else, we have learned that emergence is not a mystical concept. Theories of self-organization are rigorous, scientifically sound, and testable. Alternative hypotheses, equally rigorous, scientifically sound and testable have also been presented. Our presentation has not been exhaustive but the very fact that there exist hypotheses for emergence is an important lesson. Instead of invoking mystical forces or complex individual behavior one may look for explanations based on self-organization or its alternatives.

Self-organization is a set of mechanisms that rely on readily available, easy to use components. As such, natural selection has taken advantage of, and later modified, these components to solve a variety of problems faced by groups of organisms. Let us review the main components that form a self-organizing system.

Simple Rules of Thumb

In the examples we presented, the interactions among individuals and between individuals and the environment were largely based on simple rules of thumb requiring limited access to global information. In many cases the rules appear to require limited cognitive abilities, thus contrasting sharply with the relative complexity of the emergent collective phenomenon. One aspect of the simplicity of the rules is that they can often be presented in the form of a brief if-then statement, such as "if there is a pile of building material here, then add to it," or "if this cell contains an egg, then lay another egg in the neighboring cell." The rules of thumb are simple stimulus-response acts executed in a probabilistic manner.

Although we refer to these rules as "simple," this does not deny the enormously complex neural processing that may be required to execute the act. More explicitly, we mean that we consider the underlying sensory and motor processing as a black box, and that the behavioral rule of thumb itself can be expressed rather simply and executed without the need to acquire global information. What is most striking about these rules of thumb is that, as part of a self-organizing mechanism, they can lead to a diversity of complex collective responses without the need to explicitly code for the collective complexity in the individual genes. Complexity at the collective level arises as an emergent property of the system. Furthermore, there is nothing especially remarkable about the rules of thumb. Similar, simple behavioral rules are used by organisms in many situations that do not involve self-organization. It is how these rules of thumb are used that is important.

Multiple Interactions

Self-organization arises in systems with multiple, iterative interactions among many individuals executing the rules of thumb. In certain situations a single individual can generate a self-organized structure such as a stable foraging trail if, for example, the lifetime of the pheromone is sufficient for the individual's, trail-following events to interact with its own previous trail-laying activity. More commonly, self-organization arises when a minimal density of mutually tolerant individuals make use of the results of their own activities as well as that of others. Their ongoing interactions can modify the characteristics of the system and provide new stimuli for further interactions. This becomes an especially powerful mechanism when coupled with positive feedback.

Positive Feedback

In the social systems we have discussed, most of the positive feedback is genetically coded behavior, such as recruitment to a food source by ants. Negative feedback, in contrast, often arises as an automatic by-product of the positive feedback that leads to the depletion of building materials or to food consumption at a source.

Building behavior provided us with a clear example. In one of our models we assumed that an animal is stimulated to dig when it detects an area that was previously excavated (Deneubourg and Franks 1995). During the extraction of material from this developing nest site, the insect adds pheromone to the walls of the cavity, and the pheromone, in turn, stimulates nestmates to dig at the site. When a worker crosses a freshly dug zone, the pheromone is still there and increases the probability of digging at that site. Digging stimulates more digging—a simple case of positive feedback. However, as the pheromone evaporates into the air, and its concentration declines over time, it finally ceases to act as a stimulus. As the volume of the nest cavity increases, the density of insects decreases, and so the frequency of insect visits to the digging sites decreases. The less frequent the visits and bouts of digging (and pheromone addition), the lower the remaining quantity of pheromone and the lower the stimulus to dig. In this way, a scenario involving only a simple form of positive feedback (amplification in the form of the simple rule of thumb, "dig where another has recently dug") can lead to a self-regulating nest volume appropriate for the size of the population. The regulatory process arises simply as an interplay with physical constraints (the decrease in the density of insects and the evaporation of pheromone) with no need for individuals with an explicit overview of the cavity size, or population size or density.

Negative Feedback

Negative feedback counterbalances positive feedback and helps stabilize the collective pattern by braking and shaping the positive feedback aspects of the system. Negative feedback often takes the form of saturation, depletion, or competition. In the case of pillar formation in termites, negative feedback arises from the depletion of building materials surrounding the growing pillar as well as competition from the construction of other nearby pillars. In the example of foraging, negative feedback arises from the limited number of available foragers, satiation, consumption of the food source, crowding at the food source, or competition among food sources.

Environmental Constraints

We have tried to emphasize that environmental constraints are often an overlooked feature of self-organizing processes. These constraints can play an important role in canalizing emergent phenomena. Initial conditions, boundary conditions, and random enviromental fluctuations are involved in many self-organizing systems. Most often, part of the complexity of the structure or process is a reflection of the complexity of the environment, as in the example of the foraging patterns of army ants (Chapter 14).

Why is Self-Organization Important?

If self-organization is just another mechanism used by organisms to solve particular problems, why did we write a book about it? We briefly review here some of the reasons why we believe self-organization is important.

General Importance

The first point is that it is as important to study the proximate determinants of behavior as it is to examine the ultimate causes. Understanding the mechanisms that underlie a behavior is a necessary first step toward understanding how that behavior evolved. Evolutionary theories—optimality theory, phenotypic models, game theory, population genetics—often make assumptions that can only be justified by looking at mechanisms. For example, optimality approaches, which assume that an observed behavior has been optimized by evolution, require one to find optimal phenotypes in a given phenotypic space. Other issues related to optimality theories notwithstanding, the optimization procedure must be performed over the right ensemble of phenotypes; otherwise flawed conclusions will be drawn. Optimal foraging theory often makes predictions without knowledge of the foraging mechanism, for instance, most optimal foraging models for social insects do not take into account recruitment, which is known to be important for the overall economy of the colony.

In summary, no evolutionary stance is fully plausible without a careful examination of the potential underlying mechanisms. This is all the more necessary since evolution is often assumed to be able to take a system from one phenotypic configuration to another. Since evolution proceeds by tinkering, accumulating small changes, some barriers will not be easily overcome by variation and natural selection. Identifying such barriers is only possible through a knowledge of mechanisms.

Self-Organization and Evolutionary Theories

A key property of self-organizing systems is their ability to generate complex group-level patterns with relatively simple constituent units. Evolutionary theories often predict that animals or groups of animals should follow complex strategies, from parental investment or foraging to the partitioning of reproduction among cofoundresses. Often the question of how such complex strategies are implemented is not addressed. We believe they are sometimes implemented through self-organization. Pacala et al. (1996) showed how the evolutionarily stable allocation of foragers to food sources or, more generally, evolutionarily stable task allocation can be achieved by means of a simple self-organizing mechanism based on local interactions among insects and simple recruitment. By the modification of simple rules, through selection, evolution can lead to the implementation of complex strategies.

Furthermore, environmental parameters may act not only as boundary conditions but also as tuning parameters for the behavior of a self-organized system, and thus facilitate implementation of complex strategies responsive to environmental variations. A possible example is the swarm-raid patterns of army ants described in Chapter 14. Although it has never been rigorously analyzed, a reasonable hypothesis is that the patterns observed in the different army-ant species are adapted to the kinds of prey distribution that the ants usually encounter. Here the currency in which the level of performance of the colony can be measured may be the average energy expenditure per unit weight of prey taken back to the bivouac. The model presented in Chapter 14 informs us that some simple rules may lead to the best adapted swarm-raid patterns, given the spatial distribution of prey. The environment—the spatial distribution of prey—plays the role of a control parameter that determines the kind of self-organizing swarm-raid pattern to be expected.

In summary, self-organization permits the implementation of a variety of complex collective responses that do not necessarily reflect changes in the behavior of individuals but rather changes in the interaction between the group and its environment.

Self-Organization and Efficient Adaptions

One of the mysteries of biology is how the enormous amount of morphogenic, physiological, and behavioral complexity of an organism can be achieved with the limited amount of genetic information available within its genes. One such example is the ability of the mammalian immune system to recognize and respond to an almost limitless diversity of non-self antigens. To accomplish such feats, there must exist special mechanisms for economizing on the amount of information that must be coded in the genes. One of our fascinations with self-organization is its ability to create complexity from simplicity with remarkable economy.

The economy of coding arises in several ways. In many cases, diverse responses merely reflect interactions with the environment, not individual complexity. This was our hypothesis for the diversity of specialized patterns of army ant raids. Since environmental parameters (in contrast to parameters intrinsic to the organism, such as sensory thresholds) play a role in shaping self-organized systems, it may be possible to economize on what needs to be coded in the individual.

Another possible means of creating complexity is through a cascade of interacting mechanisms. For instance, nest construction by the fungus-growing termites, *Macrotermes*, described in Chapter 18, can be likened to a morphogenic process. Interactions among individuals, or between individuals and emergent structures, can modify the characteristics of the system and provide new conditions for the emergence of other structures in a "morphogenetic cascade." The nest develops as if it were an embryo, starting as an undifferentiated cell—the copularium, which transforms into the royal cell—and becomes progressively more complex, with many different parts.

Imagine a homogeneous medium in which structure emerges through self-organization—for example, pillars in termite nests. Once a pillar has emerged, this structure acts as a source of heterogeneity that modifies individuals' actions. The actions, in turn, create new stimuli that trigger new building actions (Deneubourg and Theraulaz 1997). Complexity unfolds progressively (Bonabeau et al. 1998b); increasingly diverse stimuli result from previous building activities, and facilitate the construction of ever more complex structures. In termites, as the nest grows, air streams develop to become new sources of stimulation for the builders. This may explain why large nests generally are more complex and diverse than smaller ones. Although our understanding of the origin of colony-level spatiotemporal patterns is far from complete, the complexity-generating mechanism that we just sketched is promising. It is general and may provide a framework to study how past actions influence and constrain future actions so that complex patterns are produced robustly and consistently.

Limits of Self-Organization

Although we discussed examples of self-organization only in bacteria, slime molds, fish, birds, and insects, self-organization can also apply to other multicellular organisms. For example, Gueron and Levin (1993) suggested that African wildebeest self-organize into large-scale migration fronts comprised of thousands of individuals. Nonetheless, in larger organisms alternative mechanisms are likely to be more efficient because higher cognitive capabilities may allow each individual to perform complex tasks on its own.

Is self-organization able to produce truly complex patterns and structures? What are the limits to the degree of complexity that mechanisms based upon self-organization can accomplish? To answer this, we must be speculative as well as imaginative. There is a wide gap between the simplicity of the problems examined in this book and the complexity of the structures produced in nature. We are far from being able to describe the mechanisms involved in the production of a *Macrotermes* nest. The behavioral rules of termites that we invoked in our models are almost certainly far simpler than those employed by real termites. Workers are likely to be sensitive to a great diversity of cues, from both social origins and the external environment. These include pheromones from other members of the society along with carbon dioxide, gravity, humidity, temperature, and air currents. This raises the question of how much more complexity and diversity might arise at the collective level if we consider a greater capacity at the individual level. If we couple this with our speculations concerning cascades of morphogenic mechanisms, there are many more degrees of freedom in what self-organization can accomplish.

The Future

Although self-organization and emergent properties are familiar terms in the fields of chemistry, physics, and certain areas of cellular and developmental biology (such as morphogenesis and neurobiology), much needs to be accomplished before these concepts achieve widespread acceptance and utility in other areas of biology. Although we do not doubt the importance of self-organization in biological systems, a current weakness of this approach is that it has been clearly identified in only a few cases. Even in those cases, experiments have seldom been performed to distinguish between hypotheses based on self-organization and alternative mechanisms. The paucity of data at both individual and colony levels prevents us from making any general assessment of the contribution of self-organization to the function of biological systems.

Toward this and, we hope that the examples and models discussed in this book have increased our knowledge of the type of experimental data that are needed to determine whether a particular system is self-organized. What is most needed now are studies that are more carefully conducted than past studies and that will allow us to distinguish between self-organization and other hypotheses for pattern-formation and decision making in specific cases.

A final important task will be to continue to clarify the relationship between self-organization and evolution, and to dispel persistent misconceptions. Some authors (e.g., Waldrop 1990; see also misconception #1, Chapter 7) have suggested that self-organization could replace natural selection, or that self-organization should be viewed as an alternative to evolution. Kauffman (1993) provided an exciting introduction to self-organization in biological systems, but while his writings have had great impact they may also have promoted some confusion, namely, that self-organization and natural selection belong in opposing camps.

Natural selection and self-organization are intimately linked, since natural selection molds the rules of interaction among the components of a living system to produce adaptive responses taking advantage, for example, of the relative simplicity of the behavioral rules needed to produce relatively complex collective responses.

Notes

Chapter 1

1. At a more microscopic level of analysis, the rhythmicity of the cluster of pacemaker cells may also be considered to be a self-organized process (Peskin 1975; Jalife 1984; Michaels et al. 1987). Also see Chapter 10. It is important therefore to indicate the desired level of description and the nature of the units involved.

Chapter 3

1. See the program Dendroctonus to explore the emergence of clustering in this system. Macintosh versions in Pascal and StarLogo can be downloaded at
   http://beelab.cas.psu.edu
2. During the following discussion and in Box 3.2 it may be helpful to refer to the programs Logistic Equation and Bifurcation Diagram. Macintosh versions in Pascal can be downloaded at http://beelab.cas.psu.edu

Chapter 6

1. The program Monte Carlo Ants allows exploration of this simulation in detail and can be downloaded at http://beelab.cas.psu.edu
2. The program Epidemic allows readers to explore this simulation in detail. It can be downloaded at http://beelab.cas.psu.edu
3. See the program Life to explore this model. A Macintosh version can be downloaded at http://beelab.cas.psu.edu
4. See the program Pattern Formation to explore this model. A Macintosh version can be downloaded at http://beelab.cas.psu.edu
5. See the program Linear CA to explore the cellular automaton model. A Macintosh version can be downloaded at http://beelab.cas.psu.edu
6. See the program DLA to explore this model. A Macintosh version can be downloaded at http://beelab.cas.psu.edu

Chapter 16

1. See the program Comb Pattern to explore this model. A Macintosh version can be downloaded at http://beelab.cas.psu.edu

References

Adam, E. 1988. Un oiseau de fer dans un fourreau de plumes: le flamant. *Terre Sauvage* 19:46–57.

Adamson, J. 1961. *Living Free*. London: Harvill.

Agladze, K., L. Budriene, G. Ivanitsky, V. Krinsky, V. Shakhbazyan, and M. Tsyganov 1993. Wave mechanisms of pattern formation in microbial populations. *Proceedings of the Royal Society of London, Series B*. 253:131–135.

Alcock, J. 1993. *Animal Behavior*. Sunderland, MA: Sinauer.

———. 1998. *Animal Behavior: An Evolutionary Approach*. 6th ed. Sunderland, MA: Sinauer.

Alexander, R. D. 1961. Aggressiveness, territoriality, and sexual behaviour in field crickets (Orthoptera: Gryllidae). *Behaviour* 17:130–223.

———. 1967. Acoustical communication in arthropods. *Annual Review of Entomology* 12:495–526.

———. 1975. Natural selection and specialized chorusing behavior in acoustical insects. In D. Pimentel, ed. *Insects, Science and Society*, New York: Academic Press.

Allee, W. C. 1931. *Animal Aggregations: A Study in General Sociology*. Chicago: University of Chicago.

———. 1942. Social dominance and subordination among vertebrates. Levels of integration in biological and social systems. *Biological Symposia* 8:139–162.

———. 1951. *Cooperation among Animals*. New York: Henry Schulman.

———. 1952. Dominance and hierarchy in societies of vertebrates. *Colloques Internationaux du CNRS* 34:157–181.

Anderson, R. S. 1990. Eolian ripples as examples of self-organization in geomorphological systems. *Earth-Science Reviews* 29:77–96.

Aoki, I. 1982. A simulation study on the schooling mechanism in fish. *Bulletin of the Japanese Society of Scientific Fisheries* 48:1081–1088.

Aron, S., J. L. Deneubourg, and J. M. Pasteels. 1988. Visual cues and trail-following idiosyncrasy in *Leptothorax unifasciatus*: an orientation process during foraging. *Insectes Sociaux* 35:355–366.

Aron, S., J. L. Deneubourg, S. Goss and J. M. Pasteels. 1990. Functional self-organisation illustrated by inter-nest traffic in the argentine ant *Iridomyrmex humilis*. In W. Alt, and G. Hoffman, eds., *Biological Motion*, Lecture Notes in Biomathematics, pp. 533–547. Berlin: Springer-Verlag.

Aron, S., J. M. Pasteels, J. L. Deneubourg, and J. L. Boeve. 1986. Foraging recruitment in *Leptothorax unifasciatus*: the influence of foraging area familiarity and the age of the nest-site. *Insectes Sociaux* 33:338–351.

Bagine, R. A. N., J. P. E. C. Darlington, P. Kat, and J. M. Ritchie. 1989. Nest structure, population structure and genetic differentiation of some morphologically similar species of *Macrotermes* in Kenya. *Sociobiology* 15:125–132.

Bagnoli, P., M. Brunelli, F. Magni, and D. Musumeci. 1976. Neural mechanisms underlying spontaneous flashing and its modulation in the firefly, *Luciola lusitanica*. *Journal of Comparative Physiology A* 108:133–156.

Bak, P. 1996. *How Nature Works: The Science of Self-Organized Criticality*. New York: Springer-Verlag.

Bak, P., K. Chen, and M. Creutz. 1989. Self-organized criticality in the "Game of Life." *Nature* 342:780–782.

Bak, P., C. Tang, and K. Wiesenfeld. 1987. Self-organized criticality: an explanation of $1/f$ noise. *Physical Review Letters* 59:381–384.

Baldwin, J. D. 1971. The social organization of a semifree-ranging troop of squirrel monkeys (*Saimiri sciureus*). *Folia Primatologica* 14:23–50.

Barchas, P. R., and S. D. Mendoza. 1984. Emergent hierarchical relationships in rhesus macaques: an application of Chase's model. In P. R. Barchas, ed., *Social Hierarchies: Essays Toward a Sociophysiological Perspective*, pp. 81–95. Wesport, CT: Greenwood Press.

Barenholz-Paniry, V., J. S. Ishay, and Z. Grossman. 1988. Rhythmic signalling and entrainment in *Vespa orientalis* larvae: Characterization of the underlying interactions. *Bulletin of Mathematical Biology* 50:661–679.

Barlow, G. W. 1974. Hexagonal territories. *Animal Behaviour* 22:876–878.

Barton, E. P., S. L. Donaldson, M. A. Ross, and J. L. Albright. 1974. Social rank and social index as related to age, body weight and milk production in dairy cows. *Proceedings of the Indiana Academy of Sciences* 83:473–477.

Beaugrand, J. P., and R. C. Zayan. 1984. An experimental model of aggressive dominance in *Xiphophorus helleri* (Pisces: Poeciliidae). *Behavioral Processes* 10:1–53.

Beckers, R. 1992. *L'auto-organisation—Une réponse alternative à la complexité individuelle?* Ph.D. Thesis, Université Paris Nord.

Beckers, R., J. L. Deneubourg, and S. Goss. 1992a. Trail laying behaviour during food recruitment in the ant *Lasius niger* (L.). *Insectes Sociaux* 39:59–72.

———. 1992b. Trails and U-turns in the selection of a path by the ant *Lasius niger*. *Journal of Theoretical Biology* 159:397–415.

———. 1993. Modulation of trail laying in the ant *Lasius niger* (Hymenoptera: Formicidae) and its role in the collective selection of a food source. *Journal of Insect Behavior* 6:751–759.

Beckers, R., J. L. Deneubourg, S. Goss, and J. M. Pasteels. 1990. Collective decision making through food recruitment. *Insectes Sociaux* 37:258–267.

Belic, M. R., V. Skarka, J. L. Deneubourg, and M. Lax. 1986. Mathematical model of honeycomb construction. *Journal of Mathematical Biology* 24:437–449.

Ben-Jacob, E., I. Cohen, and O. Shochet. 1995. Complex bacterial patterns. *Nature* 373:566–567.

Ben-Jacob, E., O. Schochet, A. Tenenbaum, I. Cohen, A. Czirok, and T. Vicsek. 1994. Generic modelling of cooperative growth patterns in bacterial colonies. *Nature* 368:46–49.

Berland, L., and P.-P. Grassé. 1951. Super-famille des Vespoidea Ashmead. In: P.-P. Grassé, ed., *Traité de Zoologie vol. 10*, pp. 1127–1174. Paris: Masson.

Bernstein, R. 1975. Foraging strategies of ants in response to variable food density. *Ecology* 56:213–219.

Bill, R. G., and W. F. Herrnkind. 1976. Drag reduction by formation movement in spiny lobsters. *Science* 193:1146–1148.

Black-Cleworth, P. 1970. The role of electrical discharges in the non-reproductive social behaviour of *Gymnotus carapo* (Gymnotidae, Pisces). *Animal Behaviour Monographs* 3:1–77.

Blackwell, P., M. Jennions, and N. Passmore. 1998. Synchronized courtship in fiddler crabs. *Nature* 391:31–32.

Blair, K. G. 1915. Luminous insects. *Nature* 96:411–415.

Bodenheimer, F. S. 1937. Studies in animal populations. II. Seasonal population-trends of the honeybee. *Quarterly Review of Biology* 12:406–425.

Boice, R., and D. W. Witter. 1969. Hierarchical feeding behaviour in the leopard frog (*Rana pipiens*). *Animal Behaviour* 17:474–479.

Bonabeau, E., G. Theraulaz, and J. L. Deneubourg. 1996. Mathematical models of self-organizing hierarchies in animal societies. *Bulletin of Mathematical Biology* 58:661–719.

———. 1998a. Latency time and absence of group effect. *Insectes Sociaux* 45:191–195.

Bonabeau, E., G. Theraulaz, J. L. Deneubourg, N. R. Franks, O. Rafelsberger, J. L. Joly, and S. Blanco. 1998b. The emergence of pillars, walls and royal chambers in termites nests. *Philosophical Transactions of the Royal Society of London—Series B, Biological Sciences* 353:1–16.

Bonner, J. T. 1967. *The Cellular Slime Molds*. 2nd Ed. Princeton: Princeton University Press.

———. 1983. Chemical signals of social amoebae. *Scientific American* 248:114–120.

Bouillon, A. 1958. Les Termites du Katanga. *Naturalistes Belges* 39:198–207.

Bourke, A. F. G. 1988. Dominance orders, worker reproduction, and queen-worker conflict in the slave-making ant *Harpagoxenus sublaevis*. *Behavioral Ecology and Sociobiology* 23:323–333.

Bovbjerg, R. V. 1956. Some factors affecting aggressive behaviour in crayfish. *Physiological Zoology* 29:127–136.

Bovbjerg, R. V., and S. L. Stephen. 1971. Behavioural changes in crayfish with increased population density. *Bulletin of the Ecological Society of America* 52:37–38.

Bradbury, J. W., and R. M. Gibson. 1983. Leks and mate choice. In P. Bateson, ed., *Mate Choice*, pp. 109–138. Cambridge: Cambridge University Press.

Breed, M. D., and G. J. Gamboa. 1977. Behavioral control of workers by queens in primitively eusocial bees. *Science* 195:694–696.

Brian, M. V. 1983. *Social insects: ecology and behavioural biology*. New York: Chapman and Hall.

Bristow, K. L., and J. A. Holt. 1987. Can termites create local energy sinks to regulate mound temperature? *Journal of Thermal Biology* 12:19–21.

Brockman, J. 1980. The control of the nest depth in a digger wasp (*Sphex ichneumoneus* L.). *Animal Behavior* 28:426–445.

Brokenshire, 1929. Through Philippine Jungles. *Missionary Herald* 125:323–325.

Brooks, C. M., and K. Koizumi. 1974. The hypothalamus and control of integrative processes. In V. B. Mountcastle, ed., *Medical Physiology*, pp. 813–836. St. Louis: C. V. Mosby.

Broom, D. M. 1975. Aggregation behaviour of the brittle-star *Ophiothrix fragilis*. *Journal of the Marine Biological Association of U K.* 55:191–197.

Bruinsma, O. H. 1979. *An Analysis of Building Behaviour of the Termite Macrotermes subhyalinus*. Ph.D. thesis, Lanbouwhogeschool te Wageningen.

Brünnich, C. 1923. A graphic representation of the oviposition of a queen-bee II. *Bee World* 4:223–224.

Buck, J. 1937. Studies on the firefly: I. The effects of light and other agents on flashing in *Photinus pyralis*, with special reference to periodicity and diurnal rhythm. *Physiological Zoology* 10:45–58.

———. 1938. Synchronous rhythmic flashing of fireflies. *Quarterly Review of Biology* 13:301–314.

———. 1988. Synchronous rhythmic flashing of fireflies. II. *Quarterly Review of Biology* 63:265–289.

Buck, J., and E. Buck. 1968. Mechanism of rhythmic synchronous flashing of fireflies. *Science* 159:1319–1327.

———. 1976. Synchronous fireflies. *Scientific American* 234:74–85.

———. 1978. Toward a functional interpretation of synchronous flashing by fireflies. *American Naturalist* 112:471–492.

———. 1980. Flash synchronization as tool and as enabler in firefly courtship competition. *American Naturalist* 116:591–593.

Buck, J., E. Buck, J. F. Case, and F. E. Hanson. 1981. Control of flashing in fireflies. V. Pacemaker synchronization in *Pteroptyx cribellata*. *Journal of Comparative Physiology A* 144:287–298.

Budrene, E. O., and H. C. Berg. 1991. Complex patterns formed by motile cells of *Escherichia coli*. *Nature* 349:630–633.

———. 1995. Dynamics of formation of symmetrical patterns by chemotactic bacteria. *Nature* 376:49–53.

Buonomici, M., and F. Magni. 1967. Nervous control of flashing in the firefly *Luciola italica* L. *Archives italiennes de Biologie* 105:323–338.

Burk, T. E. 1979. *An analysis of social behaviour in crickets*. D. Phil. thesis, University of Oxford.

Buschinger, A. 1986. Evolution of social parasitism in ants. *Trends in Ecology and Evolution* 1:155–160.

Buschinger, A. 1989. Evolution, speciation and inbreeding in the parasitic ant genus *Epimyrma* (Hymenoptera, Formicidae). *Journal of Evolutionary Biology* 2:265–283.

Camazine, S. 1991. Self-organizing pattern formation on the combs of honey bee colonies. *Behavioral Ecology and Sociobiology* 28:61–76.

Camazine, S., and J. Sneyd. 1991. A model of collective nectar source selection by honey bees: self-organization through simple rules. *Journal of Theoretical Biology* 149:547–571.

Camazine, S., K. Crailsheim, N. Hrassnigg, G. E. Robinson, B. Leonhard, and H. Kropiunigg. 1998. Protein trophallaxis and the regulation of pollen foraging by honey bees (*Apis mellifera* L.). *Apidologie* 29:113–126.

Camazine, S., J. Sneyd, M. J. Jenkins, and J. D. Murray. 1990. A mathematical model of self-organized pattern formation on the combs of honeybee colonies. *Journal of Theoretical Biology* 147:553–571.

Candland, D. K., and A. I. Leshner. 1971. Formation of squirrel monkey dominance order is correlated with endocrine output. *Bulletin of the Ecological Society of America* 52:54.

Case J. F., and J. B. Buck. 1963. Control of flashing in fireflies II. Role of central nervous system. *Biological Bulletin* 125:234–250.

Case, J. F., and L. G. Strause. 1978. Neurally controlled luminescent systems. In P. J. Herring, ed., *Bioluminescence in Action*, pp. 331–366. London: Academic Press.

Cassill, D., and W. Tschinkel. 1999. Self-organizing rules in larval feeding by worker fire ants. In C. Detrain, J. L. Deneubourg and J. M. Pasteels, eds., *Information Processing in Social Insects*. Basel: Bikhaüser Verlag.

Chase, I. D. 1974. Models of hierarchy formation in animal societies. *Behavioural Sciences* 19:374–382.

———. 1980. Social process and hierarchy formation in small groups: a comparative perspective. *American Sociological Review* 45:905–924.

———. 1982a. Dynamics of hierarchy formation: the sequential development of dominance relationships. *Behaviour* 80:218–240.

———. 1982b. Behavioral sequences during dominance hierarchy formation in chickens. *Science* 216:439–440.

———. 1985. The sequential analysis of aggressive acts during hierarchy formation: an application of the 'jigsaw' puzzle approach. *Animal Behaviour* 33:86–100.

———. 1986. Explanations of hierarchy structure. *Animal Behaviour* 34:1265–1267.

Chase, I. D., and S. Rohwer. 1987. Two methods for quantifying the development of dominance hierarchies in large groups with applications to Harris' sparrows. *Animal Behaviour* 35:1113–1128.

Chauvin, R. 1952. Sur la reconstruction du nid chez les fourmis (*Oecophylles Oecophylla longinoda* L.). *Behavior* 4:190–201.

———. 1968. *Animal societies from the bee to the gorilla*, tr. by G. Ordish. New York: Hill and Wang.

Chen, S. C. 1937a. Social modification of the activity of ants in nest-building. *Physiological Zoology* 10:420–436.

———. 1937b. The leaders and followers among the ants in nest-building. *Physiological Zoology* 10:437–455.

Clayton, D. A. 1978. Socially facilitated behavior. *The Quarterly Review of Biology* 53:373–392.

Cole, B. J. 1981. Dominance hierarchies in Leptothorax ants. *Science* 212:83–84.

———. 1991. Short-term activity cycles in ants: generation of periodicity by worker interaction. *American Naturalist* 137:244–259.

Cole, B. J., and I. Trampus. 1998. Activity cycles in ant colonies: Worker interactions and decentralized control. In C. Detrain, J. L. Deneubourg, and J. M. Pasteels, eds., *Information Processing in Social Insects*, Basel: Bikhaüser Verlag, in press.

Collias, N. E., and E. C. Collias. 1962. An experimental study of the mechanism of nest building in a weaverbird. *Auk* 79:568–595.

Collins, N. M. 1979. The nests of *Macrotermes bellicosus* (Smeathman) from Mokwa, Nigeria. *Insectes Sociaux* 26:240–246.

———. 1981. Populations, age structure and survivorship of colonies of *Macrotermes bellicosus* (Isoptera: Macrotermitinae). *Journal of Animal Ecology* 50:293–311.

Connor, F. P. 1933. Rhythmic sound produced by termites at work. *Journal of the Bombay Natural History Society* 36:1018.

Craig, W. 1916. Synchronism in the rhythmic activities of animals. *Science* 44:784–786.

———. 1917. On the ability of animals to keep time with an external rhythm. *Journal of Animal Behavior* 7:444–448.

Crick, F. 1994. *The Astonishing Hypothesis: The Scientific Search for the Soul.* New York: Charles Scribner's Sons.

Crisp, D. J., and M. Barnes. 1954. The orientation and distribution of barnacles at settlement with particular reference to surface contour. *Journal of Animal Ecology* 23:142–162.

Crisp, D. J., and P. S. Meadows. 1962. The chemical basis of gregariousness in cirripedes. *Proceedings of the Royal Society of London, Series B* 156:500–520.

Croome, D. J., and B. M. Roberts. 1981. *Airconditioning and ventilation of buildings.* (Second Edition, Volume 1). Oxford: Pergamon Press.

Crowley, M., and J. Bovet. 1980. Social synchronization of circadian rhythms in deer mice (*Peromyscus maniculatus*). *Behavioral Ecology and Sociobiology* 7:99–105.

Crutchfield, J. P., J. Farmer, N. H. Packard, and R.S. Shaw. 1986. Chaos. *Scientific American* 255:46–57.

Darchen, R. 1957. La reine d'*Apis mellifica* les ouvrières pondeuses et les constructions cirières. *Insectes Sociaux* 4:322–325.

———. 1959. Les techniques de la construction chez *Apis mellifica*. Thèse, Université de Paris.

Darling, F. 1938. *Bird Flocks and the Breeding Cycle.* London: Cambridge University Press.

Darlington, J. P. E. C. 1982. The underground passages and storage pits using in foraging by a nest of the termite *Macrotermes michaelseni* in Kajiado, Kenya. *Journal of Zoology* 198:237–247.

———. 1985. Attacks by doryline ants and termite nest defences (Hymenoptera; Formicidae; Isoptera; Termitidae). *Sociobiology* 11:189–200.

———. 1990. Populations in nests of the termite *Macrotermes subhyalinus* in Kenya. *Insectes Sociaux* 37:158–168.

Darlington, J. P. E. C., P. R. Zimmerman, and S. O. Wandiga. 1992. Populations in nests of the termite *Macrotermes jeanneli* in Kenya. *Journal of Tropical Ecology* 8:73–85.

Dawkins, R. 1982. *The Extended Phenotype.* San Francisco: W. H. Freeman.

De Block, J. W., and H. J. Geelen. 1958. The substratum required for the settling of mussels (*Mytilus edulis*). *Archives Néerlandaises Jubilee Volume*: 446–460.

DeAngelis, D. L., W. M. Post, and C. C. Travis. 1986. *Positive Feedback in Natural Systems.* Berlin: Springer Verlag.

Deleurance, E. P. 1946. Une régulation sociale à base sensorielle périphérique: l'inhibition de la ponte des ouvrières par la présence de la fondatrice chez les *Polistes* (Hyménoptères Vespidae). *Comptes Rendus de l'Académie des Sciences de Paris* 223:871–872.

———. 1957. Contribution à l'étude biologique des polistes (Hyménoptère: vespide). I. Activité de construction. *Annales des sciences naturelles, Zoologie, 11ème serie,* 93–222.

Deneubourg, J. L. 1977. Application de l'ordre par fluctuations a la description de certaines étapes de la construction du nid chez les termites. *Insectes Sociaux* 24:117–130.

Deneubourg, J. L., and N. R. Franks. 1995. Collective control without explicit coding: the case of communal nest excavation. *Journal of Insect Behavior* 8:417–432.

Deneubourg, J. L., and G. Theraulaz. 1997. Croissance et complexité des structures construites par les animaux. In G. Theraulaz and F. Spitz, eds., *Auto-organisation et comportement*, pp. 267–278. Paris: Hermès.

Deneubourg, J. L., J. C. Gregoire, and E. Le Fort. 1990a. Kinetics of the larval gregarious behaviour in the bark beetle *Dendroctonus micans*. *Journal of Insect Behavior* 3:169–182.

Deneubourg, J. L., J. M.Pasteels, and J. C. Verhaeghe. 1983. Probabilistic behaviour in ants: a strategy of errors? *Journal of Theoretical Biology* 105:259–271.

Deneubourg, J. L., G. Theraulaz, and R. Beckers. 1992. Swarm-made architectures. In F. J. Varela, and P. Bourgine, eds., *Toward a Practice of Autonomous Systems, Proceedings of The First European Conference on Artificial Life*, pp. 123–133. Cambridge: MIT Press/Bradford Books.

Deneubourg, J. L., S. Aron, S. Goss, and J. M. Pasteels. 1990b. The self-organizing exploratory pattern of the Argentine ant *Iridomyrmex humilis*. *Journal of Insect Behavior* 3:159–168.

Deneubourg, J. L., S. Goss, N. Franks, and J. M. Pasteels. 1989. The blind leading the blind: modelling chemically mediated army ant raid patterns. *Journal of Insect Behavior* 2:719–725.

Deneubourg, J. L., S. Goss, N. Franks, A. Sendova-Franks, C. Detrain, and L. Chrétien. 1990c. The dynamics of collective sorting robot-like ants and ant-like robots. In J. A. Meyer, and S. W. Wilson, eds., *From Animals to Animats*, pp. 356–363. *Proceedings of the First International Conference on Simulation of Adaptive Behaviour*. Cambridge: MIT Press.

Detrain, C., J. L. Deneubourg, and J. M. Pasteels. 1999. Decision rules and recruitment behaviour in ants. In C. Detrain, J. L. Deneubourg, and J. M. Pasteels, eds., *Information Processing in Social Insects*. Basel: Bikhaüser Verlag, in press.

Detrain, C., J. L. Deneubourg, S. Goss, and Y. Quinet. 1991. Dynamics of collective exploration in the ant *Pheidole pallidula*. *Psyche* 98:21–31.

Dewdney, A. K. 1991. Leaping into Lyapunov space. *Scientific American* 265:178–180.

DeYoung, G., P. B. Monk, and H. G. Othmer. 1988. Pacemakers in aggregation fields of *Dictyostelium discoideum*: Does a single cell suffice? *Journal of Mathematical Biology* 26:487–517.

Downing, H. A., and R. L. Jeanne. 1988. Nest construction by the paper wasp, *Polistes*: a test of stigmergy theory. *Animal Behaviour* 36:1729–1739.

———. 1990. The regulation of complex behaviour in the paper wasp, *Polistes fuscatus* (Insecta, Hymenoptera, Vespidae). *Animal Behaviour* 39:105–124.

Dreller, C., and W. H. Kirchner. 1995. The sense of hearing in honey bees. *Bee World* 76:6–17.

Driver, P. M., and D. A. Humphries. 1988. *Protean Behavior: The Biology of Unpredictability*. Oxford: Oxford University Press.

Dropkin, J. A., and G. J. Gamboa. 1981. Physical comparisons of foundresses of the paper wasp *Polistes metricus* (Hymenoptera: Vespidae). *Canadian Entomologist* 113:457–461.

Dworkin M., and D. Kaiser. 1985. Cell interactions in myxobacterial growth and development. *Science* 230:18–24.

Edelman, G. M. 1984. Cell-adhesion molecules: a molecular basis for animal form. *Scientific American* 250:118–129.

Edelstein-Keshet, L. 1994. Simple models for trail-following behavior: Trunk trails versus individual foragers. *Journal of Mathematical Biology* 32:303–328.

Edelstein-Keshet, L., and G. B. Ermentrout. 1990a. Contact response of cells can mediate morphogenetic pattern formation. *Differentiation* 45:147–159.

———. 1990b. From cell to tissue; contact mediated orientation selection in a population. In W. Alt, and G. Hoffmann, eds., *Biological Motion. Lecture Notes in Biomathematics 89*, pp. 566–576. Berlin: Springer Verlag.

———. 1990c. Models for contact-mediated pattern formation: cells that form parallel arrays. *Journal of Mathematical Biology* 29:33–58.

Edelstein-Keshet, L., Watmough, J., and G. B. Ermentrout. 1995. Trail-following in social insects: Individual properties determine population behaviour. *Behavioral Ecology and Sociobiology* 36:119–133.

Eibl-Eibesfeldt, I. 1970. *Ethology: The Biology of Behavior*, 2nd Ed. New York: Holt, Rinehart and Winston.

Ermentrout, G. B. 1991. An adaptive model for synchrony in the firefly *Pteroptyx malaccae. Journal of Mathematical Biology* 29:571–585.

Ermentrout, G. B., and L. Edelstein-Keshet. 1993. Cellular automata approaches to biological modeling. *Journal of Theoretical Biology* 160:97–133.

Ermentrout, B., J. Campbell, and G. Oster. 1986. A model for shell patterns based on neural activity. *Veliger* 28:369–388.

Evans, H. E., and M. J. West Eberhard. 1970. *The Wasps*. Chicago: University of Michigan Press.

Evans, L. T. 1951. Field study of the social behaviour of the black lizard, *Ctenosaura pectinata. Am Museum Novitiates* 1943.

———. 1953. Tail display in an iguanid lizard, *Liocephalus carinatus* coryi. *Copeia* 1:50–54.

Fabre, J. H. 1870–1910. *Souvenirs Entomologiques*. Paris: Delagrave.

———. 1889. *Souvernirs Entomologiques*. Paris: Robert Laffont.

Farb, P. 1962. *The Insects*. New York: Time.

Farine, J. P., and J. P. Lobreau. 1984. Le grégarisme chez *Dysdercus cingulatus* Fabr. (Heteroptera, Pyrrhocoridae): nouvelle méthode d'interprétation statistique. *Insectes Sociaux* 31:277–290.

Ferber, J. 1995. *Les Systèmes Multi-Agents*. Paris: InterEditions.

Fitzgerald, T. D. 1995. *The Tent Caterpillars*. Ithaca: Cornell University Press.

Focardi, S., J. L. Deneubourg, and G. Chelazzi. 1985. How shore morphology and orientation mechanisms can affect the spatial organization of intertidal molluscs. *Journal of Theoretical Biology* 112:771–782.

Foelix, R. F. 1982. *Biology of Spiders*. Cambridge: Harvard University Press.

Forrest, S. B., and P. K. Haff. 1992. Mechanics of wind ripple stratigraphy. *Science* 255:1240–1243.

Fourcassié, V., and J. L. Deneubourg. 1992. Collective exploration in the ant *Monomorium pharaonis* L. In J. Billen, ed., *Biology and Evolution of Social Insects*, pp. 369–373. Leuven: Leuven University Press.

Fowler, D. R., H. Meinhardt, and P. Prusinkiewicz. 1992. Modeling seashells. *Computer Graphics* 26:379–387.

Fraenkel, G. S., and D. L. Gunn. 1961. *The Orientation of Animals*. New York: Dover.

Francis, R. C. 1983. Experimental effects on agonistic behaviour in the paradise fish, *Macropodus opercularis*. *Behaviour* 85:292–313.

Franks, N. R., and S. Bryant. 1987. Rhythmical patterns of activity within the nests of ants. In J. Eder, and H. Rembold, eds., *Chemistry and Biology of Social Insects. Proceedings of the 10th International Congress of the International Union for the Study of Social Insects*, pp. 221–223. München: J. Peperny Verlag.

Franks, N. R., S. Bryant, R. Griffiths, and L. Hemerik. 1990. Synchronization of the behaviour within nests of the ant *Leptothorax acervorum* Fabricius–I. Discovering the phenomenon and its relation to the level of starvation. *Bulletin of Mathematical Biology* 52:597–612.

Franks, N. R., and J. L. Deneubourg. 1997. Self-organizing nest construction in ants: individual worker behaviour and the nest's dynamics. *Animal Behaviour* 54:779–796.

Franks, N. R., and L. C. Partridge. 1993. Lanchester battles and the evolution of combat in ants. *Animal Behaviour* 45:197–199.

Franks, N. R., and E. Scovell. 1983. Dominance and reproductive success among slave-making worker ants. *Nature* 304:724–725.

Franks, N. R., and A. B. Sendova-Franks. 1992. Brood sorting in ants: distributing the workload over the work-surface. *Behavioural Ecology and Sociobiology* 30:109–123.

Franks, N. R., A. Wilby, B. W. Silverman, and C. Tofts. 1992. Self-organizing nest construction in ants: sophisticated building by blind bulldozing. *Animal Behaviour* 44:357–375.

Free, J. B. 1987. Pheromones of Social Bees. London: Chapman and Hall.

Free, J. B., and Y. Spencer-Booth. 1958. Observations on the temperature regulation and food consumption of honeybees (*Apis mellifera*). *Journal of Experimental Biology* 35:930–937.

Frisch, K. von. 1967. *The Dance Language and Orientation of Bees*. Cambridge: Harvard University Press.

———. 1975. *Animal Architecture*. London: Hutchinson.

Frost, S. W. 1959. *Insect Life and Insect Natural History*. New York: Dover.

Fuchs, S. 1976a. An informational analysis of the alarm communication by drumming behavior in nests of carpenter ants (*Camponotus*, Formicidae, Hymenoptera). *Behavioral Ecology and Sociobiology* 1:315–336.

———. 1976b. The response to vibrations of the substrate and reactions to the specific drumming in colonies of carpenter ants (*Camponotus*, Formicidae, Hymenoptera). *Behavioral Ecology and Sociobiology* 1:155–184.

Galef, B. G., Jr., and L. L. Buckley. 1996. Use of foraging trails by Norway rats. *Animal Behaviour* 52:765–771.

Gardner, M. 1970. The fantastic combinations of John Conway's new solitaire game "life." *Scientific American* 223:120–123.

———. 1971. On cellular automata, self-reproduction, the Garden of Eden and the game "life." *Scientific American* 224:112–117.

Gerhardt, M., H. Schuster, and J. J. Tyson. 1990. A cellular automaton model of excitable media including curvature and dispersion. *Science* 247:1563–1566.

Gerisch, G., and B. Hess. 1974. Cyclic-AMP controlled oscillations in suspended *Dictyostelium* cells: their relation to morphogenetic cell interactions. *Proceedings of the National Academy of Sciences* 71:2118–2122.

Gervet, J. 1962. Études de l'effet de groupe sur la ponte dans la société polygyne de *Polistes gallicus* L. (Hymenoptera, Vespidae). *Insectes Sociaux* 9:231–263.

———. 1964. La ponte et sa régulation dans la société polygyne de *Polistes gallicus* L. *Annales de Sciences Naturelles et de Zoologie* 6:601–778.

Gervet, J., L. Blanc, M. Pratte, and S. Semenoff-Tian-Chansky. 1993. Experimentally induced circular dominance relationships in a polygynous *Polistes* (*Polistes dominulus* Christ) wasp colony. *Experientia* 49:599–604.

Ghent, A. W. 1960. A study of the group-feeding behaviour of larvae of the jack pine sawfly, *Neodiprion pratti banksianae* Roh. *Behaviour* 16:110–148.

Gibson, R. M., and J. W. Bradbury. 1986. Male and female strategies on sage grouse leks. In D. I. Rubenstein, and R. W. Wrangham, eds., *Ecological Aspects of Social Evolution*, pp. 379–398. Princeton: Princeton University Press.

Gilbert, S. F. 1994. *Developmental Biology*, 4th Ed. Sunderland: Sinauer.

Ginsburg, B., and W. C. Allee. 1975. Some effects of conditioning on social dominance and subordination in inbred strains of mice. In M. W. Schein, ed., *Social Hierarchy and Dominance*. Dowden: Hutchinson and Ross.

Glass, C. W., and F. A. Huntingford. 1988. Initiation and resolution of fights between swimming crabs (*Liocarcinus depurator*). *Ethology* 77:237–249.

Gleick, J. 1987. *Chaos: Making a New Science*. New York: Viking Penguin.

Goldbeter, A. 1996. *Biochemical Oscillations and Cellular Rhythms*. Cambridge: Cambridge University Press.

Goldbeter, A., and L. A. Segel. 1977. Unified mechanism for relay and oscillations of cyclic AMP in *Dictyostelium discoideum*. *Proceedings of the National Academy of Sciences* 74:1543–1547.

———. 1980. Control of developmental transitions in the cyclic AMP signaling system of *Dictyostelium discoideum*. *Differentiation* 17:127–135.

Gordon, H. R. S. 1958. Synchronous claw-waving of fiddler crabs. *Animal Behaviour* 6:238–241.

Goss, S., and J. L. Deneubourg. 1988. Autocatalysis as a source of synchronised rhythmical activity in social insects. *Insectes Sociaux* 35:310–315.

Goss, S., and J. L. Deneubourg. 1989. Self-organizing clock patterns of *Messor pergandei* (Formicidae, Myrmicinae). *Insectes Sociaux* 36:339–346.

Goss, S., S. Aron, J. L. Deneubourg, and J. M. Pasteels. 1989. Self-organized shortcuts in the Argentine ant. *Naturwissenschaften* 76:579–581.

Gössmann, C., and R. Huber. 1998. Behavioral mechanisms in the formation and maintenance of hierarchy in crayfish (*Astacus astacus*). Unpublished.

Gotwald, W. H. Jr. 1995. *Army Ants: The Biology of Social Predation*. Ithaca: Cornell University Press.

Gould, L. L., and F. Heppner. 1974. The vee formation of Canada geese. *Auk* 91:494–506.

Grassé, P.-P. 1939. La reconstruction du nid et le travail collectif chez les termites supérieurs. *Journal de Psychologie*: 370–396.

———. 1946. Sociétés animales et effet de groupe. *Experientia* 2:77–82.

———. 1959. La reconstruction du nid et les coordinations interindividuelles chez *Bellicositermes natalensis* et *Cubitermes* sp. La théorie de la stigmergie: essai d'interprétation du comportement des termites constructeurs. *Insectes Sociaux* 6:41–83.

———. 1960. La régulations automatiques du comportement des insectes sociaux et la stigmergie. *Journal de Psychologie* 57:1–10

———. 1967. Nouvelles expériences sur le termite de Müller (*Macrotermes mülleri*) et considérations sur la théorie de la stigmergie. *Insectes Sociaux* 14:73–102.

———. 1984. *Termitologia, Tome II. Fondation des Sociétés. Construction.* Paris: Masson.

Grassé, P.-P., and C. Noirot. 1949. Les termitières géantes de l'Afrique équatoriale. *Comptes Rendue de l'Académie des Sciences* 228:727–730.

———. 1958. Construction et architecture chez les termites champignonnistes (Macrotermitinae). In *Proceedings of the 10th International Congress of Entomology.* Montréal 2:515–520.

Greenfield, M. D., and I. Roizen. 1993. Katydid synchronous chorusing is an evolutionarily stable outcome of female choice. *Nature* 364:618–620.

Grégoire J. C. 1988. The greater European spruce beetle. In A. A. Berryman, ed., *Population Dynamics of Forest Insects*, pp. 455–478. New York: Plenum.

Grégoire J. C., J. C. Braekman, and A. Tondeur. 1982. Chemical communication between the larvae of *Dendroctonus micans* Kug. *Les Colloques de l'INRA, 7, Les Médiateurs chimiques* 253–257.

Gregory, R. L. 1994. DNA in the mind's eye. *Nature* 368:359–360.

Gross, M. R., and A. M. MacMillan. 1981. Predation and the evolution of colonial nesting in bluegill sunfish (*Lepomis macrochirus*). *Behavioral Ecology and Sociobiology* 8:163–174.

Gueron, S., and S. A. Levin. 1993. Self-organization of front patterns in large wildebeest herds. *Journal of Theoretical Biology* 165:541–552.

Guhl, A. M. 1968. Social stability and social inertia in chickens. *Animal Behaviour* 16:219–232.

Hahn, M., and U. Maschwitz. 1985. Foraging strategies and recruitment behaviour in the European harvester ant *Messor rufitarsis. Oecologia* 68:45–51.

Haken, H. 1977. *Synergetics: An Introduction. Nonequilibrium Phase Transitions and Self-Organization in Physics, Chemistry and Biology.* New York: Springer-Verlag.

———. 1978. *Synergetics: An Introduction.* 2nd ed. Berlin: Springer-Verlag.

Hamilton, W. D. 1964. The genetical evolution of social behaviour. I, II. *Journal of Theoretical Biology* 7:1–52.

———. 1971. Geometry for the selfish herd. *Journal of Theoretical Biology* 31:295–311.

Hansell, M. H. 1981. Nest construction in the subsocial wasp *Parischnogaster mellyi* (Saussure) Stenogastrinae (Hymenoptera). *Insectes sociaux* 28:208–216.

———. 1984. *Animal Architecture and Building Behaviour.* London: Longman.

———. 1985. The nest material of Stenogastrinae (Hymenoptera, Vespidae) and its effect on the evolution of social behaviour and nest design. *Actes Coll. Insectes Sociaux* 2:57–63.

Hanson, F. E. 1978. Comparative studies of firefly pacemakers. *Federation Proceedings* 37:2158–2164.

Hanson F. E., J. F. Case, E. Buck, and J. Buck. 1971. Synchrony and flash entrainment in a New Guinea firefly. *Science* 174:161–164.

Harris, W. V., and W. A. Sands. 1965. The Social Organization of Termite Colonies. In P. E. Ellis, ed., *Symposia of the Zoological Society of London #14, Social Organization of Animal Communities*, pp. 113–131. London: Academic Press.

Haubrich, R. 1961. Hierarchical behaviour in the South African clawed frog, *Xenopus laevis* Daudin. *Animal Behaviour* 9:71–76.

Hegh, E. 1922. *Les Termites* (Partie Générale) Brussels: Louis Desmet-Verteneuil.

Heiligenberg, W. 1977. *Studies of Brain Function, Vol. 1: Principles of Electrolocation and Jamming Avoidance*. New York: Springer Verlag.

———. 1991. *Neural Nets in Electric Fish*. Cambridge: MIT Press.

Heinrich, B. 1980. Mechanisms of body-temperature regulation in honeybees, *Apis mellifera*. I. Regulation of head temperature. *Journal of Experimental Biology* 85:61–72.

———. 1981a. Energetics of honeybee swarm thermoregulation. *Science* 212:565–566.

———. 1981b. The mechanisms and energetics of honeybee swarm temperature regulation. *Journal of Experimental Biology* 91:25–55.

———. 1981c. The regulation of temperature in the honeybee swarm. *Scientific American* 244:146–160.

———. 1985. The social physiology of temperature regulation in honeybees. In B. Hölldobler, and M. Lindauer, eds., *Experimental Behavioral Ecology and Sociobiology*, pp. 393–406. Sunderland: Sinauer.

———. 1987. Thermoregulation by individual bees. In R. Menzel, and A. Mercer, eds., *Neurobiology and Behavior of Honeybees*, pp. 102–111. Berlin: Springer-Verlag.

Heinze, J. 1990. Dominance behaviour among ant females. *Naturwissenchaften* 77:41–43.

Heinze, J., B. Hölldobler, and C. Peeters. 1994. Conflict and cooperation in ant societies. *Naturwissenchaften* 81:489–497.

Heldmann, G. 1936. Über das Leben auf Waben mit mehreren überwinterten Weibchen von *Polistes gallica* L. *Biologische Zentralblatt* 56:389–400.

Hemelrijk, C. K. 1996. Dominance interactions, spatial dynamics and emergent reciprocity in a virtual world. In P. Maes, M. J. Mataric, J.-A. Meyer, J. Pollack, and S. W. Wilson, eds., *From Animals to Animats 4: Proceedings of the Fourth International Conference on Simulation of Adaptive Behavior*, pp. 545–552. Cambridge: MIT Press.

Herrnkind, W. F., and R. McLean. 1971. Field studies of homing, mass emigration, and orientation in the spiny lobster, *Panulirus argus*. *Annals of the New York Academy of Sciences* 188:359–377.

Hess, W. N. 1920. Notes on the biology of some common Lampyridae. *Biological Bulletin* 38:39–76.

Higdon, J. J. L., and S. Corrsin. 1978. Induced drag of a bird flock. *American Naturalist* 112:727–744.

Hingston, R. W. G. 1926. The Mason wasp (*Eumenes conica*). *Journal of Bombay Natural History Society* 31:241–257.

———. 1927. The potter wasp (*Rynchium nitidulum*). *Journal of Bombay Natural History Society* 32:98–110.

Höfer, T., J. A. Sherratt, and P. K. Maini. 1995. *Dictyostelium discoideum*: cellular self-organization in an excitable biological medium. *Proceedings of the Royal Society of London, Series B* 259:249–257.

Hogeweg, P., and B. Hesper. 1983. The ontogeny of the interaction structure in bumblebee colonies: a mirror model. *Behavioral Ecology and Sociobiology* 12:271–283.

———. 1985. Socioinformatic processes: MIRROR modelling methodology. *Journal of Theoretical Biology* 113:311–330.

Höglund, J., and R. V. Alatalo. 1995. *Leks*. Princeton: Princeton University Press.

Hölldobler, B. 1971. Recruitment behaviour in *Camponotus socius*. *Zeitschrift für Vergleichende Physiologie* 75:123–142.

Hölldobler, B. and E. O. Wilson. 1978. The multiple recruitment systems of the African weaver ant *Oecophylla longinoda* (L.) (Hymenoptera: Formicidae). *Behavioral Ecology and Sociobiology* 391:19–60.

———. 1986. Nest area exploration and recognition in leafcutter ants (*Atta cephalotes*). *Journal of Insect Physiology* 32:143–150.

———. 1990. *The Ants*. Cambridge: Harvard University Press.

Horn, H. S. 1968. The adaptive significance of colonial nesting in the Brewer's blackbird *Euphagus cyanocephalus*. *Ecology* 49:682–694.

Howard, D. F., and W. R. Tschinkel. 1980. The effect of colony size and starvation on flodd flow in the fire ants, *Solenopsis invicta* (Hymenoptera: Formicidae). *Behavioural Ecology and Sociobiology* 7:293–300.

Howard, S. F. 1929. Synchronous flashing of fireflies. *Science* 70:556.

Howse, P. E. 1966. Air movement and termite behavior. *Nature* 210:967–968.

———. 1970. *Termites: a study in social behavior*. London: Hutchinson and Co.

Hubel, D. H., and T. N. Wiesel. 1977. Functional architecture of the macaque monkey visual cortex. *Proceedings of the Royal Society of London B* 198:1–59.

Huber, R., and A. Delago. 1998. Serotonin alters decisions to withdraw in fighting crayfish, *Astacus astacus*: the motivational concept revisited. *Journal of Comparative Physiology A* (in press).

Huber, R., and E. A. Kravitz 1995. A quantitative analysis of agonistic behavior in juvenile American lobsters (*Homarus americanus* L). *Brain, Behavior and Evolution* 46:72–83.

Huber, R., K. Smith, A. Delago, K. Isaksson, and E. A. Kravitz. 1997. Serotonin and aggressive motivation in crustaceans: altering the decision to retreat. *Proceedings of the National Academy of Sciences USA* 94:5939–5942.

Hunt, J. H. 1983. Foraging and morphology in ants: the role of vertebrate predators as agents of natural selection. In P. Jaisson, ed., *Social Insects in the Tropics*, pp. 83–104. Villetaneuse: Université Paris Nord.

Huth, A., and C. Wissel. 1992. The simulation of the movement of fish schools. *Journal of Theoretical Biology* 156:365–385.

Jachowsky, R. L. 1974. Agonistic behaviour of the blue crab, *Callinectes sapidus* Rathbun. *Behaviour* 50:232–253.

Jacob, F. 1982. *The Possible and the Actual*. Seattle: University of Washington.

Jaeger, H. M., and S. R., Nagel. 1992. Physics of the granular state. *Science* 255:1523–1531.

Jäger, E., and L. A. Segel. 1992. On the distribution of dominance in populations of social organisms. *SIAM Journal of Applied Mathematics* 52:1442–1468.

Jalife, J. 1984. Mutual entrainment and electrical coupling as mechanisms for synchronous firing of rabbit sino-atrial pace-maker cells. *Journal of Physiology* 356:221–243.

Janeway, C. A. 1993. How the immune system recognizes invaders. *Scientific American* 269:40–47.

Jeanne, R. L. 1970a. Descriptions of the nests of *Pseudochartergus fuscatus* and *Stelopolybia testacea*, with a note on a parasite of *S. testacea* (Hymenoptera: Vespidae). *Psyche* 77:54–69.

———. 1970b. Chemical defence of brood by a social wasp. *Science* 168:1465–1466.

———. 1975. The adaptativeness of social wasp nest architecture. *The Quarterly Review of Biology* 50:267–287.

———. 1977. Ultimate factors in social wasp nesting behavior. *Proc. 8th. Int. Cong. Int. Union Study Soc Insects.* 50:541–557.

———. 1991a. Polyethism. In K. G. Ross, and R. G. Matthews, eds., *The Social Biology of Wasps*, pp. 389–425. Ithaca: Cornell University Press.

———. 1991b. The swarm-founding Polistinae. In K. G. Ross, and R. W. Matthews, eds., *The Social Behavior of Wasps*, pp. 191–231. Ithaca: Cornell University Press.

———. 1996. The evolution of exocrine gland function in wasps. In S. Turillazi and M. J. West-Eberhard, eds., *Natural History of Paper-Wasps*, pp. 144–160. Oxford: Oxford University Press.

———. 1999. Group size, productivity, and information flow in social wasps. In C. Detrain, J. L. Deneubourg, and J. M. Pasteels, eds., *Information Processing in Social Insects*, Basel: Birkhaüser.

Johnson, C. G. 1963. Aerial migration of Insects. *Scientific American* December, p.?

Jones, R. J. 1979. Expansion of the nest of *Nasutitermes costalis*. *Insectes Sociaux* 26:322–342.

———. 1980. Gallery construction by *Nasutitermes costalis*: polyethism and the behavior of individuals. *Insectes Sociaux* 27:5–28.

Judd, T., and P. Sherman. 1996. Naked Mole Rats recruit colony mates to food source. *Animal Behaviour* 52:957–969.

Kaiser, D., and C. Crosby. 1983. Cell movement and its coordination in swarms of *Myxococcus xanthus*. *Cell Motility* 3:227–245.

Kammer, A. E., and B. Heinrich. 1974. Metabolic rates related to muscle activity in bumblebees. *Journal of Experimental Biology* 61:219–227.

Kanciruk, P., and W. Herrnkind. 1978. Mass migration of spiny lobster, *Panulirus argus* (Crustacea: Palinuridae): behavior and environmental correlates. *Bulletin of Marine Science* 28:601–623.

Kapral, R., and K. Showalter, eds. 1994. *Chemical Waves and Patterns*. Dordrecht: Kluver Academic.

Karsai, I., and G. Theraulaz. 1995. Nest Building in a Social Wasp: Postures and constraints (Hymenoptera: Vespidae). *Sociobiology* 26:83–113.

Kauffman, S. A. 1993. *The Origins of Order: Self-Organization and Selection in Evolution*. Oxford: Oxford University Press.

Kawasaki, M. 1993. Temporal hyperacuity in the gymnotiform electric fish, *Eigenmannia*. *American Zoologist* 33:86–92.

Keller, E. F. 1985. The force of the pacemaker concept in theories of aggregation in cellular slime mold. In *Reflections on Gender and Science*. New Haven: Yale University Press.

Keller, E. F., and L. A. Segel. 1970. Initiation of slime mold aggregation viewed as an instability. *Journal of Theoretical Biology* 26:399–415.

———. 1971. Travelling bands of chemotactic bacteria: a theoretical analysis. *Journal of Theoretical Biology* 30:235–248.

Keller, E. F., and G. M. Odell. 1975. Necessary and sufficient conditions for chemotactic bands. *Mathematical Bioscience* 270:309–317.

Keller, L., and H. K. Reeve. 1994. Partitioning of reproduction in animal societies. *Trends in Ecology & Evolution* 9:98–102.

Kern, F., R. Klein, E. Janssen, H. J. Bestmann, A. B. Attygalle, D. Schäfer, and U. Maschwitz. 1997. Mellein, a trail pheromone component of the ant *Lasius fuliginosus*. *Journal of Chemical Ecology* 23:779–792.

Kojima, J. 1993. A latitudinal gradient in intensity of applying ant-repellent substance to the nest petiole in paper wasps (Hymenoptera: Vespidae). *Insectes sociaux* 40:403–421.

Kojima, J., and R. L. Jeanne. 1986. Nest of Ropalidia (*Icarielle nigrescens*) and R. (I). extrema from the Philippines, with reference to the evolutionary radiation in nest architecture within the subgenus Icarielle (Hymenoptera: Vespidae). *Biotropica* 18:324–336.

Krafft, B. 1975. La tolérance réciproque chez l'araignée sociale *Agelena consociata* Denis. *Proc 6th Int. Arachnol. Congr. 1974* Amsterdam.

Krafft, B., and A. Pasquet. 1991. Synchronized and rhythmical activity during prey capture in the social spider *Anelosimus eximius* (Araneae, Theridiidae). *Insectes Sociaux* 38:83–90.

Krebs, J. R. 1974. Colonial nesting and social feeding as strategies for exploiting food resources in the great blue heron (*Ardea herodias*). *Behaviour* 51:99–134.

Krebs, J. R., and N. B. Davies, eds. 1984. *Behavioural Ecology: An Evolutionary Approach*. 2nd ed. Oxford: Blackwell.

Kruuk, H. 1964. Predators and anti-predator behaviour of the black-headed gull (*Larus ridibundus* L.). *Behaviour Supplement* 11:1–129.

Kummer, H. 1968. *Social organization of Hamadrayas baboons*. Chicago: University of Chicago Press.

Landau, H. G. 1951. On dominance relations and the structure of animal societies: I. Effect of inherent characteristics. *Bulletin of Mathematical Biophysics* 13:1–19.

Lapidus, I. R., and R. Schiller. 1976. Model for chemotactic response of a bacterial populations. *Biophysical Journal* 16:779–789.

Lauffenburger, D., M. Grady, and K. H. Keller. 1984. An hypothesis for approaching swarms of myxobacteria. *Journal of Theoretical Biology* 110:257–274.

Laurent, P. 1917. The supposed synchronal flashing of fireflies. *Science* 45:44.

Lauzeral, J., J. Halloy, and A. Goldbeter. 1997. Desynchronization of cells on the developmental path triggers the formation of spiral waves of cAMP during *Dictyostelium* aggregation. *Proceedings of the National Academy of Sciences* 94:9153–9158.

Leinaas, H. P. 1983. Synchronized moulting controlled by communication in group-living Collembola. *Science* 219:193–195.

Lemke, M., and I. Lamprecht. 1990. A model for heat production and thermoregulation in winter clusters of honey bees using differential heat conduction equations. *Journal of Theoretical Biology* 142:261–273.

Levin, S. A., and L. A. Segel. 1985. Pattern generation in space and aspect. *SIAM Review* 27:45–67.

Lewin, R. 1992. *Complexity: Life at the Edge of Chaos*. New York: Macmillan.

Lindauer, M. 1955. Schwarmbienen auf Wohnungssuche. *Zeitschrift für Vergleichende Physiologie* 37:263–324.

Lindsay, D. T. 1977. Simulating molluscan shell pigment lines and states: implications for pattern diversity. *Veliger* 24:297–299.

Linsley, E. G. 1962. Sleeping aggregations of aculeate Hymenoptera. *Annals of the Entomological Society of America* 55:148–164.

Lissaman, P. B. S., and C. A. Shollenberger. 1970. Formation flight of birds. *Science* 168:1003–1005.

Lloyd, J. E. 1966. Studies on the flash communication system of *Photinus* fireflies. *Misc. Publ. Mus. Zool. Univ. Michigan.* #130:1–95.

———. 1971. Bioluminescent communication in insects. *Annual Review of Entomology* 16:97–122.

———. 1973a. Fireflies of Melanesia: Bioluminescence, mating behavior, and synchronous flashing (Coleoptera: Lampyridae). *Environmental Entomology* 2:991–1008.

———. 1973b. Model for the mating protocol of synchronously flashing fireflies. *Nature* 245:268–270.

———. 1979. Sexual selection in luminescent beetles. In M. S. Blum, and N. A. Blum, eds., *Sexual Selection and Reproductive Competition in Insects*, pp. 293–342. New York: Academic Press.

———. 1981. Mimicry in the sexual signals of fireflies. *Scientific American* 245:110–117.

———. 1983. Bioluminescence and communication in insects. *Annual Review of Entomology* 28:131–160.

Lowe, M. E. 1956. Dominance-subordinance relationships in the crayfish *Cambarellus shufeldti. Tulane Studies in Zoology, New Orleans* 4:139–170.

Lüscher, M. 1955. Der Sauerstoffverbrauch bei Termiten und die Ventilation des Nestes bei *Macrotermes natalensis* (Haviland). *Acta Tropica* 12:289–307.

———. 1961. Air-conditioned termite nests. *Scientific American* 205:138–145.

Mac Donald, J. F. 1977. Comparative and adaptative aspects of vespine nest construction. *Proc. 8th Int. Cong. Int. Union Study Soc. Insects*, pp. 169–172.

Mackie, G. O. 1973. Coordinated behavior in hydrozoan colonies. In R. S. Boardman, A. H. Cheetham, and W. A. J. Oliver, eds., *Animal Colonies: Development and Function Through Time*, pp. 95–106. Stroudsburg: Dowden, Hutchinson & Ross.

Maeterlinck, M. 1927. *The Life of the White Ant*. London: George Allen and Unwin.

Magni, F. 1967. Central and peripheral mechanisms in the modulation of flashing in the firefly *Luciola italica* L. *Archives italiennes de Biologie* 105:339–360.

Makino, S., and S. Yamane. 1980. Heat production by the foundress of *Vespa simillima*, with description of its embryo nest (Hymenoptera: Vespidae). *Insecta Matsumurana (N S)* 19:89–101.

Mallon, E., and N. R. Franks. 2000. Ants estimate area using Buffon's needle. *Proc. Roy. Soc. Lond. (B)* 267:765–770.

Marais, E. 1937. *The Soul of the White Ant.* D New York: Dodd, Mead and Co.

Markus, M., and B. Hess. 1990. Isotropic cellular automaton for modelling excitable media. *Nature* 347:56–58.

Martiel J.-L., and A. Goldbeter. 1987. A model based on receptor desensitization for cyclic AMP signaling in *Dictyostelium* cells. *Biophysical Journal* 52:807–828.

Martinez, D. R., and E. Klinghammer. 1970. The behavior of the whale *Orcinus orca*: a review of the literature. *Z. Tierpsychol* 27:828–839.

Maruyama, M. 1963. The second cybernetics: deviation-amplifying mutual causal processes. *American Scientist* 51:164–179.

Maschwitz, U., S. Lenz, and A. Buschinger. 1986. Individual specific trails in the ant *Leptothorax affinis* (Formicidae: Myrmicinae). *Experientia* 42:1173–1174.

May, R. M. 1974. Biological populations with nonoverlapping generations: stable points, stable cycles, and chaos. *Science* 186:645–647.

———. 1976. Simple mathematical models with very complicated dynamics. *Nature* 261:459–467.

———. 1979. Flight formations in geese and other birds. *Nature* 282:778–780.

Maynard-Smith, J. 1974. The theory of games and the evolution of animal conflict. *Journal of Theoretical Biology* 47:209–221.

Maynard-Smith, J., and G. Price. 1973. The logic of animal conflict. *Nature* 246:15–18.

McBride, G. 1958. The measurement of aggressiveness in the domestic hen. *Animal Behaviour* 6:87–91.

McClintock, M. K. 1971. Menstrual synchrony and suppression. *Nature* 229:244–245.

McDermott, F. A. 1916. Flashing of fireflies. *Science* 44:610.

Medeiros, F. N. S., L. E. Lopes, P. R. S. Moutinho, P. S. Oliveira, and B. Hölldobler. 1992. Functional Polygyny, agonistic interactions and reproductive dominance in the neotropical ant *Odontomachus chelifer* (Hymenoptera, Formicidae, Ponerinae). *Ethology* 91:134–146.

Meinhardt, H. 1995. *The Algorithmic Beauty of Sea Shells.* Berlin: Springer-Verlag.

———. 1982. *Models of Biological Pattern Formation.* London: Academic Press.

Mendoza, S. D., and P. R. Barchas. 1983. Behavioral processes leading to linear status hierarchies following group formation in rhesus macaques. *Journal of Human Evolution* 12:185–192.

Meyer, J. A., and S. Wilson. 1989. *From Animals to Animats.* Cambridge: MIT Press.

———. 1991. *From Animals to Animats. II.* Cambridge: MIT Press.

Michaels, D. C., E. P. Matyas, and J. Jalife. 1987. Mechanisms of sinoatrial pacemaker synchronization: a new hypothesis. *Circulation Research* 61:704–714.

Michener, C. D. 1964. Evolution of the nest of bees. *American Zoologist* 4:227–239.

———. 1974. *The Social Behavior of Bees: A Comparative Study.* Cambridge: Harvard University Press.

Miller, K. D., J. B. Keller, and M. P. Stryker. 1989. Ocular dominance column development: analysis and simulation. *Science* 245:605–615.

Mirollo, R. E., and S. H. Strogatz. 1990. Synchronization of pulse-coupled biological oscillators. *SIAM Journal of Applied Mathematics* 50:1645–1662.

Monk, P. B., and H. G., Othmer. 1989. Cyclic AMP oscillations in suspensions of *Dictyostelium discoideum*. *Philosophical Transactions of the Royal Society of London, Series B* 323:185–224.

———. 1990. Wave propagation in aggregation fields of the cellular slime mold *Dictyostelium discoideum*. *Proceedings of the Royal Society of London, Series B* 240:555–589.

Montagner, H. 1964. Instinct et mécanismes stéréotypes chez les guêpes sociales. *Psychologie Française* 9:257–279.

Moritz, R. F. A., and E. E. Southwick. 1992. *Bees as superorganisms. An evolutionary reality*. Berlin: Springer-Verlag.

Morse, A. N. C. 1993. How do planktonic larvae know where to settle? In P. W. Sherman, and J. Alcock, eds., *Exploring Animal Behavior: Readings from American Scientist*, pp. 140–153. Sunderland: Sinauer Associates.

Morse, R. A., and T. Hooper. 1985. *The Illustrated Encyclopedia of Beekeeping*. New York: E. P. Dutton.

Mountcastle, V. B. 1974. *Medical Physiology*. St. Louis: C. V. Mosby.

Murray, J. D. 1981. A prepattern formation mechanism for animal coat markings. *Journal of Theoretical Biology* 88:161–199.

———. 1988. How the leopard gets its spots. *Scientific American* 259:80–87.

———. 1989. *Mathematical Biology. Biomathematics Vol. 19*. Berlin: Springer-Verlag.

Myerscough, M. R. 1993. A simple model for temperature regulation in honeybee swarms. *Journal of Theoretical Biology* 162:381–393.

Nagy, K. A., and J. N. Stallone. 1976. Temperature maintenance and CO_2 concentration in a swarm cluster of honeybees, *Apis mellifera*. *Comparative Biochemistry and Physiology* 55A:169–171.

Naylor, A. F. 1959. An experimental analysis of dispersal in the flour beetle, *Tribolium confusum*. *Ecology* 140:453–465.

Neill, S. R. S. J., and J. M. Cullen. 1974. Experiments on whether schooling by their prey affects the hunting behavior of cephalopods and fish predators. *Journal of Zoology* 172:549–569.

Newell, R. C. 1972. *Biology of Intertidal Animals*. New York: American Elsevier.

Newman, H. H. 1917. A case of synchronic behavior in Phalangidae. *Science* 45:44.

Nicolis, G. 1995. *Introduction to Nonlinear Science*. Cambridge: Cambridge University Press.

Nicolis, G., and I. Prigonine. 1977. *Self-organization in Non-equilibrium Systems*. New York: Wiley.

———. 1989. *Exploring Complexity*. New York: W. H. Freeman.

Nicolis, S. C., and J. L. Deneubourg. 1999. Emerging patterns and food recruitment in ants: an analytical study. *Journal of Theoretical Biology* 198:575–592.

Nijhout, H. F. 1994. *Insect Hormones*. Princeton: Princeton University Press.

Nolan, W. J. 1925. The brood-rearing cycle of the honeybee. *Bulletin of the US Department of Agriculture*. No. 1349, pp. 1–53.

Noonan, K. M. 1981. Individual strategies of inclusive-fitness-maximizing in *Polistes fuscatus* foundresses. In R. D. Alexander, and W. D. Tinkle, eds., *Natural Selection and Social Behavior*, pp. 18–44. New York: Chiron Press.

Okubo, A. 1980. *Diffusion and Ecological Problems: Mathematical Models. Biomathematics Series* Heidelberg: Springer-Verlag.

———. 1986. Dynamical aspects of animal groupings: swarms, schools, flocks, and herds. *Advances in Biophysics* 22:1–94.

Olberg, G. 1959. Das Verhalten der solitären Wespen Mitteleuropas (Vespidae, Pompilidae, Sphecidae). Berlin: Deutscher Verlag der Wissenschaften.

Oliveira, P. S., and B. Hölldobler. 1990. Dominance orders in the ponerine ant *Pachycondyla apicalis* (Hymenoptera, Formicidae). *Behavioral Ecology and Sociobiology* 27:385–393.

Omholt, S. W. 1987. Thermoregulation in the winter cluster of the honeybee, *Apis mellifera*. *Journal of Theoretical Biology* 128:219–231.

Omholt, S. W., and K. Lønvik. 1986. Heat production in the winter cluster of the honeybee, *Apis mellifera*. A theoretical study. *Journal of Theoretical Biology* 120:447–456.

Orledge, G. M. 1988. The identity of *Leptothorax albipennis*, (C.) (Hymenoptera: Formicidae) and its presence in Great Britain. *Systematic Entomology* 23:25–33.

Oster, G., and E. O. Wilson. 1978. *Caste and Ecology in the Social Insects*. Princeton: Princeton University Press.

Otte, D. 1980. On theories of flash synchronization in fireflies. *American Naturalist* 116:587–590.

Otte, D., and J. Smiley. 1977. Synchrony in Texas fireflies with a consideration of male interaction models. *Biology of Behaviour* 2:143–158.

Pacala, S. W., D. M. Gordon, and H. C. J. Godfray. 1996. Effects of social group size on information transfer and task allocation. *Evolutionary Ecology* 10:127–165.

Page, R. E. 1997. The evolution of insect societies. *Endeavour* 21:114–120.

Pálsson, E., and E. C. Cox. 1996. Origin and evolution of circular waves and spirals in *Dictyostelium discoideum* territories. *Proceedings of the National Academy of Sciences, USA* 93:1151–1155.

Pardi, L. 1942. Ricerche sui Polistini. V. La poliginia iniziale di *Polistes gallicus* (L.). *Boll. Ist. Entom. Univ. Bologna* 14:1–106.

———. 1946. Richerche sui Polistini. VII. La "dominazione" e il ciclo ovario annuale in *Polistes gallicus* (L.). *Boll. Ist. Entom. Univ. Bologna* 15:25–84.

———. 1948. Dominance order in *Polistes* wasps. *Physiological Zoology* 21:1–13.

Parr, A. E. 1927. A contribution to the theoretical analysis of the schooling behaviour of fishes. *Occasional Papers Bingham Oceanographical Colloquium* 1:1–32.

Partridge, B. L. 1981. Internal dynamics and the interrelations of fish schools. *Journal of Comparative Physiology* 144:313–325.

———. 1982. The structure and function of fish schools. *Scientific American* 246:90–99.

Partridge, B. L., T. J. Pitcher. 1979. Evidence against a hydrodynamic function for fish schools. *Nature* 279:418–419.

———. 1980. The sensory basis of fish schools: relative role of lateral line and vision. *Journal of Comparative Physiology* 135:315–325.

———. 1980. The 3-D structure of fish schools. *Behavioral Ecology and Sociobiology* 6:277–288.

Passera, L. 1984. *L'organisation sociale des fourmis*. Toulouse: Privat.

Pasteels, J. M., J. L. Deneubourg, and S. Goss. 1987. Self-organisation mechanisms in ant societies I.: the example of food recruitment. In J. M. Pasteels, and J. L. Deneubourg, eds., *From Individual to Collective Behaviour in Social Insects*. Basel: Birkhäuser Verlag.

Peairs, L. M. 1917. Synchronous rhythmic movements of fall web-worm larvae. *Science* 45:501–502.

Peskin, C. S. 1975. *Mathematical Aspects of Heart Physiology*. New York: Courant Institute of Mathematical Sciences.

Pfistner, B. 1990. A one dimensional model for the swarming behavior of myxobacteria. In W. Alt, and G. Hoffmann, eds., *Biological Motion. Lecture Notes in Biomathematics 89*, pp. 556–563. Berlin: Springer-Verlag.

Pielou, E. C. 1977. *Mathematical Ecology*. New York: John Wiley & Sons.

Pinsker, H. M. 1977a. *Aplysia* bursting neurons as endogenous oscillators. I. Phase-response curves for pulsed inhibitory synaptic input. *Journal of Neurophysiology* 40:527–543.

———. 1977b. *Aplysia* bursting neurons as endogenous oscillators. II. Synchronization and entrainment by pulsed inhibitory synaptic input. *Journal of Neurophysiology* 40:544–556.

Pitcher, T. J. 1986. The functions of shoaling behaviour. In T. J. Pitcher, ed., *The Behaviour of Teleost Fishes*, pp. 294–337. London: Croom Helm.

Pitcher, T. J., B. L. Partridge, and C. S. Wardle. 1976. A blind fish can school. *Science* 194:963–965.

Potts, W. K. 1984. The chorus-line hypothesis of manoeuvre coordination in avian flocks. *Nature* 309:344–345.

Press, W. H., B. P. Flannery, S. A. Teukolsky, and W. T. Vetterling. 1986. *Numerical Recipes: The Art of Scientific Computing*. Cambridge: Cambridge University Press.

Prestwich, G. D. 1983. The chemical defenses of termites. *Scientific American* 249:68–73.

———. 1984. Defense mechanisms of termites. *Annual Review of Entomology* 29:201–232.

Prigogine, I., and P. Glansdorf. 1971. *Thermodynamic Theory of Structure, Stability and Fluctuations*. New York: Wiley and Sons.

Pusey, A. E., and C. Packer. 1997. The ecology of relationships. In J. R. Krebs and N. B. Davies, eds., *Behavioural Ecology*, 4th ed. pp. 254–283. Oxford: Blackwell Science.

Reeve, H. K. 1991. *Polistes*. In K. G. Ross, and R. G. Matthews, eds., *The Social Biology of Wasps*, pp. 99–148. Ithaca: Cornell University Press.

———. 1992. Queen activation of lazy workers in colonies of the eusocial naked mole-rat. *Nature* 358:147–149.

Reeve, H. K., and F. L. W. Ratnieks. 1993. Queen-queen conflict in polygynous societies: mutual tolerance and reproductive skew. In L. Keller, ed., *Queen Number and Sociality in Insects*, pp. 45–85. Oxford: Oxford University Press.

Reeve, H. K., and G. J. Gamboa. 1983. Colony activity integration in primitively eusocial wasps: the role of the queen (*Polistes fuscatus*:, Hymenoptera: Vespidae). *Behavioral Ecology and Sociobiology* 13:63–74.

———. 1987. Queen regulation of worker foraging in paper wasps: A social feedback control system (*Polistes fuscatus*: Hymenoptera: Vespidae). *Behaviour* 102:147–167.

Reiswig, H. M. 1970. Porifera: sudden sperm release by tropical demospongiae. *Science* 170:538–539.

Resnick, M. 1994. Learning about life. *Artificial Life* 1:229–241.

Ribbands, C. R. 1953. *The Behavior and Social Life of Honeybees*. London: Bee Research Association.

Richard, B. 1968. La construction de barrages par les castors de France. *Revue du comportement animal* 2:1–52.

———. 1980. *Les Castors*. Balland, France.

Richard, G. 1969. Nervous system and sense organs. In K. Krishna, and F. M. Weesner, eds., *Biology of Termites*. Vol. 1, pp. 161–192. New York: Academic Press.

Richards, O. W. 1978. *The Social Wasps of the Americas, Excluding the Vespinae*. London: British Museum (Natural History).

Richards, O. W., and M. J. Richards. 1951. Observations on the social wasps of South America (Hymenoptera Vespidae). *Transactions of the Royal Entomological Society* 102, 1–170.

Richmond, C. A. 1930. Fireflies flashing in unison. *Science* 71:537–538.

Rissing, S. W., and J. Wheeler. 1976. Foraging responses of *Veromessor pergandei* to changes in seed production. *Pan-Pacific Entomologist* 52:63–72.

Rivault, C., and A. Cloarec. 1998. Cockroach aggregation: discrimination between strain odours in *Blattella germanica*. *Animal Behaviour* 55:177–184.

Roos, W., C. Scheidegger, and G. Gerisch. 1977. Adenylate cyclase oscillations as signals for cell aggregation in *Dictyostelium discoideum*. *Nature* 266:259–61.

Roos, W., V. Nanjundiah, D. Malchow, and G. Gerisch. 1975. Amplification of cyclic-AMP signals in aggregating cells of *Dictyostelium discoideum*. *FEBS Letters* 53:139–142.

Roper, T. J., and J. Ryon. 1977. Mutual synchronization of diurnal activity rhythms in groups of Red wolf/coyote hybrids. *Journal of Zoology* 182:177–185.

Rose, G. A. 1993. Cod spawning on a migration highway in the north-west Atlantic. *Nature* 366:458–461.

Röseler, P. F. 1985. Endocrine basis of dominance and reproduction in polistine paper wasps. *Fortschrifte. Zool.* 31:259–272.

———. 1991. Reproductive competition during colony establishment. In K. G. Ross, and R. G. Matthews, eds., *The Social Biology of Wasps*, pp. 309–335. Ithaca: Cornell University Press.

Röseler, P. F., and I. Röseler. 1989. Dominance of ovariectomized foundresses of the paper wasp *Polistes gallicus*. *Insectes Sociaux* 36:219–234.

Röseler, P. F., I. Röseler, and A. Strambi. 1980. The activity of corpora allata in dominant and subordinated females of the wasp *Polistes gallicus*. *Insectes Sociaux* 27:97–107.

Röseler, P. F., I. Röseler, and A. Strambi. 1985. Role of ovaries and ecdysteroids in dominance hierarchy establishment among foundresses of the primitively social wasp, *Polistes gallicus*. *Behavioral Ecology and Sociobiology* 18:9–13.

———. 1986. Studies of the dominance hierarchy in the paper wasp, *Polistes gallicus* (L.) (Hymenoptera, Vespidae). *Monitore Zool Ital* 20:283–290.

Röseler, P. F., I. Röseler, A. Strambi, and R. Augier. 1984. Influence of insect hormones on the establishment of dominance hierarchies among foundresses of the paper wasp *Polistes gallicus*. *Behavioral Ecology and Sociobiology* 15:133–142.

Rosen, R. 1981. Pattern generation in networks. *Progress in Theoretical Biology* 6:161–209.

Rouland, C., F. Lenoir, and M. Lepage. 1991. The role of the symbiotic fungus in the digestive metabolism of several species of fungus-growing termites. *Comparative Biochemistry and Physiology* (A) 99:657–663.

Russell, M. J., G. M. Switz, and K. Thompson. 1980. Olfactory influences on the human menstrual cycle. *Pharmacology Biochemistry and Behavior* 13:737–738.

Ruttner, F. 1973. Races of bees. In Dadant and Sons, eds., *The Hive and the Honey Bee*, pp. 20–38. Hamilton: Dadant and Sons.

Sakagami, S. H., R. Ohgushi, and D. W. Roubik. 1990. Biology of three Vespa species in central Sumatra (Hymenoptera, Vespidae). In *Natural History of Social Wasps and Bees in Equatorial Sumatra*, pp. 113–124. Sapporo: Hokkaido University Press.

Sander, L. M. 1986. Fractal growth processes. *Nature* 322:789–793.

———. 1987. Fractal growth. *Scientific American* 256:94–100.

Saussure, H. de. 1853–1858. *Etudes sur la famille des vespides. 2. Monographie des guêpes sociales, ou de la tribu des vespiens*. Paris: Masson.

Schaap, P. 1986. Regulation of size and pattern in cellular slime molds. *Differentiation* 33:1–16.

Schein, M. W., and M. H. Fohrman. 1955. Social dominance relationships in a herd of dairy cattle. *British Journal of Animal Behavior* 3:45–55.

Schjelderup-Ebbe T. 1913. Hönsenes stemme. *Bidrag til hönsenes psykologi. Naturen* 37:262–276.

———. 1922. Beiträge zur sozialpsychologie des haushuhns. *Zeitschrifte Psychologie* 88:225–252.

Schneirla, T. C. 1949. Problems in the environmental adaptation of some New-World species of doryline ants. *Anales del Instituto de Biología, Universidad de México* 20:371–384.

———. 1956. The army ants. In *Annual Report of the Smithsonian Institution*, 4232:379–406. Waschington, D.C.: Smithsonian Institution.

Schöner, G., and J. A. S. Kelso. 1988. Dynamic pattern generation in behavioral and neural systems. *Science* 239:1513–1520.

Schremmer, F., L. März, and P. Simonberger. 1985. Chitin im Speichel der papier-wespen (soziale Faltenwespen, Vespidae): *Biologie, Chemismus, Feinstruktur. Mikroskopie* 42:52–56.

Seeley, T. D. 1978. Life history strategy of the honey bee, *Apis mellifera. Oecologia* 32:109–118.

———. 1982. Adaptive significance of the age polyethism schedule in honeybee colonies. *Behavioral Ecology and Sociobiology* 11:287–293.

———. 1983. The ecology of temperate and tropical honeybee societies. *American Scientist* 71:264–272.

———. 1985. *Honeybee Ecology*. Princeton: Princeton University Press.

———. 1989a. Social foraging in honey bees: how nectar foragers assess their colony's nutritional status. *Behavioral Ecology and Sociobiology* 24:181–199.

———. 1989b. The honey bee colony as a superorganism. *American Scientist* 77:546–553.

———. 1995. The Wisdom of the Hive. Cambridge: Harvard University Press.

Seeley, T. D., S. Camazine, and J. Sneyd. 1991. Collective decision-making in honey bees: how colonies choose among nectar sources. *Behavioral Ecology and Sociobiology* 28:277–290.

Seeley, T. D., and B. Heinrich. 1981. Regulation of temperature in the nests of social insects. In B. Heinrich, ed., *Insect Thermoregulation*. New York: John Wiley and Sons.

Seeley, T. D., and R. Morse. 1976. The nest of the honey bee (*Apis mellifera* L.). *Insectes Sociaux* 23:495–512.

Seeley, T. D., and P. K. Visscher. 1988. Assessing the benfits of cooperation in honeybee foraging: search costs, forage quality, and competitive ability. *Behavioral Ecology and Sociobiology* 22:229–237.

Sendova-Franks, A. B., and N. R. Franks. 1993. Task allocation in ant colonies within variable environments. (A study of temporal polyethism: experiment) *Bulletin of Mathematical Biology* 55:75–96.

———. 1994. Social resilience in individual worker ants and its role in division of labour. *Proceedings of the Royal Society of London Series (B)* 256:305–309.

———. 1995a. Division of labour in crisis management: task allocation during colony emigration in the ant *Leptothorax unifasciatus* (L.) *Behavioral Ecology and Sociobiology* 36:269–282.

———. 1995b. Spatial relationships within nests of the ant *Leptothorax unifasciatus* (L.) and their implications for the division of labour. *Animal Behaviour* 50:121–136.

———. 1995c. Demonstrating new social interactions in ant colonies through randomization tests: separating seeing from believing. *Animal Behaviour* 50:1683–1696.

Shaffer, B. M. 1975. Secretion of cyclic AMP induced by cyclic AMP in the cellular slime mould *Dictyostelium discoideum. Nature* 255:549–552.

Shapiro, J. A. 1988. Bacteria as multicellular organisms. *Scientific American* 258:82–89.

———. 1992. Concentric rings in *E. coli* colonies. In L. Rensing, ed., *Oscillations and Morphogenesis*, pp. 297–310. New York: Marcell Dekker.

Shaw, E. 1962. The schooling of fishes. *Scientific American* 206:128–138.

———. 1970. Schooling in fishes: critique and review. In L. R. Aronson, E. Tobach, D. S. Lehrman, and J. S. Rosenblatt, eds., *Development and Evolution of Behaviour*, pp. 452–480. San Francisco: Freeman.

Sherman A., and J. Rinzel. 1991. Model for synchronization of pancreatic β-cells by gap junction coupling. *Biophysical Journal* 59:547–559.

Sherman, A., J. Rinzel, and J. Keizer. 1988. Emergence of organized bursting in clusters of pancreatic beta-cells by channel sharing. *Biophysics Journal* 54:411–425.

Shimkets, L. J. 1990. Social and developmental biology of the myxobacteria. *Microbiological Reviews* 54:473–501.

Shimkets, L. J., and D. Kaiser. 1982. Induction of coordinated movement of *Myxococcus xanthus* cells. *Journal of Bacteriology* 152:451–461.

Silvertown, J., S. Holtier, J. Johnson, and P. Dale. 1992. Cellular automaton models of interspecific competition for space-the effect of pattern on process. *Journal of Ecology* 80:527–533.

Simon, H. A. 1981. *The Sciences of the Artificial*. Cambridge: MIT Press.

Simpson, J. 1961. Nest climate regulation in honeybee colonies. *Science* 133:1327–1333.

Sinclair, A. R. E., and M. Norton-Griffiths. 1979. *Serengeti: Dynamics of an Ecosystem*. Chicago: University of Chicago.

Sismondo, E. 1990. Synchronous, alternating, and phase-locked stridulation by a tropical katydid. *Science* 249:55–58.

Skarka, V., J. L. Deneubourg, and M. R. Belic. 1990. Mathematical model of building behavior of *Apis mellifera. Journal of Theoretical Biology* 147:1–16.

Slater, P. J. B. 1986. Individual differences and dominance hierarchies. *Animal Behaviour* 34:1264–1265.

Smith, A. P. 1978. An investigation of the mechanisms underlying nest construction in the mud wasp *Paralastor* sp. (Hymenoptera: Eumenidae). *Animal Behaviour* 26:232–240.

Smith, H. M. 1935. Synchronous flashing of fireflies. *Science* 82:151–152.

Sorenson, A. A., T. M. Busch, and S. Bradleigh Vinson. 1985. Control of food influx by temporal subcastes in the fire ant, *Solenopsis invicta*. *Behavioural Ecology and Sociobiology* 17:191–198.

Southwick, E. E. 1983. The honey bee cluster as a homeothermic superorganism. *Comparative Biochemistry and Physiology* 75A:641–645.

———. 1985. Bee hair structure and the effect of hair on metabolism at low temperature. *Journal of Apicultural Research* 24:144–149.

Southwick, E. E., and R. F. A. Moritz. 1987. Social synchronization of circadian rhythms of metabolism in honeybees (*Apis mellifera*). *Physiological Entomology* 12:209–212.

Southwick, E. E., and J. N. Mugaas. 1971. A hypothetical homeotherm: the honeybee hive. *Comparative Biochemistry and Physiology* 40A:935–944.

Spradberry, J. P. 1973. *Wasps: an account of the biology and natural history of solitary and social wasps*. Seattle: University of Washington Press.

Starr, C. K. 1991. The nest as the locus of social life. In K. Ross, and R. Matthews, eds., *The Social Biology of Wasps*, pp. 520–539. Ithaca: Comstock Publ. Assoc.

Steinbock, O., H. Hashimoto H., and S. C. Müller. 1991. Quantitative analysis of periodic chemotaxis in aggregation patterns of *Dictyostelium discoideum*. *Physica D* 49:233–239.

Stephens, D. W., and J. R. Krebs. 1986. *Foraging Theory*. Princeton: Princeton University Press.

Stevens, A. 1990. Simulations of the gliding behavior and aggregation of myxobacteria. In W. Alt, and G. Hoffmann, eds., *Biological Motion. Lecture Notes in Biomathematics 89*, pp. 548–555. Berlin: Springer-Verlag.

Stickland, T., and N. R. Franks. 1994. Computer image analysis provides new observations of ant behaviour patterns. *Proceedings of the Royal Society of London (B)* 257:279–286.

Strausfeld, N. J. 1976. *Atlas on an Insect Brain*. Berlin: Springer-Verlag.

Strogatz, S. H., and I. Stewart. 1993. Coupled oscillators and biological synchronization. *Scientific American* 269:102–109.

Strogatz, S. H., R. E. Mirolio, and P. C. Matthews. 1992. Coupled nonlinear oscillators below the synchronization threshold: Relaxation by generalized Landau damping. *Physical Review Letters* 68:2730–2733.

Stuart, A. M. 1967. Alarm, defense, and construction behavior relationships in termites (Isoptera). *Science* 156:1123–1125.

———. 1969. Social Behavior and Communication in Biology of Termites. In K. Krishna, and F. M. Weesner, eds., *Biology of Termites*, Vol. 1, pp. 193–232. New York: Academic Press.

———. 1972. Behavioral regulatory mechanisms in the social homeostasis of termites (Isoptera). *American Zoologist* 12:589–594.

————. 1957. Communication and recruitment in *Monomorium pharaonis*. *Animal Behaviour* 5:104–109.

Sudd, J. H. 1963. How insects work in groups. *Discovery* 26:15–19.

Swindale, N. V. 1980. A model for the formation of ocular dominance stripes. *Proceeding of the Royal Society of London, Series B* 208:243–264.

Tang Y., and H. G. Othmer. 1994. A G protein-based model of adaptation in *Dictyostelium discoideum*. *Mathematical Biosciences* 120:25–76.

Tanner, W. L. 1930. Plant lice pumping in unison. *Science* 72:560.

Theraulaz, G. 1991. *Morphogenèse et Auto-Organisation des Comportements dans les Colonies de guêpes Polistes dominulus* (C.). Ph.D. Dissertation, Université de Provence.

Theraulaz, G., and E. Bonabeau. 1995a. Coordination in distributed building. *Science* 269:686–688.

————. 1995b. Modelling the collective building of complex architectures in social insects with lattice swarms. *Journal of Theoretical Biology* 177:381–400.

Theraulaz, G., E. Bonabeau, and J. L. Deneubourg. 1995. Self-organization of hierarchies in animal societies: the case of the primitively eusocial wasp *Polistes dominulus* (C.). *Journal of Theoretical Biology* 174:313–323.

Theraulaz, G., and J. Gervet. 1992. Les performances collectives des sociétés d'insectes. *Psychologie Française* 37:7–14.

Theraulaz, G., J. Gervet, and S. Semenoff-Tian-Chansky. 1991a. Social regulation of foraging activities in *Polistes dominulus* (C.): a systemic approach to behavioural organization. *Behaviour* 116:292–320.

Theraulaz, G., S. Goss, J. Gervet, and J. L. Deneubourg. 1991b. Task differentiation in *Polistes* wasp colonies: a model for self-organizing groups of robots. In J. A. Meyer, and S. W. Wilson, eds., *From Animals to Animats: Proceedings of the First International Conference on Simulation of Adaptive Behaviour*, pp. 346–355. Cambridge: MIT Press.

Theraulaz, G., J. Gervet, B. Thon, M. Pratte, and S. Semenoff-Tian-Chansky. 1992. The dynamics of colony organization in the primitively eusocial wasp *Polistes dominulus* (C.). *Ethology* 91:177–202.

Theraulaz, G., M. Pratte, and J. Gervet. 1989. Effects of removal of α-individuals from a *Polistes dominulus* (C.) wasp society: changes in behavioural patterns resulting from hierarchical changes. *Actes des Colloques Insectes Sociaux* 5:169–179.

————. 1990. Behavioural profiles in *Polistes dominulus* (C.) wasp societies: a quantitative study. *Behaviour* 113:223–250.

Thierry, B. 1985. Patterns of agonistic interactions in three species of macaques (*Macaca mulatta, M. fascicularis, M. tonkeana*). *Aggressive Behavior* 11:223–233.

Thorpe, W. H. 1963. *Learning and Instinct in Animals*. London: Methuen.

Tomchik, K. J., and P.N. Devreotes. 1981. Adenosine $3',5'$-monophosphate waves in *Dictyostelium discoideum*. A demonstration by isotope dilution fluorography. *Science* 212:443–446.

Traniello, J. F. A. 1977. Recruitment behaviour, orientation, and the organization of the foraging in the carpenter ant *Camponotus pennsylvanicus*. *Behavioural Ecology and Sociobiology* 2:61–79.

Traniello, J. F. A., and S. K. Robson. 1995. Trail and territorial communication in social insects. In W. J. Bell, and R. Cardé, eds., *The Chemical Ecology of Insects* vol. 2 pp. 241–246. London: Chapman and Hall.

Treherne, J. E., and W. A. Foster. 1981. Group transmission of predator avoidance in a marine insect: the Trafalgar Effect. *Animal Behaviour* 29:911–917.

Tschinkel, W. R. 1993. Sociometry and sociogenesis of colonies of the fire ant *Solenopsis invicta* during one annual cycle. *Ecological Monographs* 63:425–427.

Tucker, V. A. 1969. Wave-making by whirligig beetles (Gyrinidae). *Science* 166:897–899.

Turillazi, S., M. T. Marino-Piccioli, L. Hervatin, and L. Pardi. 1982. Reproductive capacity of single foundress and associated foundress females of *Polistes gallicus* (L.) (Hymenoptera Vespidae). *Monitore Zoologico Italiano* 16:75–88.

Turillazi, S., and L. Pardi. 1977. Body size and hierarchy in polygynic nests of *Polistes gallicus* (L.) (Hymenoptera, Vespidae). *Monitore Zoologico Italiano* 16:75–88.

Turillazzi, S. 1989. The origin and evolution of social life in the Stenogastrinae (Hymenoptera, Vespidae). *Journal of Insect behavior* 2:649–661.

Turillazzi, S., and A. Ugolini. 1979. Rubbing behavior in some European Polistes (Hymenoptera: Vespidae). *Monitore Zoologico Italiano (Nuova Serie)* 13:129–41.

Turing, A. 1952. The chemical basis for morphogenesis. *Philosophical Transactions of the Royal Society of London* 237:37–72.

Tyler, S. J. 1972. The behaviour and social organisation of the new forest ponies. *Animal Behavior Monographs* 48:223–233.

Tyson, J. J., and J. D. Murray. 1989. Cyclic AMP waves during aggregation of *Dictyostelium* amoebae. *Development* 106:421–426.

Tyson, J. J., K. A. Alexander, V. S. Manoranjan, and J. D. Murray. 1989. Spiral waves of cyclic AMP in a model of slime mold aggregation. *Physica D* 34:193–207.

Uvarov, B. P. 1928. *Grasshoppers and Locusts.* London: Imperial Bureau of Entomology.

Van de Poll, N. E., F. De Jonge, H. G. Van Oyen, J. Van Pett. 1982. Aggressive behaviour in rats: effects of winning and losing on subsequent aggressive interactions. *Behavioural Processes* 7:143–155.

Van der Kloot, W. G., and C. M. Williams. 1953a. Cocoon construction by the cecropia silkworm I. The role of the external environment. *Behaviour* 5:141–156.

Van der Kloot, W.G., and C. M. Williams. 1953b. Cocoon construction by the cecropia silkworm II. The role of the internal environment. *Behaviour* 5:157–174.

Van Honk, C., and P. Hogeweg. 1981. The ontogeny of the social structure in a captive *Bombus terrestris* colony. *Behavioral Ecology and Sociobiology* 9:111–119.

Vehrencamp, S. 1983. A model for the evolution of despotic versus egalitarian societies. *Animal Behaviour* 31:667–682.

Velarde, M. G., and C. Normand. 1980. Convection. *Scientific American* 243:93–108.

Velthuis, H. H. W. 1976. Egg laying, aggression and dominance in bees. *Proceedings of the XVth International Congress of Entomology*, pp. 436–449.

Verhaeghe, J. C. 1982. Food recruitment in *Tetramorium impurum* (Hymenoptera: Formicidae). *Insectes Sociaux* 29:65–85.

Verhaeghe, J. C., and J. L. Deneubourg. 1983. Experimental study and modelling of food recruitment in the ant *Tetramorium impurum* (Hym. Form.). *Insectes Sociaux* 30:347–360.

Visscher, K. P., and S. Camazine. 1998. The mystery of swarming honey bees: from individual behaviors to collective decisions. In C. Detrain, J. L. Deneubourg, and J. M. Pasteels, eds., *Information Processing in Social Insects*. In press. Basel: Bikhaüser Verlag.

Visscher, K. P., and T. D. Seeley. 1982. Foraging strategy of honeybee colonies in a temperate deciduous forest. *Ecology* 63:1790–1801.

Vogel, S. 1981. *Life in Moving Fluids: The Physical Biology of Flow*. Boston: Willard Grant Press.

Waldrop, M. M. 1990. Spontaneous order, evolution, and life. *Science* 247:1543–1545.

Walker, T. J. 1969. Acoustic synchrony: Two mechanisms in the snowy tree cricket. *Science* 166:891–894.

Warburton K., and J. Lazarus. 1991. Tendency-distance models of social cohesion in animal groups. *Journal of Theoretical Biology* 150:473–488.

Watmough, J., and S. Camazine. 1995. Self-organized thermoregulation of honey bee clusters. *Journal of Theoretical Biology* 176:391–402.

Watmough, J., and L. Edelstein-Keshet. 1995a. A one dimensional model of trail propagation by army ants. *Journal of Mathematical Biology* 33:459–476.

———. 1995b. Modelling the formation of trail networks by foraging ants. *Journal of Theoretical Biology* 176:357–371.

Wehner, R., 1970. Etudes sur la construction des cratères au-dessus des nids de la fourmi *Cataglyphis bicolor*. *Insectes Sociaux* 17:124–133.

Wehner, R., A. C. Marsh, and S. Wehner. 1992. Desert ants on a thermal tightrope. *Nature* 357:586–587.

Wenzel, J. W. 1989. Endogenous factors, external cues, and accentric construction in *Polistes annularis* (Hymenoptera: Vespidae). *Journal of Insect Behaviour* 2:679–699.

———. 1991. Evolution of nest architecture. In K. Ross, and R. Matthews, eds., *The Social Biology of Wasps*, pp. 480–519. Ithaca: Comstock Publ. Assoc.

———. 1996. Learning, behavioural programs, and higher-level rules in construction of *Polistes*. In S. Turillazzi and M. J. West-Eberhard, eds., *Natural History of Paper-Wasps*, pp. 58–74. Oxford: Oxford University Press.

Werner, B. T., and B. Hallet. 1993. Numerical simulation of self-organized stone stripes. *Nature* 361:142–145.

West-Eberhard, M. J. 1969. The social biology of polistine wasps. *Miscellaneous publications, Museum of Zoology, University of Michigan* 140:1–101.

———. 1981. Intragroup selection and the evolution of insect societies. In R. D. Alexander, and D. W. Tinkle, eds., *Natural Selection and Social Behavior: Recent Research and New Theory*, pp. 3–17. New York: Chiron.

Wheeler, W. M. 1917. The synchronic behavior of Phalangidae. *Science* 45:189–190.

Wiehs, D. 1973. Hydromechanics of fish schooling. *Nature* 241:290–291.

Wiener, N. 1961. *Cybernetics or Control and Communication in The Animal and The Machine*, 2nd ed. Cambridge: MIT Press.

Williams, G. C. 1966. *Adaptation and Natural Selection*. Princeton: Princeton University Press.

Wilson, D. S., and E. Sober. 1989. Reviving the superorganism. *Journal of Theoretical Biology* 136:337–356.

Wilson, E. O. 1962. Chemical communication among workers of the fire-ant *Solenopsis saevissima* (Fr. Smith). *Animal Behaviour* 10:134–164.

———. 1971. *The Insect Societies*. Cambridge: Harvard University Press.

———. 1975. Sociobiology: *The New Synthesis*. Cambridge: Harvard University Press.

Wilson, E. O., W. H. Bossert, and F. E. Regnier. 1969. A general method of estimating threshold concentrations of odorant molecules. *Journal of Insect Physiology* 15:597–610.

Wilson, E. O., and B. Hölldobler. 1988. Dense heterarchies and mass communication as the basis of organization in ant colonies. *Trends in Ecolology and Evolution* 3:65–68.

Winfree, A. T. 1967. Biological rhythms and the behavior of populations of coupled oscillators. *Journal of Theoretical Biology* 16:15–42.

———. 1972. Spiral waves of chemical activity. *Science* 175:634–636.

———. 1980. *The Geometry of Biological Time*. New York: Springer-Verlag.

———. 1984. The prehistory of the Belousov-Zhabotinski oscillator. *Journal of Chemical Education* 61:661–663.

———. 1987. *The Timing of Biological Clocks*. New York: Scientific American Library.

Winston, M. 1987. *The Biology of the Honey Bee*. Cambridge: Harvard University Press.

Woodward, D. E., R. Tyson, M. R. Myerscough, J. D. Murray, E. O. Budrene, and H. C. Berg. 1995. Spatio-temporal patterns generated by *Salmonella typhimurium*. *Biophysical Journal* 68:2181–2189.

Yagi, N. 1926. The cocooning behaviour of a saturnian caterpillar *Dictyoploca japonica*; a problem in analysis of insect conduct. *Journal of Experimental Zoology* 46:245–259.

Yates, F. E., A. Garfinkel, D. O. Walter, and G. B. Yates. 1987. *Self-Organizing Systems: The Emergence of Order*. New York: Plenum Press.

Young, D. A. 1984. A local activator-inhibitor model of vertebrate skin patterns. *Mathematical Bioscience* 72:51–58.

Zusman, D. R. 1984. Cell-cell interactions and development in *Myxococcus xanthus*. *Quarterly Review of Biology* 59:119–138.

Index

Illustrations are indicated by page references in boldface and followed by *f*. Tables are indicated by page references followed by *t*. References from text boxes are indicated by page references followed by *b*.